Textbook of
Biochemistry for Nurses

Textbook of
Biochemistry for Nurses

Second Edition

Ashok Kumar J.
Prof. & Head Dept. of Biochemistry
A.J. Institute of Medical Sciences
Kuntikana, Mangalore
Karnataka

TECHSAR™
NEW DELHI

Published and Printed by

Techsar Pvt. Ltd.
4435-36/7, nsari Road, Daryaganj
New Delhi–110 002 (India)
E-mail: tech_info@techsarworld.com
Website: www.techsarworld.com

ISBN: 978-93-81141-28-1

Reprint 2025

Foreword

I have a great pleasure in writing this foreword for "Textbook of Biochemistry for Nurses". I congratulate Dr. Ashok Kumar for his maiden venture.

There are several books on Biochemistry, but this book is unique in its own way. The special needs of nursing students are amply met in this book.

Dr. Ashok Kumar is a very popular teacher. He has a great interest in teaching and transferring knowledge, which he continues to do by participating in various programmes. He has been rated as one of the best teachers in our institution.

Dr. Ashok has taken tremendous pain while preparing this book, focusing attention on basics as well as advanced aspects, yet keeping simplicity and readability that would help any beginner and expert alike to enjoy reading and understanding.

I am sure that this textbook is going to be very useful not only for the students of nursing, but also for the students of physiotherapy, and allied health sciences. This book is also of immense help to undergraduate medical students.

I am confident that this book will guide them throughout their carrier. The book reminds me a principle of "small things add to perfection, but perfection is not a small thing". I congratulate Dr. Ashok Kumar for his fine contribution and wish him more successful ventures in years to come.

I strongly recommend this book to all the students.

Dr. B. Sanjeev Rai (M.D., FIAP, MBA)
Dean
Father Muller Medical College
Father Muller Road
Mangalore–575002
Karnataka

Preface to the Second Edition

I am happy to present the second edition of *Textbook of Biochemistry for Nurses*. It gives a sense of satisfaction that this book has been well received by the students and teachers. As per the suggestions received, the second edition of this book is designed as per the syllabus of nursing course including nutrition.

Many of the reactions and reaction sequences are rewritten to make it clear to the readers and illustrative diagrams have been added. Explanations are simpler. New chapters on gene expression and various topics in nutrition are added.

I am grateful to almighty God for guiding me in this venture. I am thakful to Dr. Sarawathi Udupa, Professor of Biochemistry who has gone through this book and given her valuable suggestions when preparing the second edition. Also I thank my teachers Prof. Vivian D'Souza, Prof. Ashalatha V. Rao, and all my teachers for their encouragement and support. Help rendered by my wife, encouragement of my parents and cooperation of my children to get this book materialized is solicited. I hope this book would help nursing students in understanding the subject of biochemistry and nutrition. Kindly feel free to give your suggestions at drashokkumarj@gmail.com.

Ashok Kumar J.

Preface to the First Edition

After teaching nursing students for a decade, I realized the need for a textbook in Biochemistry that could provide the necessary information to achieve their academic goals. This book has been designed to cater to the academic needs of the nursing students. An earnest effort has been made to present the subject in simple words. I hope that the information present in this book would be helpful to the nurses throughout their career.

A sound knowledge of biochemistry is essential to understand the pathophysiology of disease, its diagnosis, treatment, and follow up. Since the nursing community works closely in association with clinicians dedicated to patient care, it is important for them to be aware of the biochemical aspects of human diseases. Thus, in this book, the clinical applications of biochemical knowledge have been mentioned wherever necessary.

I am grateful to the almighty God for guiding me in all my endeavours. I would like to thank I.K. International Publishing House Pvt. Ltd. for their sincere efforts and cooperation in the publication of this book.

I am thankful to the Director, Administrators and Dean of Father Muller Medical College, and my colleagues for their support and encouragement. Constant encouragement and help rendered by my wife Dr. Kavitha, and cooperation of my children, Aakanksha and Aashlesh, to get this book materialized is solicited.

I request all my readers to e-mail their sincere comments at drashokkumarj@yahoomail.com.

Ashok Kumar J.

Contents

1

Cell and Cell Membrane

INTRODUCTION

Cell is the structural and functional unit of an organism. Cells have life. To sustain life, energy is required. Living cells have the ability to harness energy from the surrounding environment. Several biochemical reactions are responsible for the production or extraction of energy. Based upon the differences with respect to the microscopic structure and composition, cells can be grouped into:

1. Prokaryotic cells
2. Eukaryotic cells.

Prokaryotes lack distinct membrane-bound subcellular structures; mainly, the nuclear membrane is absent. Eukaryotes have nuclear membrane surrounding the central mass of nucleus (Fig. 1.1).

CELL MEMBRANE

Cell is enclosed by a membrane called plasma membrane which is made up of lipids, proteins and carbohydrates. Cell membrane separates the cell content from the exterior and protects the cell. It also encloses the subcellular organelles and regulates the passage of molecules across it.

Membrane Lipids

Lipids present in the cell membrane are amphipathic lipids. Amphipathic lipids have both hydrophilic and hydrophobic part. Amphipathic lipids present in cell membrane are:

(a) **Phospholipids:** Phospholipids are phosphate containing lipids. Phospholipids present in cell membrane are glycerophospholipids (Fig 1.2), and sphingophospholipids. Glycerophospholipids contain glycerol as the alcohol group and sphingophospholipids contain sphingosine as alcohol.

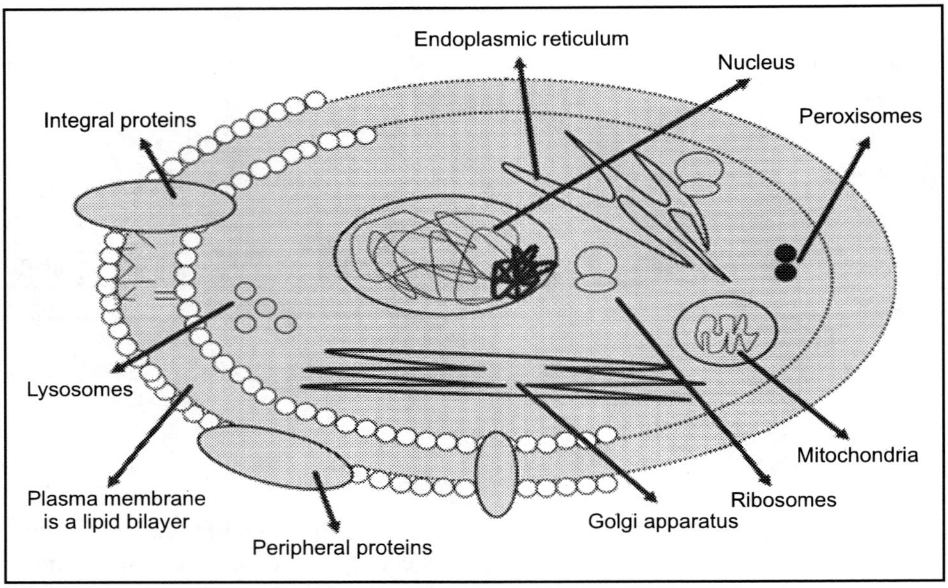

Fig. 1.1 Eukaryotic cell.

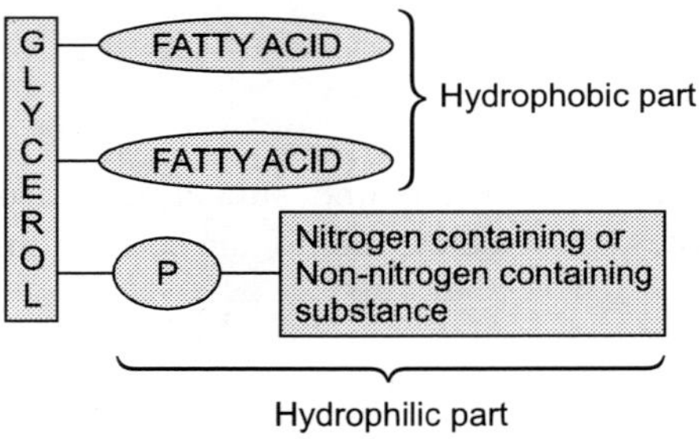

Fig. 1.2 Phospholipid.

Glycerophospholipids present in cell membrane are phosphatidylcholine, phosphatidylethanolamine, phosphatidylserine, phosphatidylinositol, etc. Cardiolipin which is a glycerophospholipid present in inner mitochondrial membrane. Sphingomyline is a sphingophospholipid. It is mainly present in myelin sheath of neurons.

(b) **Glycolipids:** These are lipids containing sugar, e.g., cerebrosides and gangliosides. (Fig. 1.3).

(c) **Cholesterol:** Cholesterol is a sterol containing cyclopentanoperhydrophenanthrene ring. The hydroxyl group of the cholesterol is hydrophilic in nature and the bulky ring structure and the hydrocarbon chain are hydrophobic in nature, so cholesterol is an amphipathic lipid. (Fig.1.4)

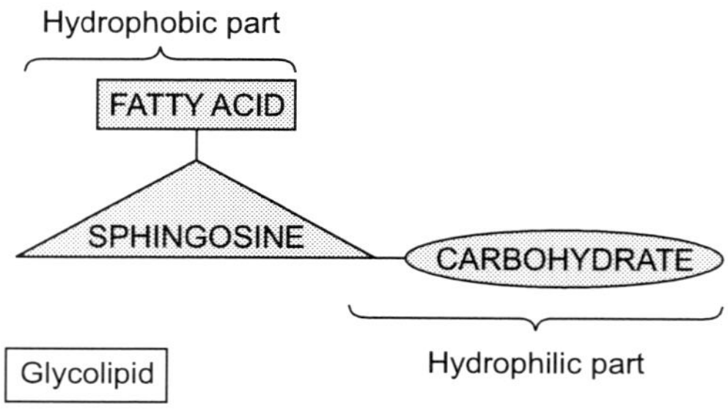

Fig. 1.3 Glycolipid.

Cholesterol can be esterified with fatty acid to form cholesteryl ester. In cholesteryl ester the hydroxyl group of cholesterol and the carboxyl group of fatty acids form ester bond. This makes cholesteryl ester hydrophobic. Cholesteryl ester is not present in the cell membrane.

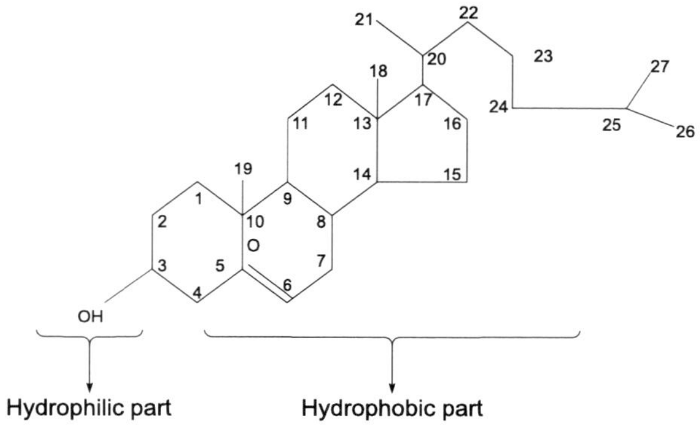

Fig. 1.4 Cholesterol.

Membrane Proteins

Cell membrane contains proteins. Some proteins which are deeply fitted into the cell membrane and can be separated from the membrane using detergents or fat solvents damaging the membrane are called **integral proteins**. Some other proteins remain attached to the surface of the membrane loosely and can be easily separated without damage to the membrane, these are called **peripheral proteins**.

Membrane Carbohydrates

Carbohydrates present in the membrane are attached to the lipids and proteins, e.g., glycoproteins and glycolipids.

STRUCTURE OF THE CELL MEMBRANE (FLUID MOSAIC MODEL)

(Fluid mosaic model was proposed in 1972 by Singer and Nicolson)

Fig. 1.5

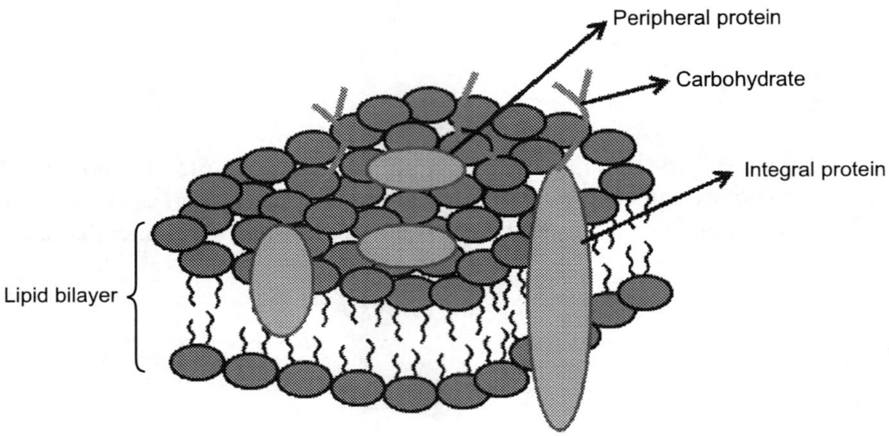

Fig. 1.6 Fluid mosaic model of cell membrane.

Phospholipids (amphipathic lipid) have hydrophilic head and hydrophobic tail (Fig.1.5). In the cell membrane phospholipids are arranged as bilayer (Fig. 1.6). Outer phospholipid layer is called outer leaflet and inner layer is called inner leaflet of the phospholipid bilayer.

In lipid bilayer the hydrophobic groups of the lipids face each other and form hydrophobic core. Hydrophilic group faces the external aqueous environment. Membranes contain integral and peripheral proteins. Some of the integral proteins travel the whole thickness of the membrane. In these proteins hydrophobic amino acids occupy the core of the membrane.

Lipid bilayer shows lateral movement of its lipid components so it is fluid in nature. Phospholipids contain two fatty acids of which one is unsaturated fatty acid. Fluidity of the cell membrane increases as the 'Cis' unsaturated fatty acid content of the cell membrane increases. Cholesterol content of the cell membrane alters the fluidity of the membrane.

Cell membrane shows asymmetry. The lipid components of the individual leaflets of the lipid bilayer are different. Phospholipid containing choline (e.g., phosphatidylcholine) is mainly located in the outer layer. Phospholipids containing amino groups (e.g., phosphatidylserine, phosphatidylethanolamine, etc.) are located in the inner layer. Distribution of protein and carbohydrates also differs between the inner and outer layer of the lipid bilayer.

Plasma membrane is the cell membrane which encloses the cell. It separates and protects the cell content from the surrounding environment. Plasma membrane controls the passage of various molecules across it (semipermeable). Some small molecules can freely cross the membrane, e.g., urea, carbon dioxide, water, etc.

Membranes are also present inside the cell. It divides the cell into different compartments called subcellular organelles. The lipid, protein and carbohydrate composition of the membranes of the subcellular organelle differ from each other. Each subcellular organelle has its own composition and a set of biochemical functions.

TRANSPORT ACROSS THE MEMBRANE

Cell membrane is selectively permeable to various molecules. There are different mechanisms by which constant exchange of molecules and information can take place between the membrane-enclosed compartment and the exterior.

Depending upon the direction and number of molecules transported, transport systems can be described as follows:

1. **Uniport:** Transports one type of molecule in both directions, e.g., glucose transporter in RBC.
2. **Cotransport:** Transport of one molecule depends upon the simultaneous or sequential transport of another molecule, e.g., Na^+ amino acid cotransport.
3. **Symport:** Transports two types of molecules in the same direction, e.g., Na^+glucose symport.
4. **Antiport:** Transport of two types of molecules in opposite direction, e.g., Na^+/K^+ ATPase.

Different mechanisms of transport across the membrane are:

1. **Diffusion**
   (a) Passive diffusion (simple diffusion)
   (b) Facilitated diffusion
      Ion channels
      - Ligand gated ion channels
      - Voltage gated ion channels

      Ionophores
      - Channel forming
      - Mobile ion carriers
2. **Active transport**
3. **Vesicular transport**
   (a) Endocytosis
   (b) Exocytosis

1. Diffusion

Movement of molecules from one side of the membrane to other side along the concentration gradient (from higher concentration to lower concentration) is called diffusion. Diffusion does not require energy in the form of ATP and at one point it reaches equilibrium.

Diffusion across the membrane without the help of carrier protein in the membrane is called **simple diffusion** e.g., transport of O_2, CO_2, and urea, etc. If diffusion across the membrane requires a carrier protein (transporter), it is called facilitated diffusion, e.g., glucose transport (GLUT) in RBCs and transport of amino acids.

Facilitated diffusion (Fig.1.7) requires the change in the conformation of the transport proteins as shown below. These conformations are called ping pong conformation.

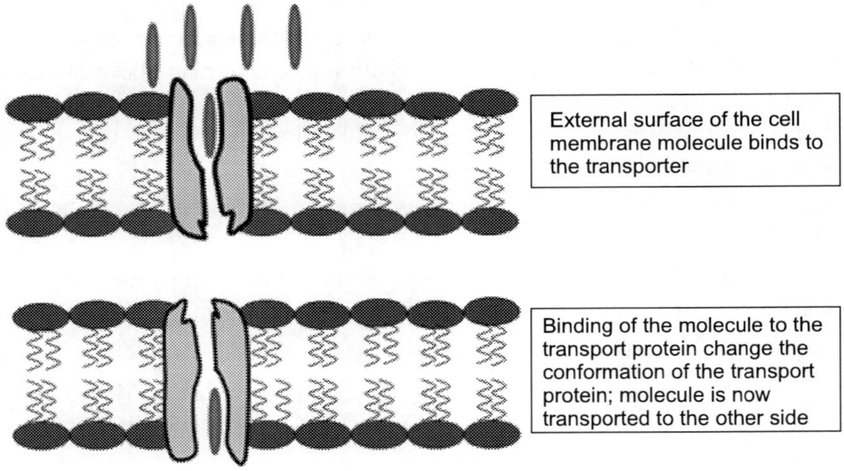

Fig. 1.7 Facilitated diffusion.

Ion Channels

Ion carrier proteins are present in the membrane. It opens in response to stimulus. When it opens, it allows the passage of specific ion along its concentration gradient. Since it opens when there is a specific stimulus, they are called **gated channels**.

When ion channels open in response to change in the membrane potential, they are called **voltage gated channels,** e.g., sodium and potassium channels of the neurons. If the opening of the channels depends upon the binding of a signal molecule (intracellular or extracellular), ion channel is called **ligand gated ion channel,** e.g., binding of acetylcholine to the post-synaptic receptors will open the sodium channels.

Ionophores

Ionophores increase the membrane permeability to certain ions. If they form a channel in the membrane to transport ions from one side of the membrane to another, they are called **channel forming ionophores,** e.g., gramicidin. When the ionophores carry ions from one side of the membrane to another, they are called **mobile ionophores**.

2. Active Transport

Molecules are transported against the concentration gradient (from lower concentration to higher concentration) with the help of transporter by utilizing energy in the form of ATP.

(a) **Primary active transport** is active transport where energy is primarily used by the transporter itself, e.g., Na + K + ATPase (Fig. 1.8).

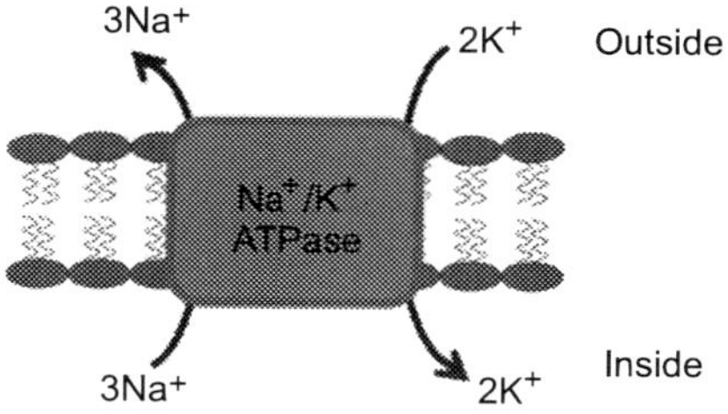

Fig. 1.8 Sodium potassium pump.

(b) **Secondary active transport** is an active transport mechanism where the energy is not utilized by the primary transporter protein but by another transport protein, which is also required for the transport of the molecule, for example, absorption of glucose in the intestine.

3. Vesicular Transport

(a) Endocytosis It is a transport process by which large molecules are transported across the membrane. The cell internalizes macromolecules to form vesicles.

There are two types:

(i) Pinocytosis (cell drinking) – uptake of fluids by the cell.
(ii) Phagocytosis (cell eating)–uptake of macromolecules, bacteria, etc. by the neutrophils or macrophages.

(b) Exocytosis It is a process by which large molecules are released from the cell e.g., release of neurotransmitters, insulin from the beta cells, etc.

Osmosis

Osmosis is the movement of water from solution having lower concentration of solute to the solution having higher concentration of solute across the membrane which separates them.

Force required to prevent the movement of the water across the membrane in this situation is osmotic force (osmotic tension). Osmotic pressure is the osmotic tension due to proteins. It is expressed as mm of Hg, e.g., osmotic pressure of plasma protein is 25 mm of Hg.

SUBCELLULAR ORGANELLES

Cytoplasm (Cytosol)

It is the matrix of the cell. It contains all the subcellular organelles. Water is the major component of cytosol containing different biomolecules. Cytoplasm is rich in protein. A number of metabolic reactions take place in cytosol.

Cytoplasm contains all the cofactors and energy required for protein synthesis. **Fatty acid synthesis**, **cholesterol synthesis**, **glycogen metabolism**, **glycolysis** takes place in cytosol.

Mitochondria

It is the powerhouse of the cell. It is enclosed by two different membranes called inner mitochondrial membrane and outer mitochondrial membrane. The inner membrane is folded so as to increase the surface area. Space between the two membranes is called intermembranous space (Fig. 1.9). The mitochondria has got its own DNA which codes for certain proteins of mitochondria.

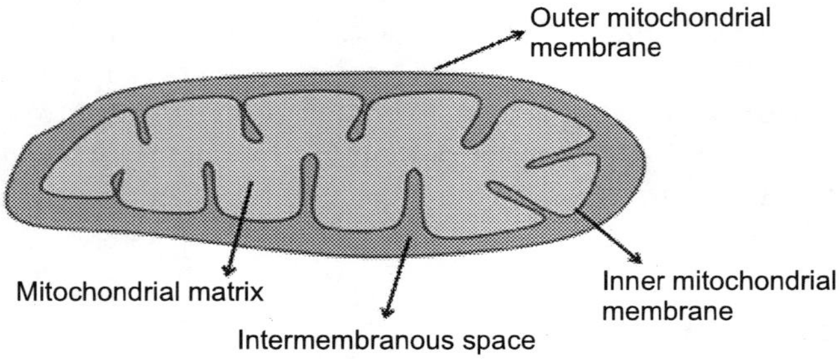

Fig. 1.9 Mitochondria.

Oxidative metabolism of various biomolecules takes place in the mitochondria, leading to the production of energy in the form of ATP.

Citric acid cycle, **fatty acid oxidation** and **oxidative phosphorylation** to produce ATP (through electron transport chain) take place in mitochondria.

Nucleus

Nucleus contains the hereditary material, deoxyribonucleic acid. Genetic material (DNA) is enclosed by a membrane called **nuclear membrane**. Nuclear membrane has two layers. The space between the two layers is called perinuclear cisterna.

There are **pores present on the nuclear membrane**. Exchange of material between cytoplasm and nucleoplasm is possible through these pores in addition to other mechanisms of transport across the nuclear membrane.

The information present in the **genetic material** is responsible for the structure and function of the cell.

Ribosome

Ribosome is a ribonucleoprotein. Ribosome may be present in the free form in the cytosol (free ribosome), or may be bound to the surface of endoplasmic reticulum. Ribosome has got two subunits called larger subunit and smaller subunit (Fig. 1.10). During protein synthesis both subunits are associated with each other to form ribosome.

Larger subunit has got two sites called A site (Aminoacyl site) and P site (Peptidyl site). Ribosomes are required for synthesis of proteins.

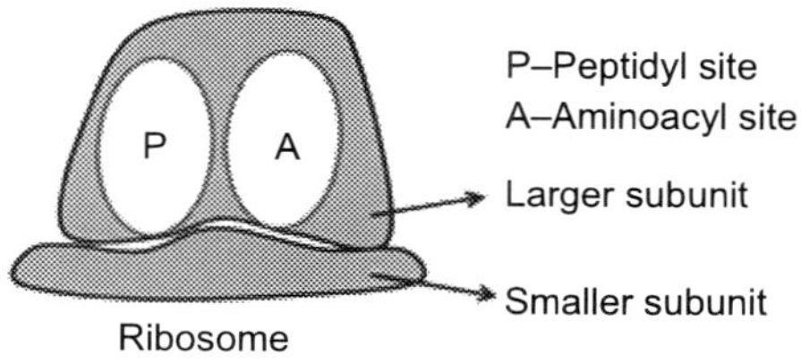

Fig. 1.10 Ribosome.

Endoplasmic Reticulum

It is a network of membranes present inside the cell. Endoplasmic reticulum with ribosome on its surface is called **rough endoplasmic reticulum** (Fig. 1.12), while the endoplasmic reticulum lacking ribosome is called **smooth endoplasmic reticulum** (Fig. 1.11).

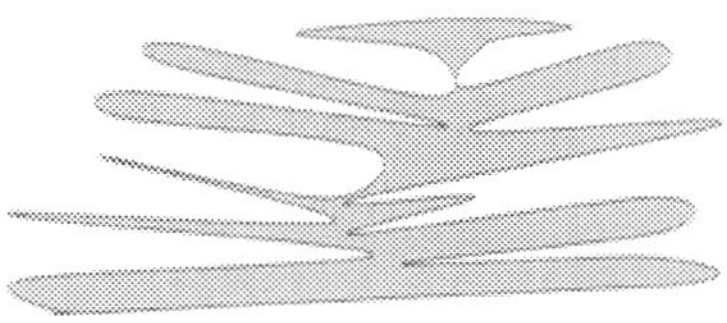

Fig. 1.11 Smooth endoplasmic reticulum.

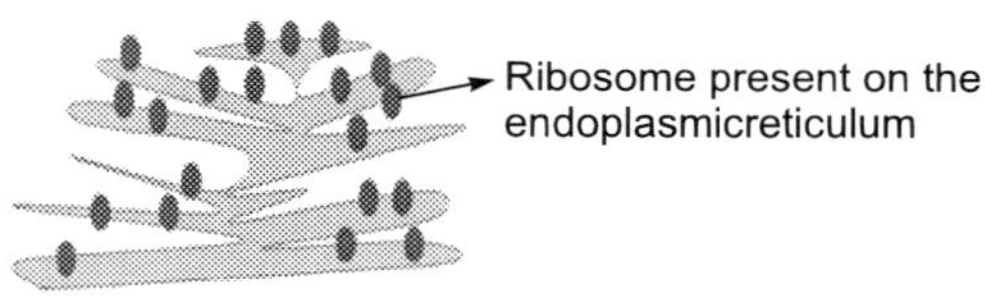

Fig. 1.12 Rough endoplasmic reticulum.

It contains various **enzymes required for the detoxification of toxic materials**. It is important in the metabolism of lipid and bilirubin. Endoplasmic reticulum of muscle has a special role, as it is **capable of concentrating calcium in the cell**. The proteins which are secreted by the cells and membrane proteins are synthesized and processed in the rough endoplasmic reticulum, and sent to golgi apparatus for further processing.

Golgi Apparatus

It is a network of membranes, mainly involved in the **addition of carbohydrate moiety (glycosylation)** to the newly-synthesized membrane proteins or secretory proteins.

Lysosomes

Lysosomes contain a variety of enzymes (hydrolases) for the **destruction of biomolecules**. It has an **acidic pH**. It is mainly involved in the **intracellular digestion** of biomolecules.

Peroxisomes

Peroxisomes are membrane-bound organelles derived from endoplasmic reticulum. It contains enzymes to **destroy the toxic substance hydrogen peroxide (H_2O_2)**.

Cytoskeleton

Cytoskeleton is the three-dimensional meshwork of protein filaments which gives shape to the cell. Three different types of cytoskeletons of the cell are microtubules actin filament (microfilament) and intermediate filaments.

Microtubules are made up of proteins called tubulins. Cilia, spindles of the cells contain microtubules. Microtubules are made up of contractile proteins so that they can contract. This helps in the ciliary movement as seen in the respiratory tract and contraction of the spindles during the mitotic division of the cell.

Microfilaments contain contractile proteins (actin and less amount of myosin) so they can contract. Phagocytosis is possible because of the pseudopodium formation, which requires microfilaments.

Intermediary filaments run between nucleus and cell membrane forming a scaffold which helps to suspend cytoplasmic materials.

Tight Junctions

Tight junctions (Fig. 1.13) are the areas where two adjacent cells are associated and their membranes joined together. In this area, there is no intercellular space. It forms an impermeable barrier to fluid and other substances. It is also called zona occludens.

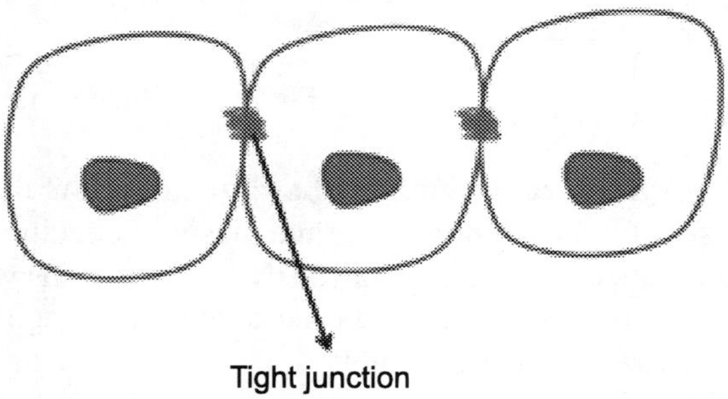

Fig. 1.13

Tight junctions are made up of various transmembrane proteins. The major types are the claudins and occludins. As the protein content of the tight junction increases, they become more impermeable to ions and water.

Tight junctions help to hold together the cells. It prevents the passage of ions and molecules through the intercellular space. Substances must enter the cell to pass through the tissue. They prevent the movement of integral membrane proteins between the apical and basolateral surface of the cell. This helps to segregate certain functions between the apical surface and the basolateral surface of the cells.

Tight junctions between the epithelial cells of the distal convoluted tubules prevent the passage of ions and molecules. The epithelia of the proximal convoluted tubules have leaky junctions.

QUESTIONS

1. Draw a labelled diagram of eukaryotic cell. List the important functions of any four subcellular sites.
2. Discuss the transport mechanisms across the membrane.
3. What is active transport? Give any two examples.
4. Discuss about the structure and functions of plasma membrane.
5. Name important subcellular organelles and mention their functions.

2

Acids, Bases and Buffers

INTRODUCTION

Electrolytes when dissolved in water, split up into two or more electrically charged particles. In 1887, Arrhenius explained this property of electrolytes, and hence, it is known as **Arrhenius theory of electrolyte dissociation**.

The charged particles, which are formed when the electrolyte is dissolved in water, are called ions, and this process is called ionization. The process of ionization is reversible. These ions may have positive or negative charge. The total number of positive charges is equal to the total number of negative charges in the solution. Since they are charged, they are responsible for carrying the electric current through the solution.

The extent of ionization is given by the ratio of the total number of dissociated molecules to the number of molecules dissolved. The degree of ionization can be increased by diluting the solution in case of weak electrolytes.

ACIDS AND BASES

Brønsted and Lowry defined acids as substances which are able to donate protons, and bases as substances which accept protons.

$$BH \longrightarrow B^- + H^+$$

In the above example, BH is an acid because it donates proton and B^- is an anion liberated by the deprotonation of the acid. B^- behaves like a base, so it is called **conjugate base**.

Acids can be classified into strong acids and weak acids.

1. **Strong acids** get dissociated almost completely, e.g., hydrochloric acid, sulphuric acid. This is because the conjugate bases of these acids are very weak (have less affinity for the proton).
2. **Weak acids** get dissociated partially, e.g., acetic acid, carbonic acid. This happens because the conjugate bases of these acids are strong (have greater affinity for proton).

Since the dissociation of the weak acids is partial, the equilibrium constant for the dissociation reaction of the weak acid (BH) can be written as follows:

$$BH \longrightarrow B^- + H^+$$

By applying the law of mass action at equilibrium

$$Ka = \frac{[B^-][H^+]}{[BH]} \quad \boxed{\text{Ka is dissociation constant}}$$

The equilibrium constants for ionization reactions are commonly called ionization or dissociation constant (Ka).

The above equation can be rearranged as:

$$[H^+] = \frac{Ka \times [BH]}{[B^-]}$$

By multiplying the above equation by –1 and taking logarithm of both sides, the following expression can be derived:

$$-[H^+] = -\left[\frac{Ka \times [BH]}{[B^-]}\right]$$

$$-\log_{10}[H^+] = -\log_{10} Ka - \log_{10}\frac{[BH]}{[B^-]}$$

$$-\log_{10}[H^+] \text{ is pH}$$

As the hydrogen ion concentration increases, the pH of the solution decreases. As the hydrogen ion concentration decreases, pH will increase. Hydrogen ion concentration and pH are reciprocally related.

$$pH = pKa + \log_{10}\frac{[B^-]}{[BH]}$$

$$pH = pKa + \log_{10}\frac{[\text{Conjugate base}]}{[\text{Acid}]}$$

This expression is called **Henderson-Hasselbalch equation**. If the conjugate base and acid concentration is the same, then **pH = pKa**. The value of pKa is lower for strong acids and higher for weak acids.

BUFFERS

Buffers are solutions that **resist changes in the pH** of a when acid or alkali is added. The buffer system has an acid component and salt component. Buffers are prepared by mixing a weak acid and its salt with a strong base or by mixing weak base and its salt with a strong acid.

Mechanism of Buffer Action

When a strong acid is added, protons are scavenged by the salt component of the buffer system. Similarly, when an alkali is added, acid component will get deprotonated and the alkali combines with the proton to form water.

Acetate buffer is made up of acetic acid and sodium acetate.

Acetic acid / Sodium acetate

CH_3COOH / CH_3COONa

(a) When acid like HCl is added to the acetate buffer system, CH_3COONa (sodium acetate) gets converted to CH_3COO^- (acetate). Acetate combines with the protons released by the strong acids to give acetic acid, which is a weak acid, and the change in pH is very small.

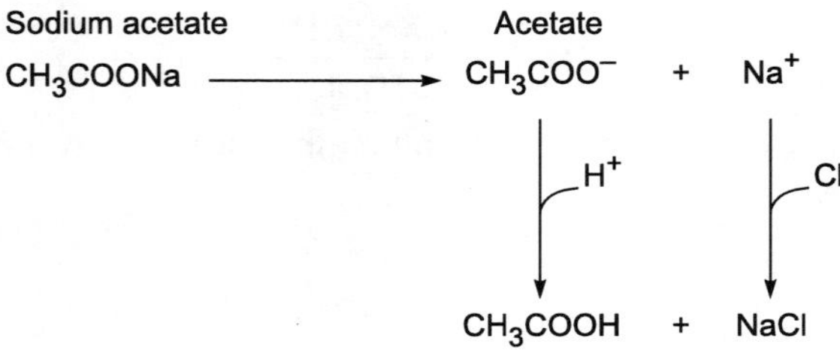

(b) When a strong alkali like NaOH is added, OH^- combines with the H^+ released by the CH_3COOH to form water. Na^+ combines with acetate to form sodium acetate.

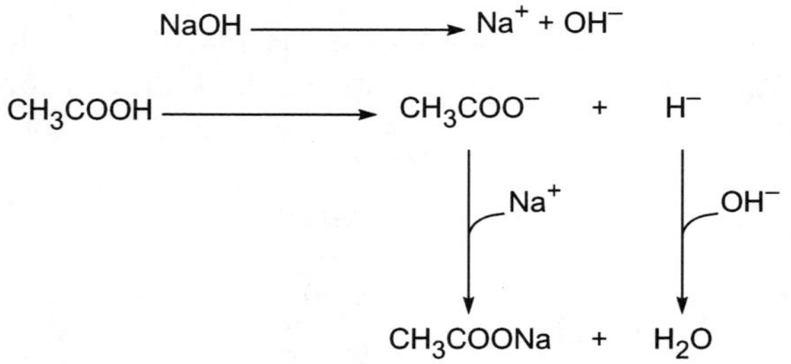

According to Henderson-Hasselbalch equation, when the concentration of the salt component and the acid component of the buffer system is equal, pH of the solution equals pKa. At this pH, buffer can respond equally to both added acid as well as the alkali.

Buffering Capacity

Capacity of the buffer to resist the change in the pH of a solution when an acid or alkali is added is called buffering capacity. It is estimated by calculating the **amount of acid or alkali required to change the pH of one litre of buffer by one unit.**

Buffering capacity depends upon the following factors:

1. **The concentration of the acid and base component of the buffer**
   As the concentration of acid and base component of the buffer increases, buffering capacity of the buffer also increases.
2. **The pH of the Buffer**
   Buffer can act best at pH = pKa, and its buffering range is about one pH unit above or below the pKa value.

If pH is close to pKa of the buffer, buffering capacity of the buffer will be more. If the pH increases or decreases from this particular pH, buffering capacity of the buffer will also decrease.

Buffer Systems Present in Human Body

Buffer systems are present in the intracellular and extracellular fluids. They help to maintain the pH of the body fluids by neutralizing the excess acids produced by the body.

Major buffer systems present in the human body are:

1. Bicarbonate buffer (HCO_3^-/H_2CO_3)
2. Phosphate buffer (HPO_4^-/H_2PO_4)
3. Protein buffer (Haemoglobin present in RBCs and plasma proteins)—The imidazole ring of the histidine residue of proteins is mainly responsible for buffering action.
4. Ammonia buffer (NH_3/NH_4^+)

QUESTIONS

1. What are acids? Give any two examples. Mention the differences between a weak acid and a strong acid.
2. What are buffers? Give two examples. Explain the mechanism of action of buffer.
3. Describe the Henderson-Hasselbalch equation and discuss its significance.
4. Write a note on buffering capacity.
5. What is ionization? Explain.

3

Chemistry of Proteins

INTRODUCTION

Proteins are abundant in the biological system. They have got diverse functions. Proteins contain carbon, hydrogen, oxygen, and nitrogen. Some proteins contain phosphorus and/or sulphur. Nitrogen is the characteristic element present in proteins. It is present to an extent of 16%.

Building blocks (monomeric subunit) of proteins are called **amino acids**. Proteins present in the biological system are made up of a ubiquitous set of 20 amino acids (standard amino acids). In proteins amino acids are linked to each other by **peptide bonds**.

Proteins differ from each other with respect to their amino acid sequence. The amino acid sequence determines the structure of the protein. The function of the protein depends upon its structure (Fig. 3.1).

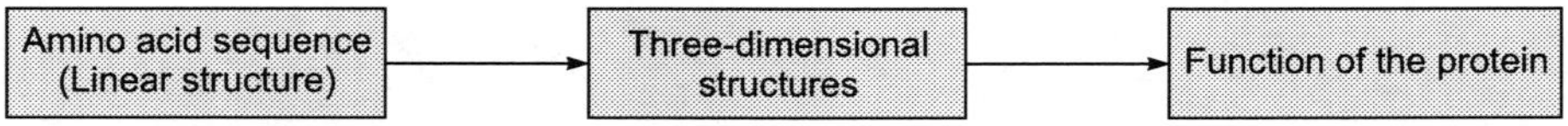

Fig. 3.1

Almost all the life processes depend on proteins. They are the end products of gene expression (Fig. 3.2). Genes are portions of DNA, which contain the information regarding the amino acid sequence of a specific protein in the form of nucleotide sequence. The processes involved in decoding this information, present in the gene, is called **gene expression**.

Fig. 3.2 Gene expression.

In nature there are about 300 different amino acids. The proteins present in biological systems contain 20 different amino acids, for which codons are present in the genetic code. These 20 amino acids are called **standard amino acids**.

FUNCTIONS OF PROTEINS

1. **Proteins having Structural Function:** Proteins are the major structural components of cell membranes, cell organelles, cytoskeletons, and also form a part of the extracellular matrix, e.g., collagen and elastins.
2. **Proteins with Catalytic (Enzyme) Function:** Most of the biocatalysts are proteins in nature e.g., glucokinase, fructokinase, etc.
3. **Proteins Acting as Transporters:** Proteins help in the transport of molecules between tissues or across the membranes. Haemoglobin transports oxygen and carbon dioxide between lungs and tissues; transferrin transports iron between the tissues; Na^+/K^+ ATPase transports Na^+ and K^+ transport glucose across the cell membrane. Glucose transporters transport glucose across the membrane.
4. **Proteins with Storage Function**: Proteins have got storage function, e.g., myoglobin stores oxygen in skeletal and cardiac muscles and ferritin stores iron.
5. **Proteins with Defence Function:** Proteins help to defend the body against the invading antigens. For example, IgG, IgM, IgD, etc.
6. **Some Hormones are Proteins:** For example, insulin, glucagon, growth hormone, etc.
7. **Some Receptors are Proteins:** Ligands bind to their respective receptors and bring about specific action, e.g., insulin binds to insulin receptor present on the cell membrane of extrahepatic tissues; steroid hormones bind to receptors present inside the cell.
8. **Proteins are Required for Locomotion (Contractile Proteins):** Proteins have made the locomotion of an organism possible, e.g., actin and myosin of the skeletal muscle.
9. **Proteins Required for Nutrition:** Proteins present in the body have nutritional importance, e.g., albumin.

AMINO ACIDS

Amino acids are small biomolecules. They are amino derivatives of weak organic acids. Amino acids contain an amino group, a carboxyl group, and a side chain group (Fig. 3.3). Amino acids differ from each other with respect to their side chain group.

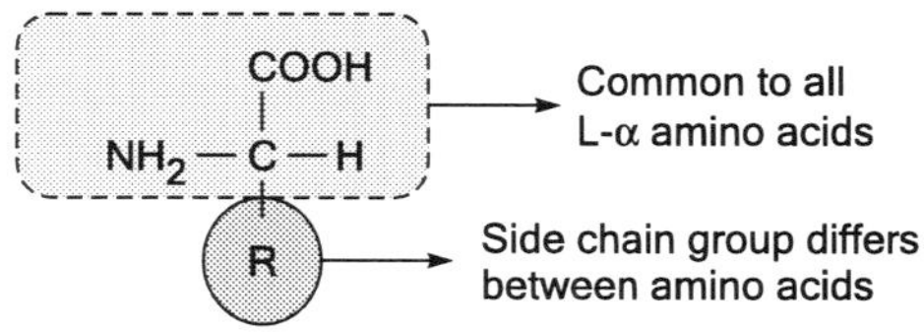

Fig. 3.3 General structure of amino acid.

The carbon atom next to the carboxyl group is called alpha carbon atom. In α-amino acids amino group is attached to α-carbon atom. If the amino group is attached to the β-carbon atom, the amino acid is called β-amino acid. All standard amino acids are α-amino acids. The 20 different standard amino acids differ from each other with respect to their side chain group (Table 3.1).

Table 3.1: Standard amino acids

Glycine	Phenylalanine	Threonine	Aspartic acid
Alanine	Tyrosine	Serine	Glutamine
Valine	Tryptophan	Methionine	Asparagine
Leucine	Histidine	Cysteine	Arginine
Isoleucine	Proline	Glutamic acid	Lysine

Classification of Amino Acids

Based on the Structure

1. **Aliphatic side chain**—Glycine, Alanine, Valine, Leucine, Isoleucine.
2. **Hydroxyl group**—Serine, Threonine.
3. **Aromatic side chain**—Phenylalanine, Tyrosine,Tryptophan.
4. **Sulphur-containing**—Cysteine, Methionine.
5. **Acidic amino acids and their amides**—Glutamic acid, Glutamine, Aspartic acid, Asparagine.
6. **Basic amino acids**—Lysine, Arginine, Histidine.
7. **Heterocyclic side chain**—Histidine, Tryptophan
8. **Imino acid**—Proline

Based on the Nutritional Requirement

1. **Essential amino acids:** Some amino acids are not synthesized in the body, so they should be supplemented in the daily diet. These are called essential amino acids. Essential amino acids are phenylalanine, valine, threonine, tryptophan, isoleucine, methionine, leucine and lysine.
2. **Semi-essential amino acids:** Some amino acids are synthesized in the body, but they are not sufficient to support the growth and development, so they should be supplied in the daily diet. These are called semi-essential amino acids. Semi-essential amino acids are histidine, arginine.
3. **Non-essential amino acids:** Amino acids which are synthesized in the human body are called non-essential amino acids.

Proteins containing sufficient amount of all the essential amino acids are called **complete proteins**. For example, albumin.

Based on Polarity

1. **Non-polar amino acids**

   They are hydrophobic in nature. For example, alanine, leucine, valine, isoleucine, methionine, phenylalanine, tryptophan, and proline.
2. **Polar amino acids**

   They are hydrophilic in nature.They are able to form hydrogen bond with water.

Based upon the presence or absence of charge on the amino acids, they can be further classified into:

(a) **Polar amino acids without charge on the side chain**
They have –OH, –SH, and $-NH_2$ group.
For example, glycine, cysteine, serine, threonine, glutamine, asparagine, and tyrosine.

(b) **Polar amino acids with charge on side chain group**

- Some have positive charge on side chain group.
  For example, lysine, arginine and histidine.
- Some have negative charge on the side chain group.
  For example, aspartic acid and glutamic acid.

Based on Metabolic Fate

1. Glucogenic amino acids
2. Ketogenic amino acids
3. Glucogenic and ketogenic amino acids

Glucogenic Amino Acids: During the metabolism of amino acids, their carbon skeleton is converted into glucose by gluconeogenic pathway, e.g., alanine, aspartate, glycine and methionine are glucogenic amino acids.

Ketogenic Amino Acids: Amino acids, whose carbon skeleton is converted into acetoacetate (ketone body) and acetyl CoA are called ketogenic amino acids e.g., leucine and lysine are ketogenic amino acids.

Glucogenic and Ketogenic Amino Acids: If carbon skeleton of amino acid is converted into glucose as well as fat, it is called glucogenic and ketogenic amino acid, e.g., isoleucine, phenylalanine, tryptophan, and tyrosine.

Functions of Amino Acids

1. Monomeric subunits of proteins and peptides.
2. Some amino acids can be converted into glucose by gluconeogenic pathway.
3. Specific amino acids can give rise to specialized biological products in the body:
   (a) Tyrosine: Hormones—adrenalin and thyroxin.
   (b) Tyrosine: Pigment—melanin.
   (c) Tryptophan: Nicotinamide Adenine Dinucleotide (NAD^+).
   (d) Glycine, arginine, and methionine are required for the synthesis of creatine.
   (e) The biologically important amine "histamine" is formed from histidine.
   (f) Glycine is required for the synthesis of "heme".
   (g) Glycine, aspartate, and glutamine are required for the synthesis of nucleotides.

Non-protein Amino Acids (Non-standard Amino Acids)

Amino acids not present in the proteins but present in the cell are called non-standard or non-protein amino acids, e.g., citrulline, ornithine, γ-amino butyric acid, β-alanine, etc.

Derived Amino Acids

These are the amino acids which may be present in the protein. They are modified form of the standard amino acids. These amino acids are formed after the protein is synthesized (translation) the process called **post-translational modification**.

For example, g-carboxy glutamate present in certain proteins involved in the blood coagulation is modified from glutamate. Hydroxylproline and hydroxylysine of collagen are modified froms proline and lysine respectively.

Properties of Amino Acids

(a) Colourless.
(b) Crystallizable.
(c) Most of them all soluble in water.
(d) Some are:

tasteless—leucine.
sweet—glycine, alanine, valine.
bitter—arginine, isoleucine.

All amino acids, except glycine, contain asymmetric carbon atom. Amino acids exist in D and L isomeric forms. D-amino acid (Fig. 3.4) has its alpha NH_2 and H oriented around its asymmetric alpha carbon in the same way as the –OH and H oriented around the asymmetric middle carbon of D-glyceraldehyde.

The corresponding L-amino acid (Fig. 3.4) bears the NH_2 and the H on its alpha carbon atom in the same way as the –OH and H on the second carbon atom of L-glyceraldehyde. All proteins of animals generally contain L-amino acids.

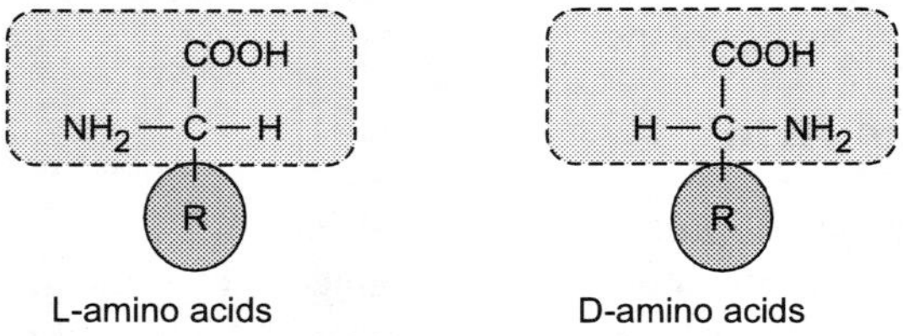

Fig. 3.4 D- and L-amino acid.

Charge Property

Depending upon the surrounding pH, amino acids can act as acids or bases. So, they are called **ampholytes**.

- At acidic pH, the amino and carboxyl groups are present in the protonated form. The net charge present on the amino acid is +ve.
- As the pH goes towards alkaline side, the proton from the carboxyl group gets dissociated and it acquires –ve charge. Now the amino acid has one +ve charge and one –ve charge. So the net charge present on the amino acid is zero. In this form, the amino acids are called **zwitter ions**.

- As the pH further moves towards alkaline side amino group gets deprotonated and the amino acid will have net –ve charge (Fig. 3.5).

The pH at which the amino acids are present in the zwitter ion form is called isoelectric pH (pI).

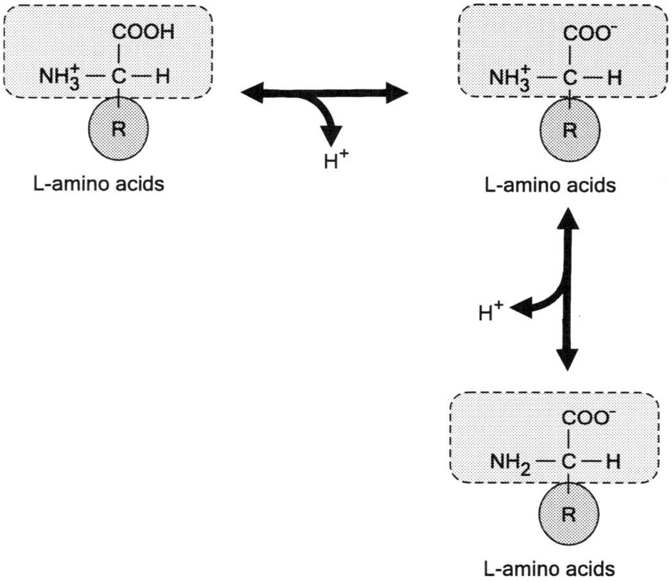

Fig. 3.5 Charge property of the amino acids.

PEPTIDE BOND

It is a covalent bond formed between the alpha amino group of an amino acid and alpha carboxyl group of an adjacent amino acid (Fig. 3.6). All the atoms involved in peptide bond are in the same plane (coplanar). Peptide bond can exist in "*trans*" configuration.

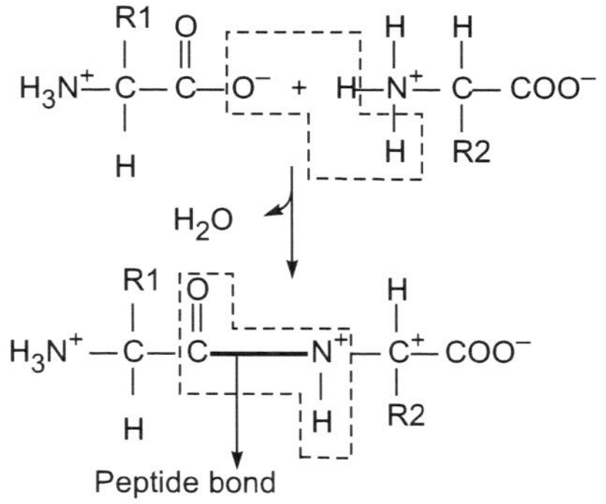

Fig. 3.6 Peptide bond.

Characteristics of Peptide Bond

1. **Partial double bond**
   The carbon to nitrogen (C–N) bond length (1.33 Å) is approximately halfway between that found for carbon to nitrogen single bond length (1.45 Å) and carbon to nitrogen double bond length (1.25 Å).
2. **Restriction for free rotation around the bond**
   Since peptide bond is a partial double bond, there is a restriction for the free rotation around the bond.
3. **Participates in hydrogen bond formation**
   Peptide bond nitrogen and carbonyl oxygen act as hydrogen donor and hydrogen acceptor respectively when hydrogen bond is formed.

The covalent bonds adjacent to the peptide bond have the liberty to rotate freely around their axis. These covalent bonds are the

(a) Carbon to carbon single bond, between the alpha carbon atom and carboxylic group (psi ψ)
(b) Carbon to nitrogen bond, between the alpha carbon atom and the amino nitrogen (phi Φ).

The rotational angles of psi ψ and phi ϕ bonds are responsible for the conformation of the polypeptide chain.

PEPTIDES

Amino acids are linked to one another by peptide bonds to form peptides, polypeptides, or proteins. If the number of amino acids present is ten or less than ten, the polymer of amino acid is called peptide. Peptides are further named as dipeptide, tripeptide, depending upon the number of amino acids present.

1. **Dipeptide:** Two amino acids linked by one peptide bond.
2. **Tripeptide:** Three amino acids linked by two peptide bonds.
3. **Tetrapeptide:** Four amino acids linked by three peptide bonds.
4. **Pentapeptides:** Contains five amino acids.
5. **Decapeptide:** Peptides containing ten amino acids.

Peptides of Biological Importance

(a) Occur in different sizes;
(b) Are required for normal physiological functions;
(c) Occur in almost all systems of the body.

1. **Glutathione**
   Glutathione is a **tripeptide**. It is made up of three amino acids (glutamate, cysteine, and glycine; Gamma-glutamylcysteinyl glycine). Glutathione is abbreviated as **G-SH** (G: Glutathione; SH: Sulfhydryl group). **Sulfhydryl group of cysteine is the functional group**. Reversible oxidation and reduction of G-SH is important for many of its biological functions. Oxidized form of Glutathione is written as G-S-S-G and reduced is G-SH

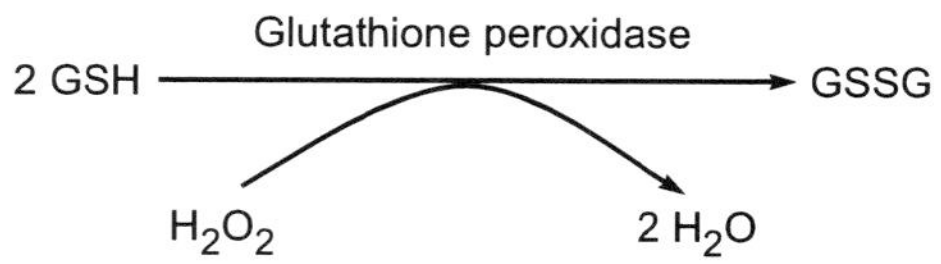

Glutathione donates its hydrogen to detoxify hydrogen peroxide as shown above.

2. **Thyrotropin releasing hormone**
   Thyrotropin releasing hormone (TRH) secreted by hypothalamus is required for the release of thyroid stimulating hormone (TSH) from anterior pituitary gland. It is a tripeptide.
3. **Angiotensin II and angiotensin I**
   They are peptides present in lungs and cardiovascular system, which bring about vasoconstriction.

CLASSIFICATION OF PROTEINS

1. Proteins are classified based upon their functions as:

(a) Structural proteins
(b) Catalytic proteins
(c) Transport proteins
(d) Storage proteins
(e) Regulatory proteins
(f) Genetic proteins
(g) Contractile proteins
(h) Nutrient proteins
(i) Defence proteins

2. Classification Based Upon the Composition

(a) ***Simple proteins:*** These are the proteins, which are made up of only amino acids. For example, serum albumin.

(b) ***Conjugated proteins:*** These are the proteins, which contain nonprotein portion (prosthetic groups) in addition to amino acids.

1. Chromoproteins, e.g., haemoglobin.
2. Glycoproteins, e.g., plasma proteins.
3. Lipoproteins, e.g., low density lipoprotein, chylomicron.
4. Phosphoproteins, e.g., casein.

(c) ***Derived proteins:*** These are the proteins, which are derived from some other proteins, e.g., collagen.

3. Classification Based on the Shape

(a) Globular proteins

The polypeptide chain is folded compactly. The hydrophilic groups of the constituent amino acids are exposed to the exterior, and the hydrophobic groups of the amino acids are inside the core of the protein, which is devoid of water. So, these proteins are soluble in the aqueous phase. Most of the plasma proteins are globular proteins. These proteins are easily digestible.

(b) Fibrous proteins

Polypeptide chains of these proteins are extended. This arrangement of the polypeptide chain gives it tensile strength and mechanical properties. They are usually components of structural proteins. Fibrous proteins are resistant to digestion.

STRUCTURE OF PROTEIN

For better understanding of the structure of protein, it has been described at several levels of complexity, which are termed different levels of structural organization.

Four different levels of structural organization have been defined. They are:

1. Primary structure
2. Secondary structure
3. Tertiary structure
4. Quaternary structure

Primary Structure

The primary structure of protein refers to the number, nature and amino acid sequence of the protein. The amino acids present in the polypeptide or protein are linked to one another by peptide bonds. Each amino acid present in the protein or polypeptide is called **residue** or **moiety**.

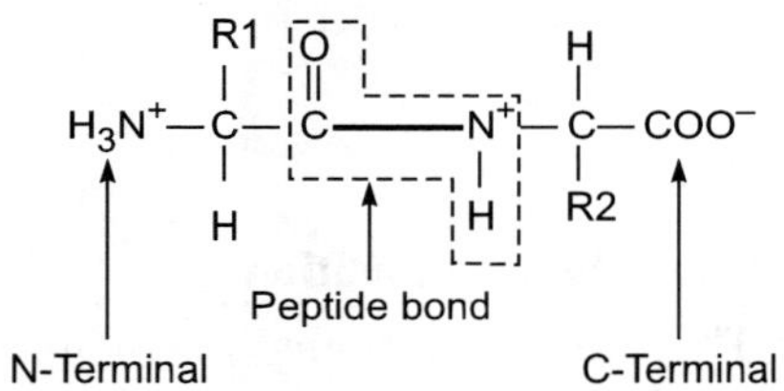

Fig. 3.7 N and C terminals.

At one end of the protein, amino group of an amino acid is free; it is not involved in the peptide bond formation. This end is called **amino terminal end** -(N-terminal) (i.e., the end bearing the amino acid residue with the free amino group). Amino terminal is always written on the left-hand side (is the first amino acid of the protein chain). End of the protein chain containing free carboxyl group is called **carboxy-terminal end** (C-terminal) (Fig. 3.7). It is written on the right-hand side.

Primary structure of the protein is mainly due to the covalent bonds formed between the adjacent amino acids. Primary structure is stabilized by peptide bond.

Primary structure is genetically determined. Remaining levels of structural organization (secondary, tertiary) of proteins are due to various non-covalent interactions is called higher levels of structural organization.

Secondary Structure

The portion of the polypeptide chain where all the phi and psi bond angles are same will exhibit regular repeating structure. Folding of the protein in regular and repeating manner is the secondary level of structural organization.

There are some common types of secondary structures that are stable and occur widely in proteins. These structures were proposed by Pauling and Corey in 1951.

1. Alpha helix
2. Beta-pleated sheets.

Alpha helix

It is a rod-like structure formed due to the tight coiling of polypeptide chain around the long axis (Fig. 3.8). The side chains of the amino acid residues of the polypeptide chain extend outwards from the central axis. These side chains are perpendicular to the helix axis. In alpha helix, the peptide bond planes are almost parallel to the helix axis.

Alpha helix is stabilized by hydrogen bonds (non-covalent bond). Hydrogen bond is formed between the carbonyl group of one peptide bond and the amino group of another peptide bond situated four amino acid residues away. The hydrogen of the amino group is shared between the peptide bond nitrogen and the carbonyl oxygen.

Except the first and the last peptide bonds of the polypeptide or proteins, all other peptide bonds are involved in the hydrogen bond formation. These hydrogen bonds are almost parallel to the helix axis. The side chains of the amino acid residues are projected outside the core of the helix. Alpha helix may be right-handed or left-handed helix.

Dimensions

1. Diameter of the helix is 0.5 nm.
2. Each turn travels a distance of 0.54 nm along the helix axis (pitch of the helix).
3. Each turn of alpha helix contains 3.6 amino acids.
4. Adjacent amino acids travel a distance of 0.15 nm along the helix axis.

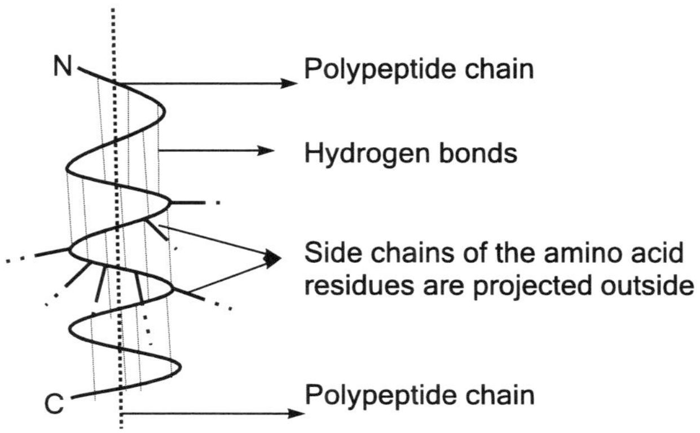

Fig. 3.8 Alpha helix.

Beta-pleated sheets

It is the most common secondary structure, where the fully extended polypeptide chains line up side by side to form a sheet-like structure (Fig. 3.9). The side chains of the amino acids are projected above and below the plane of the sheet.

Beta-pleated sheets are formed by separate polypeptide chains or a single polypeptide chain folding back on itself. About two to five adjacent strands of polypeptides together form these structures. Like alpha helix, beta-pleated sheets are stabilized by interchain hydrogen bonds.

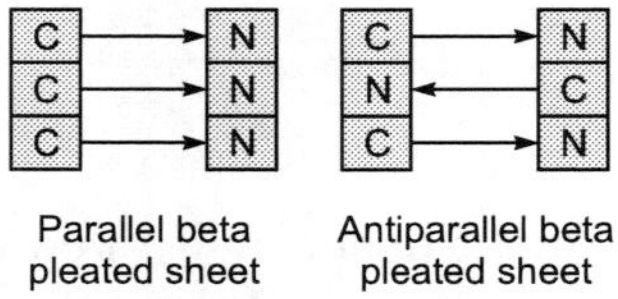

Fig. 3.9 Beta-pleated sheets.

When the adjacent polypeptide chains run in opposite directions (amino terminal of one chain corresponds to the carboxyl terminal of another chain), the structures formed are termed **antiparallel beta-pleated sheets** (Fig. 3.9). When they run in the same direction (amino terminal of one polypeptide chain corresponds to the amino terminal of the other chain), they are called **parallel beta-pleated sheets** (Fig. 3.9). Beta-pleated sheets are found in proteins present in spider web, fibre silk, claw of the reptiles, etc.

Tertiary Structure

Polypeptide chain with secondary structure that folds or twists on itself is called tertiary level of structural organization of proteins (Fig. 3.10). At this level of structural organization, the amino acid residues, which are situated far from each other in the primary structure, are brought close to one other.

Tertiary structure of the protein is due to a variety of non-covalent bonds like hydrophobic interactions between the side chains of hydrophobic amino acids; electrostatic interaction between the side chains of oppositely charged side chains of polar amino acids; hydrogen bonds; and van der Waals interaction.

The polypeptide chain folds into multiple compact semi-independent regions called **domains**. Each domain has about 150 amino acids. In the proteins having multiple domains, different domains are connected to one another by a stretch of polypeptide chain lacking regular secondary structure. The region between two domains is called cleft or **interdomain interface**. This is the place where the active sites of the enzymes usually reside. In multifunctional proteins, each domain performs different functions.

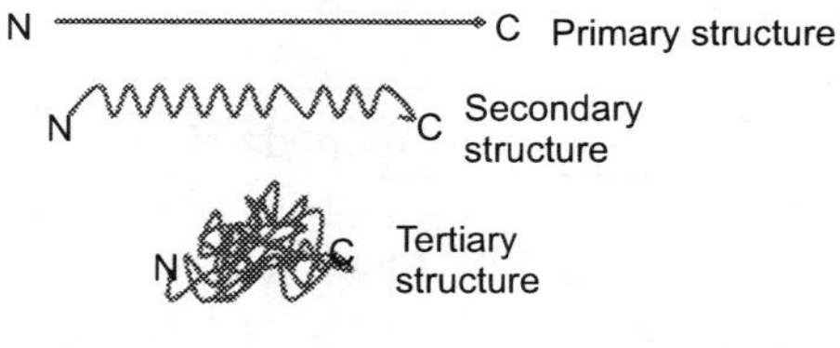

Fig. 3.10 Different levels of structural organization of proteins.

Quaternary Structure

Proteins containing more than one subunit exhibits quaternary level of structural organization. Each polypeptide chain of the proteins having the tertiary structure is called subunit or protomer (Fig. 3.11).

Fig. 3.11 Quaternary structure of protein.

The protomers interact with each other at the interface by forming non-covalent interactions like electrostatic interactions, hydrogen bonding, etc. It also involves the disulphide bond between the cysteine residues of different protomers.

For example, haemoglobin has four subunits, so it exhibits quaternary level of structural organization.

DENATURATION OF PROTEINS

Denaturation of the proteins can be described as the loss of biological as well as physical and chemical properties. Physical or chemical agents that bring about denaturation are called **denaturating agents**.

Denaturating Agents

1. ***Physical agents:*** X-rays, UV rays, violent shaking, ultrasound waves, etc.
2. ***Chemical agents:*** Acids, alkali, urea, etc.

   During denaturation, quaternary, tertiary, and secondary structure of the proteins are lost, but the primary structure remains intact. In some cases, if the denaturating agent is removed, the protein regains its original structure. This type of denaturation is called reversible denaturation.

Following are the changes in the protein following denaturation:

1. Biological activity of the protein is lost
2. Solubility of the protein is decreased
3. Increased susceptibility to hydrolysis by proteolytic enzymes
4. Loss of crystallizability.

COLOUR REACTIONS OF PROTEINS

The side chain groups of different amino acids can react with chemical reagents under the test conditions to form coloured products. These reactions are called colour reactions of proteins. With the help of colour reactions, presence of different amino acids in the protein solution can be made out.

(a) **Ninhydrin Test**
Add ninhydrin reagent to the protein solution and mix well. Heat this mixture. Upon heating it will give purple colour (Ruhemann's purple). Amino acids give yellow colour. This indicates the presence of α-amino acid.

(b) **Xanthoproteic Test**
To one mL of protein solution, add a few drops of concentrated nitric acid. Mix well and boil it for a few minutes. Yellow colour will appear. Divide the content in the test tube into two equal halfs. Add 40% NaOH to one half. Colour will change to orange. Aromatic side chain of the amino acid undergoes nitration. The nitrated product has yellow colour in the acidic medium and orange colour in the alkaline medium. This indicates the presence of tyrosine and tryptophan.

(c) **Millions Test (Cole's Test)**
To one mL of protein, add one mL of mercuric sulphate. Boil for 30 seconds and add 1% sodium nitrite. If proteins contain tyrosine, the hydroxyphenyl group reacts with mercuric sulphate and sodium nitrate to give red colour or red coloured precipitate. It is due to the formation of nitrated mercuric phenolate ion. This indicates the presence of tyrosine.

(d) **Aldehyde test**
To one mL of protein solution, add a drop of formalin and mercuric sulphate. Mix well. Now hold the test tube in a slanting position and add concentrated sulphuric acid slowly along the side of the test tube. A purple-coloured ring appears at the junction of the two solutions, indicating the presence of tryptophan in the protein. A violet-coloured complex is formed when the oxidized product of indole nucleus of tryptophan reacts with the aldehyde.

(e) **Sakaguchi's Test**
To two mL of protein solution, add 2 drops of 40% NaOH, 4 drops of a naphthol, and a few drops of bromine water. Red colour indicates the presence of gaunido group of arginine.
Gaunido group of arginine reacts with the alkaline hypobromite to give a bright red-coloured complex.

(f) **Pauley's Test**
Take equal volume (0.5 mL) of sodium nitrite and sulphanilic acid in a test tube 24 (diazotized sulphanilic acid). After keeping it for one minute, add one mL of protein solution. Mix well. Add one mL of 10% sodium carbonate to make the solution alkaline. Red colour indicates the presence of histidine and tyrosine.
Diazotized sulphonilic acid reacts with the imidazole ring of histidine, or hydroxy phenyl ring of tyrosine to form a red-coloured complex under alkaline condition.

(g) **Sulphur Test**
Mix equal volumes of protein solution and 40% sodium hydroxide. Boil it for a minute and then cool. Add one drop of lead acetate. Black or brown colour indicates the presence of cysteine and cystine.
When protein is boiled with a strong alkali, sulphur from the side chains of cysteine and cystine is released. It reacts with the lead acetate to form lead sulphide, which is black or brown in colour.

(h) Molisch Test (***See under Chemistry of Carbohydrates pp. 59***)

If Molisch test is answered by the given protein solution, it indicates that the protein is a glycoprotein.

(i) Organic Phosphorus Test

Take 5 mL of caseine (milk protein). Add 0.5 mL of 40% NaOH and boil it for few minutes. Cool under tap water and add concentrated nitric acid. Mix well and add a pinch of solid ammonium molybdate. If protein contains phosphate group, it gives a yellow colour (canary yellow).

When protein is boiled with the strong alkali, organic phosphate is converted to inorganic phosphate. It reacts with the ammonium molybdate to form canary yellow-coloured ammonium phosphomolybdate.

ELECTROPHORESIS

Electrophoresis is a separative procedure first used by Tiselius in 1937 which was called moving boundary electrophoresis. By this technique charged particles get separated depending upon the net charge present on the particle, molecular mass and shape of the particle.

Positively charged particles are called cations and negatively charged particles are called anions. Cations move towards cathode and anions move towards anode.

Electrophoresis technique can be used to separate amino acids, proteins, nucleic acids, etc. Charge present on the molecule depends on the pH of the surrounding environment. When pH of the medium is above the isoelectric pH, net charge present on the molecule is –ve. If pH of the medium is decreased below the isoelectric pH, then the net charge present on the molecule is + ve.

Types of electrophoresis

1. Horizontal electrophoresis
2. Vertical electrophoresis

Apparatus and Reagent

1. **Electrophoresis Tank**

   It is fitted with electrodes and holds electrophoresis buffer.

2. **Electrophoresis Buffer**

   Electrophoresis buffer is required to conduct electricity. The pH and the ionic strength of the buffer affect the separation of the molecule.

3. **Support Media**

   The support media used in horizontal electrophoresis are of different type, e.g., filter paper, agarose gel, agar gel, starch gel, etc. In case of vertical electrophoresis the support media mainly used is polyacrylamide. The pore size of the support media is determined by the concentration of supporting media. When the concentration is more pore size is less and vice versa.

In clinical biochemistry laboratory serum protein electrophoresis is done frequently. The buffer used is barbitone buffer of pH 8.6 and the support media is agarose gel. At pH 8.6 serum proteins are negatively charged and will move towards anode and proteins get separated depending upon the charge and molecular mass. After the separation of the proteins it is fixed on the support media and later stained to visualize different bands.

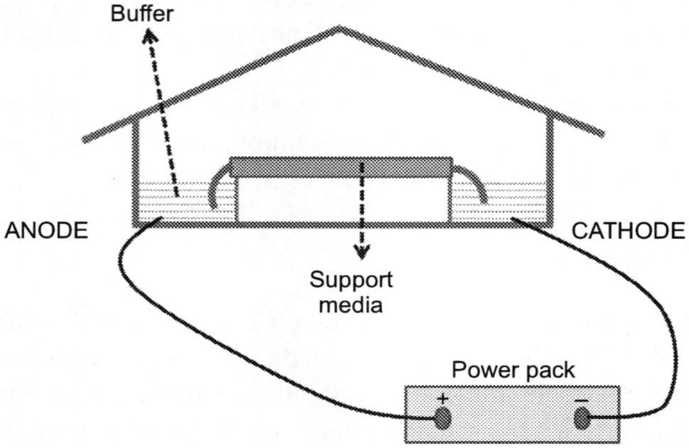

IMMUNOELECTROPHORESIS

Electrophoresis and reaction between antigen and antibody is used to detect antigen.

CAPILLARY ELECTROPHORESIS

Capillary tubes with 100 to 200 cm in length and bore size of about 50 to 100 microns are used to separate molecules. Separation can be done within a few minutes.

CHROMATOGRAPHY

It is a separative procedure used to separate most of the biological substances including proteins, lipids, nucleic acids and carbohydrates.

In chromatography separation takes place based on the ability of the substance to get distributed between the stationary phase and mobile phase. Stationary phase may be solid or liquid. Mobile phase may be liquid or gas.

Based on the mechanism used to achieve the separation of the molecules, there are different types of chromatography.

1. **Partition chromatography**

   The separation of the molecules based on their ability to partition between mobile and stationary phase. It is most commonly used to separate amino acids or peptides. It may be partitioning between liquid-liquid or gas-liquid.

   Paper chromatography is an example of partition chromatography. It is a liquid-liquid partition chromatography. Stationary phase is liquid (water) and the mobile phase is liquid (contains organic solvent).

When this mixture of mobile and stationary phase is allowed to ascend or descend over the filter paper, water forms hydrogen bonding with the cellulose of the filter paper and its movement is retarded. So water becomes the stationary phase. Organic solvent acts as a mobile phase. Based on the relative solubility of the molecules in the water and organic solvent they get separated on the filter paper.

2. **Adsorption chromatography**
   Separation takes place based on the adsorption on the surface of the solid stationary phase. Mobile phase is liquid or gas.
3. **Ion exchange chromatography**
   Separation depends upon the electrostatic attraction between the oppositely charged groups present on the resin inside the column and the molecule to be separated.
4. **Gel filtration**
   Porous gel is packed in the column. Molecules which are smaller in size will enter the gel through the pores whereas a larger molecule remains outside and move fast in the column.
5. **Affinity chromatography**
   Separation based on the high affinity of specific proteins for specific chemical groups, e.g., affinity between the coenzyme and enzyme can be used to separate enzymes, affinity between biotin and avidin, etc.

QUESTIONS

Long Essay

1. What are proteins? Discuss in detail about different levels of structural organization of proteins.
2. What are amino acids? Write the classification of amino acids based upon their solubility, nutritional requirement, and metabolic fate.
3. Discuss about the acid base property of amino acids, and add a note on isoelectric pH.

Short Essay

1. Write the classification of proteins based upon their functions.
2. Define isoelectric pH.
3. What do you mean by denaturation of proteins? Give any four examples of denaturating agents. List the changes that take place in the properties of protein after denaturation.
4. Discuss the functions of plasma proteins.
5. Describe the secondary structure of protein.
6. Write a short note on the primary structure of protein.
7. Discuss the primary structure of protein.
8. Write a note on chromatography.
9. Write a note on electrophoresis.

4

Tissue Proteins

INTRODUCTION

Plasma is a clear, pale yellow supernatant fluid, which emerges out when blood mixed with anticoagulant is centrifuged. If blood is allowed to clot, the supernatant which emerges out is called serum.

Plasma is the fraction of blood without cells, whereas serum is blood without cells, fibrinogen, and other components required for the clotting of blood. Both plasma and serum contain proteins in addition to various organic and inorganic constituents.

Serum total protein content is 6 to 8 g/dL. Total protein is made up of albumin and globulins. Plasma contains fibrinogen in addition to albumin and globulin.

SEPARATION OF SERUM PROTEINS

Precipitation by Sodium Sulphite

Total protein is estimated in the serum sample. Portion of the serum sample is treated with sodium sulphite. Sodium sulphite precipitates globulin. The estimation of protein in the supernatant gives the amount of albumin.

Serum total protein – Serum albumin = Serum globulin.

Electrophoresis

Serum protein can be separated by a technique called electrophoresis. It is a technique where biomolecules get separated depending upon the net charge present on the molecules and their density.

Serum proteins have a net negative charge under the conditions of electrophoresis. They move towards the anode. The proteins having highest charge and smallest size move fast in the electrophoretogram (Fig. 7). Since serum contains different proteins with different sizes and charge intensities; they form different bands of proteins.

The protein that moves very fast and furthest is albumin. Following the albumin, group of proteins having little less charge and large size forms next faint band called alpha-1 globulins.

Alpha-1 gloublin band is followed by alpha-2 globulin band, beta band and at the point of the application of the sample, it is the gamma band containing gamma globulins.

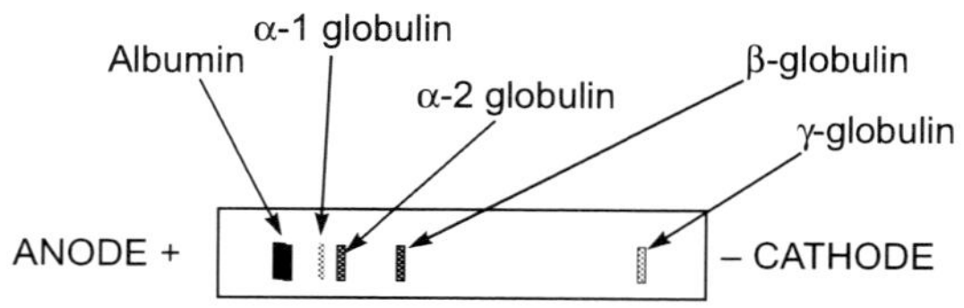

FUNCTIONS OF PLASMA PROTEINS

1. **Maintains the colloidal osmotic pressure of the blood**
   Albumin is mainly responsible for 75 to 80% of the colloidal osmotic pressure of the blood. If the plasma albumin level is reduced due to any disease, colloidal osmotic pressure of the blood will decrease and water will escape to the interstitial space. Accumulation of water in the interstitial space is called **oedema**.
2. **Transporter**
   Albumin as well as other proteins present in the plasma help in the transport of various substances in the circulation. Hydrophobic substances require a carrier molecule, which can make them more water soluble and facilitate their transport. Plasma proteins mainly help to transport these hydrophobic substances.
3. **Nutritive function**
   When energy is needed to be derived from the proteins, plasma proteins are catabolized in addition to muscle proteins.
4. **Buffering action**
   The histidine residue of plasma proteins has pKa value close to the pH of the blood. Hence, plasma proteins are good buffers.

COLLAGEN

Collagen is a fibrous protein present in the extracellular matrix, bone, cartilage, tendons, teeth and blood vessels. It is the most abundant protein present in the human body.

Synthesis and structure of collagen

Collagen synthesis takes place in fibroblast, osteoblast and chondroblasts. Polypeptide chains of collagen are called α-chains. The hydroxylation of the proline and lysine to form hydroxyproline and hydroxylysine takes place after the synthesis of polypeptide chain. Some of the hydroxylysine residues are glycosylated with glucose or glucosyl galactose.

Three α-chains wind around each other to form triple helix called procollagen. Procollagen is secreted from these cells to extracellular matrix. Procollagen has got amino terminal and carboxy terminal extension peptides. These extension peptides are cleaved by procollagen peptidase to form tropocollagen.

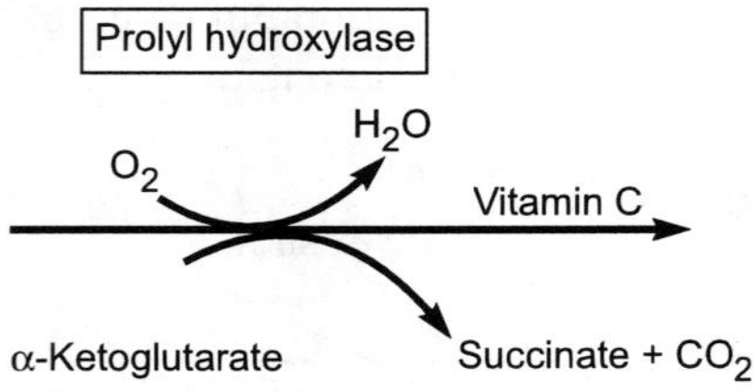

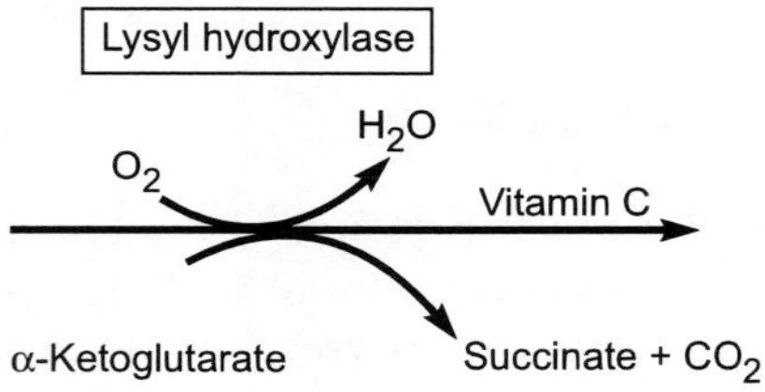

Tropocollagen has got 3 polypeptide chains. Individual polypeptide chains are called α-chains. Each α-chain contains about 1000 amino acids. In these polypeptide chains every third residue is glycine. The sequence can be written as

-Gly-X-Y-Gly-X-Y-Gly-X-Y-

X and Y are other amino acids. Most of the X position is occupied by proline and Y position is occupied by hydroxyproline. Some of the Y positions are occupied by lysine and hydroxylysine.

Tropocollagen is 300 nm long and has 1.4 nm diameter. Tropocollagens are later assembled in a staggered pattern where each tropocollagen overlapping its neighbouring tropocollagen by a length approximately three quarters of its length, forms collagen fibrils.

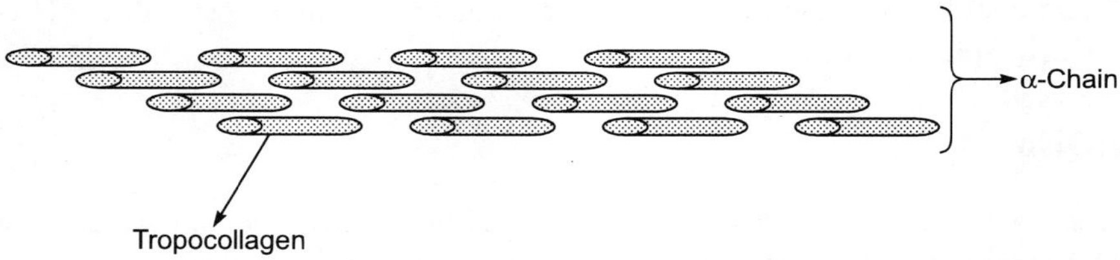

In collagen fibrils some of the lysyl and hydroxylysyl residues are oxidized by lysyl oxidase to form aldehydes. These reactive aldehydes form covalent cross-links with neighbouring lysine or hydroxylysine residues.

Vitamin C is required for hydroxylation of proline and lysine. These hydroxylated amino acids are required for the cross link formation. In deficiency of vitamin C, there is defect in the hydroxylation reaction, so the tensile strength of the collagen is reduced. It gives rise to disease called scurvy. In scurvy fragility of capillary blood vessels due to defective collagen formation is responsible for bleeding gum and bruises on the limbs as a result of subcutaneous extravasation of blood.

ELASTIN

Elastin is the connective tissue protein and has high tensile strength. They are found in the lungs, walls of the large arteries and elastic ligaments. It is synthesized from a precursor called tropoelastin. Elastin is rich in proline and lysine but contains only a little hydroxyproline. Elastin does not contain hydroxylysine.

The side chain of the lysine oxidized by lysyl oxidase enzyme to form allysyl residue. Allysyl side chain and lysine forms covalent crosslink called **desmosine**. This gives elastin tensile strength.

KERATINS

Keratins are proteins present in skin, nail and hair, etc. In keratins the polypeptide chains are rich in cysteine so they are held together by disulphide bonds.

MYOSIN

Muscle cell contains various proteins required for muscle contraction. Myosin is one of them. It constitutes about 55% of the muscle protein.

Myosin has six polypeptide chains. Two are called heavy chains and four are called light chains. The heavy chains are coiled around each other. Each heavy chain has globular head portion attached to its amino terminal.

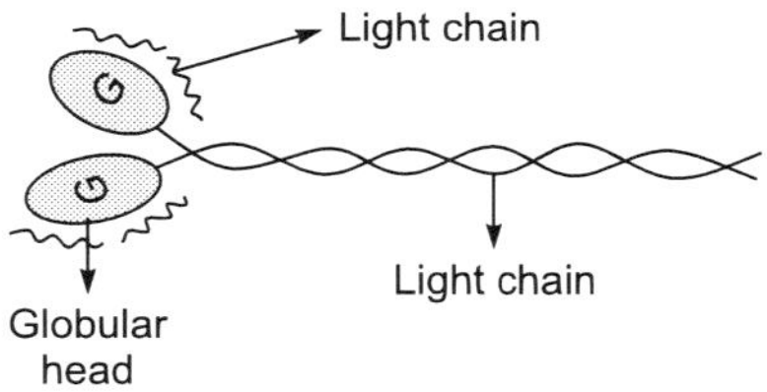

Two light chains each are attached to the globular head portions. Light chains are of two types. One is called the essential light chain and the other is called regulatory light chain.

LENS PROTEIN

Protein present in the lens is crystalline. There is no turnover of crystalline. In diabetes mellitus these proteins are glycated. Glycation of the crystalline makes it more susceptible for oxidation and aggregation which makes it opalescence and cataract.

QUESTIONS

1. Write the electrophoritogram of serum proteins and label the bands.
2. Functions of plasma proteins.
3. Elastin.
4. Myosin.

5

Enzymes

INTRODUCTION

Enzymes are **biocatalysts**. They are proteins in nature and are thermolabile. They increase the rate of reactions by thousand to millionfold without undergoing any physical or chemical change. Some of the enzymes are ribonucleic acids. Ribonucleic acids (RNAs) having the catalytic activity are called **ribozymes,** e.g., small nuclear RNA.

To date, plenty of enzymes have been identified and separated from the biological system, and the reactions catalyzed by them have been demonstrated. Enzymes have got both diagnostic and therapeutic value. The branch of biochemistry dealing with the study of enzymes is known as **enzymology**.

SUBSTRATE

Substrate is the substance on which the enzyme acts and modifies it into a product. During an enzyme catalyzed reaction, enzyme (E) binds to the substrate (S) and forms enzyme-substrate complex (ES). Enzyme modifies the substrate to form product (P). The product is still bound to the enzyme (EP). Later, product and enzyme are released in their free form.

$$E + S \longrightarrow ES \longrightarrow E + P$$

ACTIVE SITE

The substrate binds to a specific site on the enzyme, which is called active site of the protein. Substrates that have three-dimensional shapes complementary to the shape of the active site are able to bind to it. The three-dimensional shape of the active site is determined by the tertiary structure of the protein.

The part of the active site to which the substrate binds is called substrate **binding site**. On binding of the substrate to the active site, side chain groups of the amino acid residues of the active site are involved in making or breaking the covalent bonds in the substrate, to convert it into products. This part of the active site is called **catalytic site**.

ENZYME-SUBSTRATE COMPLEX

The enzyme and substrate interact at the active site, leading to the formation of enzyme-substrate complex. Enzyme and substrate are held together mainly by non-covalent interactions like hydrogen bond, electrostatic interaction, van der Waals forces, etc.

Different models have been proposed to explain the enzyme-substrate complex formation. Two of them are discussed in this section.

Fisher's Lock and Key Model

In this model, enzyme and substrate are compared to lock and key. The substrate fits into the active site of the enzyme in the same way as the key fits into the lock.

According to lock and key model, each enzyme has a preformed active site for the substrate. These enzymes can bind to substrates, which have complementary shape as that of an active site (Fig. 5.1). This model explains active site as a preformed rigid structure. It explains the absolute specificity but not the broad specificity.

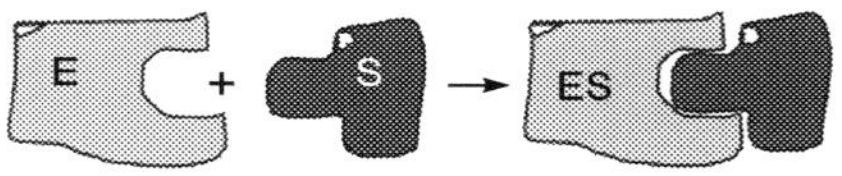

Fig. 5.1 Lock and key model.

Since the lock and key model is not able to explain the allosteric mechanism, it is not accepted.

Koshland's Induced Fit Model

According to the Koshland's induced fit theory, the active site is not preformed. As the substrate approaches the enzyme and superficially interacts with it, the three-dimensional shape of the enzyme is altered (Fig. 5.2).

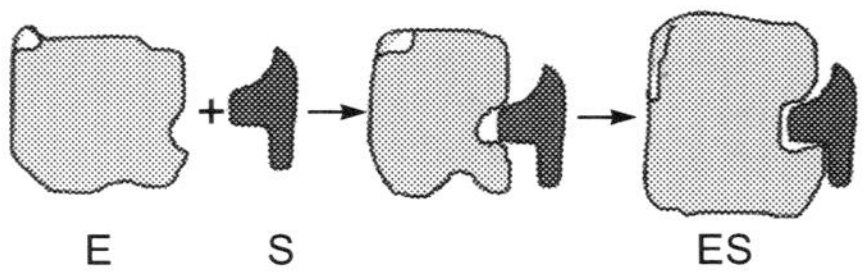

Fig. 5.2 Induced fit model.

When the shape of the active site is altered, it becomes complementary to the substrate or substrates for that enzyme. Then the substrate interacts with the enzyme, leading to the formation of enzyme-substrate complex.

According to the induced fit model, active site is flexible. This theory can also explain the allosteric mechanism of enzyme action.

SPECIFICITY

Enzymes exhibit specificity. They can act on one particular substrate, or a group of substrates, or one type of stereoisomeric form of the substrate. This depends upon the region of the substrate that forms the complementary fit with the active site.

Absolute specificity

An enzyme is absolutely specific for a given substrate. It cannot act on any substance having structural resemblance to the substrate. For example, urease acts only on urea. The test done to detect urea is called specific urease test, because urease enzyme has absolute specificity for urea.

Group specificity

Enzymes involved in the digestion of proteins can act only on the peptide bonds, where amino acids having a specific group are present in their side chains. For example, chymotrypsin and pepsin can act on peptide bonds, where the aromatic amino acids are involved in the peptide bond formation.

Bond specificity

Enzymes can act only on one type of bond. For example, glycosidic bond is acted upon by glycosidase; peptide bond is acted upon by peptidases.

Stereospecificity

Enzymes are specific only to one type of the stereoisomeric form of the substrate, and they can form only one stereoisomeric form of the product. For example, enzymes involved in the metabolism of glucose can act only on the D glucose, not on L glucose. The enzyme succinate dehydrogenase can form fumarate; it cannot form the geometric form "maleate" (malic acid).

Reaction specificity

A given substance can be a substrate for enzymes catalyzing different types of reactions. For each type of reaction, there are specific enzymes.

COENZYMES

All enzymes are proteins. The enzymes which are complex proteins contain prosthetic groups in addition to the protein portion. The organic molecule which is thermostable and absolutely necessary for the action of enzyme is called coenzyme. If it is an inorganic metal ion, it is called activator.

The protein portion of the enzyme devoid of coenzyme is inactive, and it is called apoenzyme. Apoenzyme combines with the coenzyme to form holoenzyme, which is the active form of the enzyme.

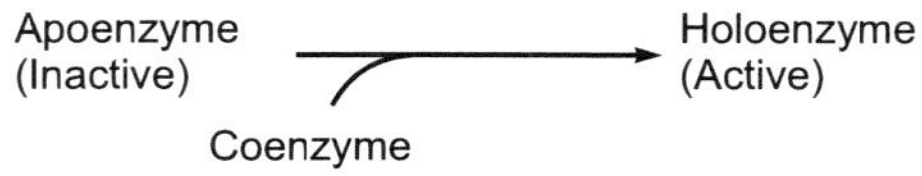

Most of the coenzymes are derivatives of water-soluble vitamins. Water-soluble vitamins are converted in our body to form coenzymes (Table 5.1).

Table 5.1: Vitamins and their coenzyme forms

Vitamins	*Coenzyme forms*
Thiamine	Thiamine pyrophosphate (TPP)
Niacin	Nicotinamide-adenine dinucleotide (NAD^+)
	Nicotinamide-adenine dinucleotide phosphate ($NADP^+$)
Riboflavin	Flavin mononucleotide (FMN)
	Flavin adenine dinucleotide (FAD)
Biotin	Biotin
Pyridoxine	Pyridoxal phosphate (PLP)
Pantothenic acid	CoA SH
Vitamin B12	Methyl cobalamin and adenosyl cobalamin

Coenzymes act as a second substrate in enzyme catalyzed reactions. The change, which takes place in the substrate by the action of enzymes, is exactly counterbalanced by the coenzyme. For example, in case of oxidation-reduction reaction catalyzed by lactate dehydrogenase, lactate is oxidized to pyruvate. The hydrogen removed from the lactate during oxidation is accepted by the coenzyme NAD^+ of the enzyme. When the substrate is oxidized, the coenzyme is reduced and it exactly counterbalances the change that takes place in the substrate.

Enzymes involved in oxidation-reduction, transfer of groups, require coenzymes. The enzymes belonging to the group "hydrolases" do not require coenzymes.

CLASSIFICATION OF ENZYMES

There are thousands of different enzymes. In an effort to systematically identify a specific enzyme, the International Union of Biochemists and Molecular Biologists (IUBMB) has classified enzymes into six different groups, depending upon the type of reaction they catalyze.

Class 1: Oxidoreductase

Oxidation-reduction reaction is catalyzed by this group of enzymes.

$$SH_2 + A \longleftrightarrow S + AH_2$$

Example : Lactate dehydrogenase

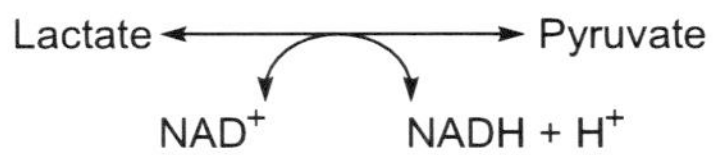

Lactate is oxidized to pyruvate; simultaneously, the oxidized form of the coenzyme is reduced. Other examples for oxidoreductases are glucose-6-phosphate dehydrogenase, xanthine oxidase, phenylalanine hydroxylase, alcohol dehydrogenase, etc.

Class 2: Transferase

Tranferases are a group of enzymes, which are involved in the transfer of groups between substrates (group transferred is other than hydrogen).

$$S\text{-}G + A \longrightarrow S + A\text{-}G$$

Example: Hexokinase

Glucose + ATP ⟶ Glucose-6-phosphate + ADP

Hexokinase catalyzes a reaction in which a high energy phosphate group is transferred from ATP to glucose, leading to the formation of glucose-6-phosphate.

Transaminases are a group of enzymes, which are involved in the transfer of amino group of an amino acid to α-ketoacid.

Serine hydroxymethyl transferase, glucose-1-phosphate uridyl transferase, etc., are other examples of transferases.

Class 3: Hydrolases

Hydrolases are involved in the hydrolytic cleavage of covalent bonds like peptide bond, glycosidic bond, ester linkage, etc.

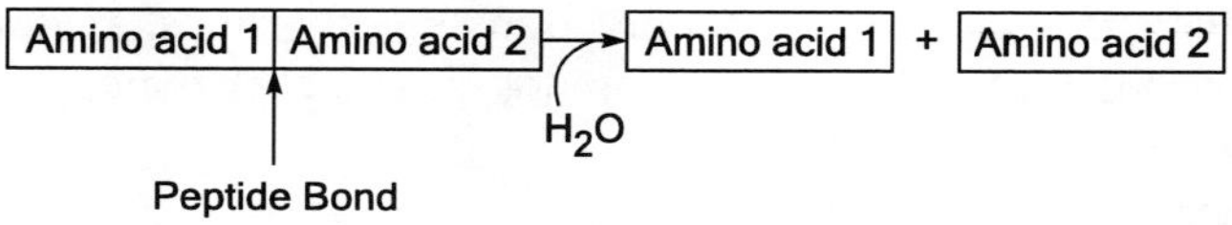

Most of the enzymes involved in the digestion of carbohydrates, proteins and lipids are hydrolases, for example, amylase, sucrase, maltase, trypsin, chymotrypsin, lipase, etc.

Class 4: Lyases

Lyases are a group of enzymes, which are involved in breaking the covalent bond by mechanisms other than hydrolysis, and can form new covalent bond without using energy.

For example,

1. HMG CoA lyase is the enzyme of ketogenesis.

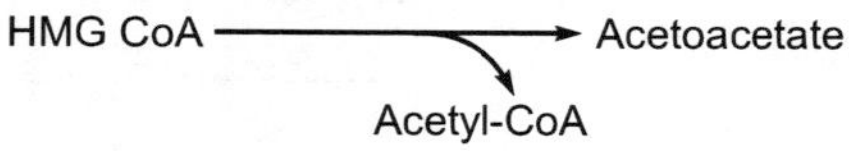

2. Aldolase is the enzyme of glycolytic pathway.

$$\text{Fructose 1,6 bisphosphate} \longrightarrow \text{Glyceraldehyde-3-phosphate} + \text{Dihydroxy acetonephosphate}$$

Other examples, citrate synthase, aromatic amino acid decarboxylase, fumarase, etc. are examples of lyases.

Class 5: Isomerase

Isomerases are a heterogeneous group of enzymes. They catalyze isomerization of different types.

$$\underset{\text{Substrate}}{S} \longrightarrow \underset{\text{Isomeric form of the substrate}}{S'}$$

For example,

1. Phosphohexose isomerase.

$$\text{Glucose-6-phosphate} \longleftrightarrow \text{Fructose-6-phosphate}$$

2. Phosphotriose isomerase.

$$\text{Glyceraldehyde-3-phosphate} \longleftrightarrow \text{Fructose-6-phosphate}$$

Epimerase, racemase, isomerase, and mutase come under the class isomerase.

Class 6: Ligases

Ligases are involved in joining two molecules to form a new compound by utilizing energy in the form of ATP.

Enzyme Commission Number (EC): Each enzyme is given a four-digit code number. This number is called enzyme commission number (EC). First digit of the number indicates the class to which the enzyme belongs. Second digit indicates the subclass, and the third digit indicates the sub-subclass. Fourth digit of EC is specific for a given enzyme.

Rate of the Enzyme Catalyzed Reaction: Enzymes do not alter the equilibrium constant of the reaction, but help to reach the equilibrium fast.

For example,

$$S \overset{E}{\longleftrightarrow} P$$

The rate at which the substrate is converted to product is called velocity of the reaction. Enzymes increase the rate of spontaneous reactions. The action of the enzyme is affected by various factors.

FACTORS AFFECTING THE ENZYME CATALYZED REACTION

1. Enzyme concentration
2. Substrate concentration
3. Temperature
4. pH
5. Product concentration
6. Activators of enzyme
7. Inhibitors of enzyme

Enzyme Concentration

As the concentration of the enzyme (E) increases, the velocity (V) of the reaction also increases, provided the substrate availability should be unlimited (Fig. 5.3).

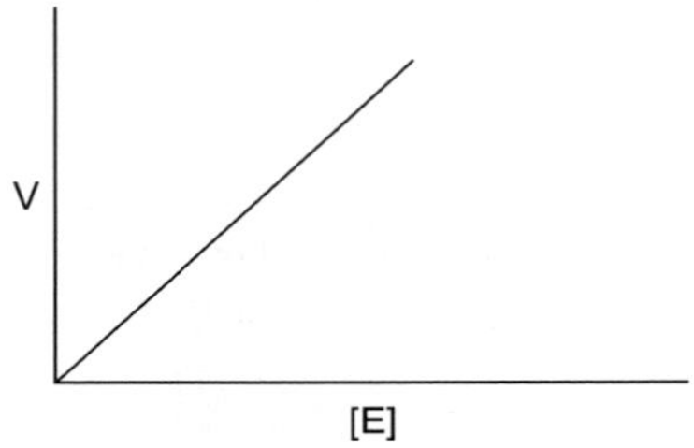

Fig. 5.3 Enzyme concentration.

Substrate Concentration

If the concentration of the substrate is gradually increased by keeping all other factors constant, velocity of the reaction is directly proportional to the concentration of the substrate in the initial phase. This phase of the enzyme catalyzed reaction is known as "**first order reaction**".

As the concentration of the substrate further increases, velocity also increases, but it is not directly proportional to the concentration of the substrate (**second order reaction**).

When the substrate concentration reaches maximum, the velocity of the reaction also reaches maximum. This velocity is called **maximum velocity (V_{max})**. This phase of the enzyme catalyzed reaction is known as "**zero order reaction**" (Fig. 5.4).

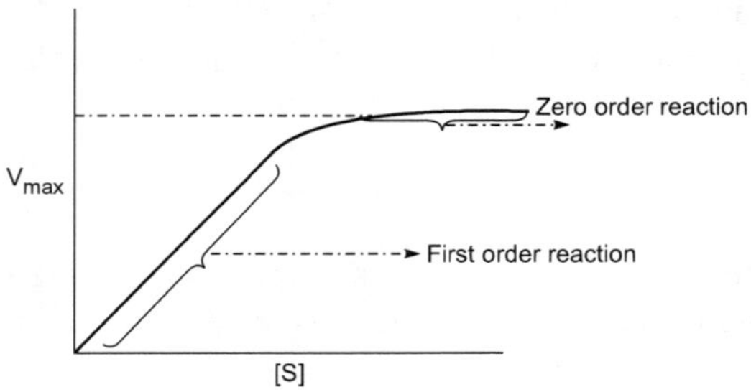

Fig. 5.4 Substrate concentration.

The relationship between the substrate concentration and the velocity of reaction has been derived by Michaelis and Menten. The following equation is called Michaelis-Menten equation:

$$V_0 = \frac{V_{max}[S]}{[S] + Km}$$

V_0 = initial velocity; [S] is the substrate concentration; Km = Michaelis-Menten constant.

When the relationship between the substrate concentration and the velocity is plotted on a graph, the plot is hyperbolic as shown in Fig. 5.4.

Km is the concentration of substrate required to reach half V_{max} (half of the maximum velocity). Km indicates the affinity of the enzyme for its substrate. If Km value is more for a given substrate, it indicates that the enzyme has less affinity for that substrate. Km value is less when the enzyme has maximum affinity for its substrate.

If an enzyme has many substrates, Km value helps to identify the natural substrate for the enzyme. The natural substrate is the one for which enzyme has got least Km value.

Temperature

Change in the temperature of an enzyme catalyzed reaction can affect the reaction rate. Initially, as the temperature increases, the reaction rate also increases, and at one particular temperature, it reaches maximum. This temperature is called **optimum temperature** (Fig. 5.5).

If the temperature is further increased, the rate of reaction starts decreasing slowly. This is due to the gradual denaturation of enzyme proteins, as the temperature goes above the optimum temperature.

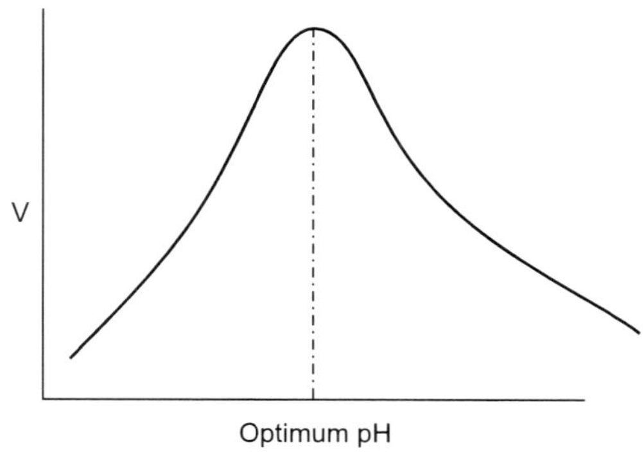

Fig. 5.5 Temperature.

pH

The velocity of enzyme catalyzed reaction is affected by the change in pH of the reaction mixture. As the pH increases, the velocity of enzyme catalyzed reaction increases and reaches a maximum at certain pH. The pH at which the enzyme catalyzed reaction has a maximum velocity is called optimum pH (Fig. 5.6). If the pH is further increased, the velocity of the reaction starts decreasing slowly due to the change in the charge property of the side chains of certain amino acids residing in the active site of the enzyme.

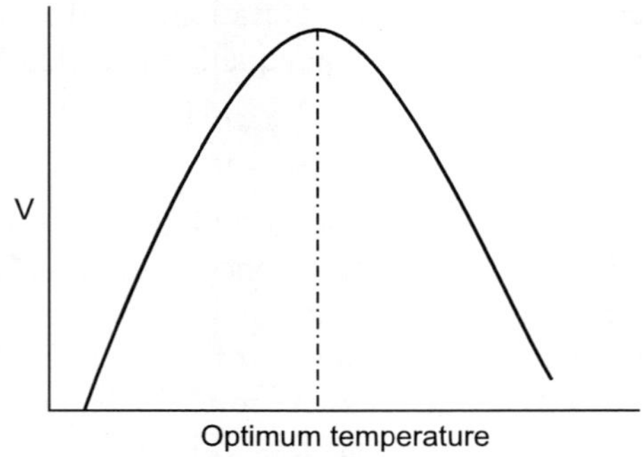

Fig. 5.6 pH.

Product Concentration

If the product of the enzymatic reaction gets accumulated, it can decrease the rate of the enzymatic reaction. It can even reverse the reaction, leading to the formation of original substrate.

Activators of Enzyme

Enzyme activity can be increased by the presence of certain substances or metal ions. The metal ions that increase the activity of enzymes are called metal cofactors. Metal ions like Ca^{2+}, Mg^{2+}, Cu^{2+}, and Zn^{2+} etc., are required by different enzymes.

Enzymes like carbonic anhydrase and carboxy peptidase require Zn^{2+} as the cofactor. Hexokinase and other ATP requiring kinases need Mg^{2+} as the cofactor. Manganese is required for the activity of enolase. Chloride is required for amylase. Copper requiring enzymes are tyrosinase, catalase, and peroxidase. Lecithinase and lipase require calcium. Xanthine oxidase requires molybdenum.

INHIBITION OF ENZYME ACTIVITY

Substances that decrease the enzyme activity are called inhibitors, and the process is known as inhibition. Enzyme activity has to be inhibited during the process of regulation of metabolic pathway.

Following are the different types of inhibition mechanisms:

1. Competitive inhibition
2. Non-competitive inhibition
3. Uncompetitive inhibition
4. Suicide inhibition
5. Allosteric inhibition

Competitive Inhibition

Competitive inhibition is a reversible type of inhibition. The inhibitor resembles the substrate

(substrate analogue). Since the inhibitor is a substrate analogue, it competes with the substrate to bind to the active site of the enzyme. Since there is a competition between the substrate and the inhibitor for the active site, this type of inhibition is called competitive inhibition (Fig. 5.7).

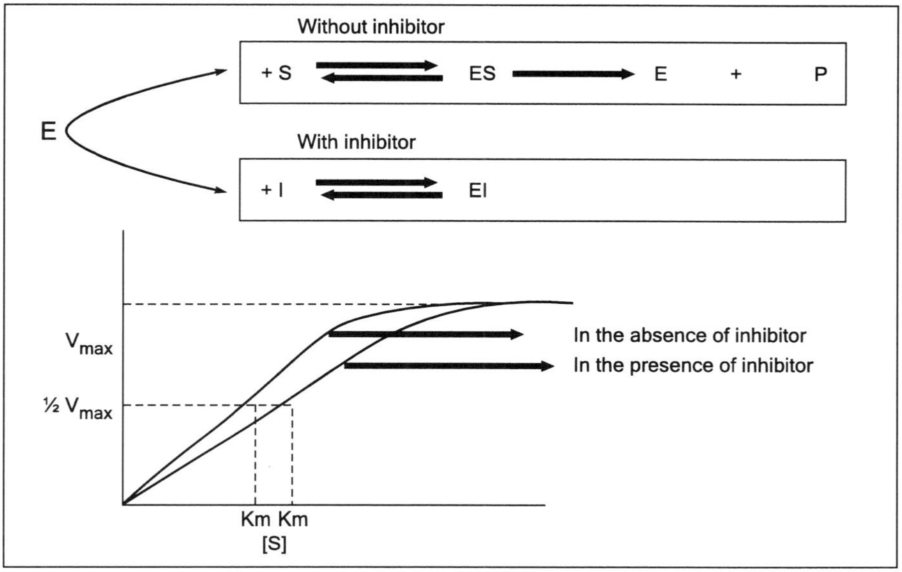

Fig. 5.7 Competitive inhibition.

Inhibition by the inhibitor can be reversed by increasing the concentration of the substrate. In competitive inhibition, **Km increases and V_{max} is not altered**. Inhibitor forms a complex with the enzyme, which is called enzyme inhibitor complex.

A classical example of competitive inhibition is the inhibition of succinate dehydrogenase by malonic acid. Malonic acid has a structural resemblance to succinate. So, it competes with succinate for the active site on the enzyme succinate dehydrogenase. If the concentration of malonic acid is more, the rate of the enzymatic reaction will be drastically reduced.

The concept of competitive inhibition can be used for therapeutic purposes.

1. Antimetabolite like methotrexate competitively inhibits the enzyme dihydrofolate reductase, which is required for the synthesis of tetrahydrofolate. So, cells are depleted of tetrahydrofolate, which is required for the synthesis of DNA. This affects replication, and the cell proliferation is decreased. This drug can be used in the treatment of cancer.
2. The antimicrobial agent sulphonamides are structural analogues of paraaminobenzoic acid. They can inhibit the enzyme pteroid synthase and hence inhibit the folic acid formation. Bacterial cell wall synthesis requires folic acid. So, sulphonamides inhibit bacterial cell proliferation.
3. Trimethoprim, another antimicrobial agent, is a competitive inhibitor of dihydrofolate reductase.
4. Methanol poisoning is treated by giving ethanol, because ethanol acts as a competitive inhibitor of alcohol dehydrogenase, and the end product of the methanol metabolism

(formic acid) is reduced. Formic acid accumulation is responsible for blindness seen in case of methanol poisoning.

Non-Competitive Inhibition

Non-competitive inhibition reactions can be reversible or irreversible (Fig. 5.8). The inhibitor is not a structural analogue of the substrate.

So, the inhibitor binds to a site on the enzyme other than the active site. Inhibitor can bind to both free enzyme and enzyme-substrate complex, and make the enzyme inactive.

In non-competitive inhibition, **the Km is not altered but V_{max} is increased**. If the inhibitor binds to the free enzyme, it forms enzyme-inhibitor complex, and if it binds to the enzyme-substrate complex, it forms enzyme-substrate-inhibitor complex. It is not possible to overcome the non-competitive inhibition by increasing the substrate concentration.

Examples of non-competitive inhibition:

1. Inhibition of the enzyme glyceraldehyde-3-phosphate dehydrogenase by iodoacetate.
2. Inhibition of acetyl choline esterase by diisopropylflurophosphate (DFP).
3. Inhibition of enalose by fluoride.
4. Trypsin inhibitor secreted by the roundworm inhibiting the enzyme, trypsin. Due to this inhibition, the roundworm is protected against the proteolytic digestion by the enzymes in the intestine.

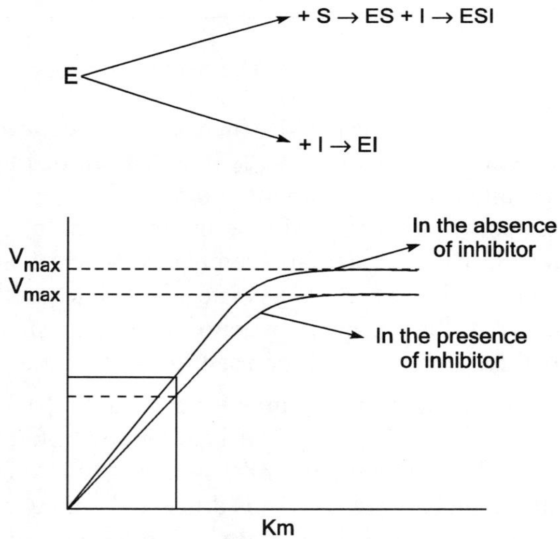

Fig. 5.8 Non-competitive inhibition.

Uncompetitive Inhibition

In case of uncompetitive inhibition, the inhibitor can bind only to the enzyme substrate complex. In this type of inhibition, both V_{max} and Km are reduced and it is rarely seen in biological system.

Suicide Inhibition

Suicide inhibition is an irreversible type of inhibition. The inhibitor is a substrate analogue, which binds to the active site of the enzyme. Once the inhibitor occupies the active site, it is acted upon by the enzyme. Action of the enzyme on the inhibitor converts it into a potent inhibitor, and it occupies the active site of the enzyme permanently, thereby inhibiting the enzyme irreversibly. The inhibition cannot be reversed by increasing the concentration of the substrate. Suicide inhibition is also called mechanism-based inhibition.

For example:

1. Allopurinol is a drug used to treat chronic gout. Gout is a clinical condition, where uric acid crystals are accumulated in the joint and soft tissues due to increased uric acid level in the blood. Xanthine oxidase is the enzyme involved in the formation of uric acid from the purines (Fig. 5.9).

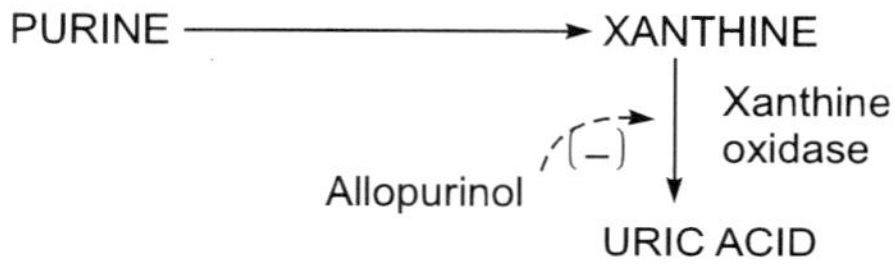

Fig. 5.9 Suicide inhibition.

When allopurinol is acted upon by the enzyme xanthine oxidase, it converts allopurinol to alloxanthine, which is a powerful inhibitor of the enzyme, and xanthine oxidase is permanently inhibited. Inhibition is effective till the new enzyme is synthesized by the cell.

2. Aspirin is a suicide inhibitor of the enzyme cylooxygenase of prostanoid biosynthetic pathway.
3. Difluromethyl ornithine is a suicide inhibitor of ornithine decarboxylase.

Allosteric Inhibition

The activity of allosteric enzymes can be increased or decreased when allosteric modulators or effectors bind to the enzyme. The allosteric effectors, which increase the activity of the enzyme, are called **positive allosteric effectors** or modifiers, and the effectors that decrease the activity of the enzyme are called **negative allosteric effectors**.

Most of the enzymes catalyzing the regulated steps of metabolic pathway are allosteric enzymes. These enzymes are called rate limiting enzymes or regulated enzymes of the metabolic pathway.

For example:

1. Aspartate transcarbamoylase is the enzyme of pyrimidine biosynthesis, catalyzing the second reaction of the pathway by which carbamoyl phosphate and aspartate are converted to aspartate transcarbamoylase (Fig. 5.10).

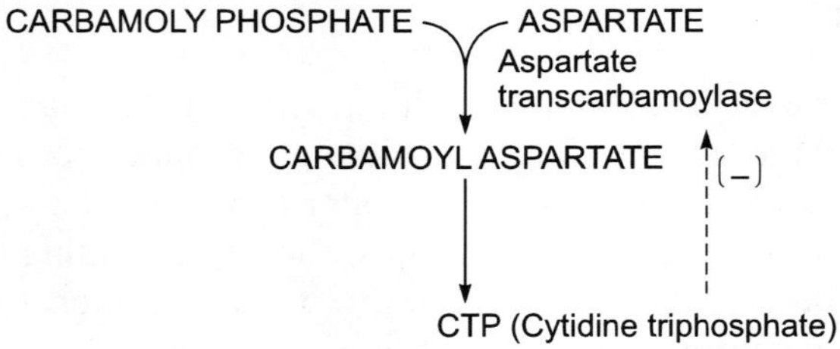

Fig. 5.10 Allosteric inhibition.

The end product of the metabolic pathway, cytidine triphosphate (CTP), is the negative modifier of the enzyme aspartate transcarbamoylase. The inhibition of the regulated enzyme by the end product of the pathway is called **feedback inhibition**. Feedback inhibition is always allosteric inhibition.

2. Acetyl CoA carboxylase is inhibited by palmitate.
3. Pyruvate carboxylase is inhibited by ADP.
4. Phosphofructokinase is inhibited by citrate and ATP.

REGULATION OF ENZYME ACTIVITY

Activity of the enzymes is well regulated, so that they can be activated only when they are required. Regulation of the enzyme activity is the means by which the metabolic pathways are regulated. Regulation of the rate of the metabolic pathway helps to prevent the depletion of the substrate in the cell and to utilize the resources available economically.

There are different modes by which the enzymes of metabolic pathway can be regulated. These can be summarized as follows:

1. Covalent modification
2. Allosteric regulation
3. Synthesis and degradation
4. Activation of the proenzyme
5. Isoenzymes

Covalent Modification

Certain groups are added to the enzyme proteins by forming covalent bond, or are removed by hydrolysis of covalent bonds leading to alteration in the structure of the enzyme. This leads to increase or decrease in the activity of the enzymes. This process is known as covalent modification.

Enzymes are covalently modified by **phosphorylation** or **dephosphorylation**. In addition to this, other means of covalent modifications are acetylation and deacetylation; methylation and demethylation, etc.

Glycogen synthase is the enzyme of glycogen synthesis, and glycogen phosphorylase is the enzyme of glycogen breakdown. On phosphorylation, glycogen synthase becomes inactive and glycogen phosphorylase becomes active. So, when glycogen synthesis is active, glycogen breakdown is inhibited and vice versa. It helps to avoid futile cycle.

Allosteric Regulation

The activity of allosteric enzymes is regulated by the small molecules called allosteric modifiers or effectors.

Allosteric modifiers bind to the allosteric site (site other than the active site), which may be situated far from the active site. The effector molecules bind to the regulatory site and bring about the structural changes in the enzyme. Because of these structural changes, activity of the enzyme might increase or decrease. The substrate saturation curve in case of allosteric enzymes is sigmoid in shape (Fig. 5.11).

The allosteric regulators which increase the activity of the enzyme are called positive allosteric effectors. Those which decrease the activity of the enzymes are called negative allosteric effectors.

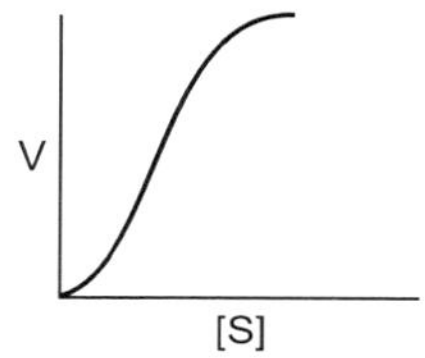

Fig. 5.11 Allosteric regulation.

Synthesis and Degradation of the Enzymes

Enzyme activity can be controlled by increasing or decreasing the concentration of the enzymes by altering the rate of synthesis or degradation. The enzymes whose concentration is always kept constant are called **constitutive enzymes**. The rate of synthesis of the regulated enzymes is constantly altered. They have a very short half-life, since their action is required only for a short period of time.

The process of increasing the synthesis of an enzyme is called **induction**. The molecules, which are responsible for this increase in the synthesis by acting at the gene level, are called **inducers**. Similarly, molecules which decrease the synthesis of an enzyme, are called **repressors**, and the process is called **repression**.

For example, the key enzymes of gluconeogenesis—pyruvate carboxylase, phosphoenolpyruvate carboxykinase, fructose 1,6 bisphosphatase, and glucose-6-phosphatase are induced by glucagon, whereas insulin acts as a repressor.

Activation of the Proenzyme

Enzymes that are synthesized and secreted in the inactive form are called **proenzymes (zymogens)**. The proenzymes are activated only when they reach the site of action or whenever their activity is required. Activation of the proenzyme takes place by the hydrolysis of a few or sometimes even one peptide bond due to change in the pH of the surrounding environment, or by the action of other enzymes.

For example, digestive enzymes like pepsinogen, trypsinogen, chymotrypsinogen, etc., are the inactive form of the enzymes pepsin, trypsin, and chymotrypsin respectively. Pepsinogen is activated to pepsin by the change in the surrounding pH. The pepsin can further activate the remaining pepsinogen to pepsin. During the clotting of blood, a cascade of zymogen activation is involved.

Isoenzymes

Different physical forms of the same catalytic activity are called isoenzymes. They differ from each other with respect to the antigenic property, electrophoretic mobility, heat stability, and sensitivity to the inhibitors. They catalyze the same reaction and differ from one another with respect to Km, V_{max} and regulatory properties.

The different molecular forms may be found in different individuals in the population, or in the same individual at different tissues, or in the same tissue at different subcellular sites. Most of the enzymes, which exist in multiple molecular forms, are oligomeric (have more than one subunit).

For example, creatine kinase (CK) is an enzyme which converts creatine to creatine phosphate by using phosphate group from ATP.

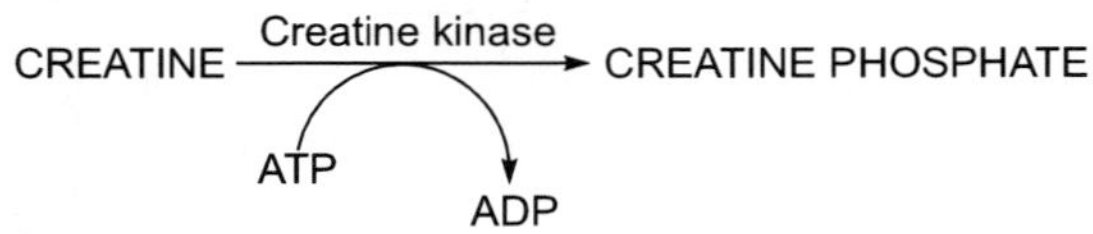

Creatine kinase is a dimeric protein. It has two subunits called M subunit and B subunit. Different subunits are predominantly synthesized in different tissues. The subunit M is synthesized mainly in the muscle, and the subunit B is mainly synthesized in the brain. These subunits can unite in different combinations to form dimeric enzyme protein. The resulting isomeric forms are as follows:

BB → CK1 → Rich in brain and central nervous system
MB → CK2 → Specific for cardiac muscle
MM → CK3 → Rich in skeletal muscle

There are three different forms of creatine kinase—CK1, CK2, and CK3.

Lactate dehydrogenase is another example

It is a tetrameric protein (having four subunits) made up of two different types of subunits—H (heart) subunit and M (skeletal muscle) subunit. These two subunits combine at different combinations to form the tetrameric lactate dehydrogenase as follows:

HHHH → LD1
HHHM → LD2
HHMM → LD3
HMMM → LD4
MMMM → LD5

LDH has five different molecular forms—LD1, LD2, LD3, LD4 and LD5.

DIAGNOSTIC ENZYMOLOGY

Enzymes have been used to diagnose the diseases. The branch of enzymology that deals with the diagnosis of diseases is called diagnostic enzymology. The enzyme concentration is measured in the freshly-collected serum. Serum has two kinds of enzymes:

1. Functional enzymes
2. Non-functional enzymes

Functional Enzymes

These are the enzymes which have a definite function in the blood. These enzymes are normally synthesized in different tissues and enter circulation to discharge their respective duties there.

Non-functional Enzymes

These enzymes do not have any function in the blood. They are the intracellular enzymes of different tissues. They have definite functions in the tissues where they exist. They leak out from the cells of these tissues during the daily wear and tear (turnover of cells). They are responsible for the low activity of these enzymes in the blood of normal individuals.

The enzyme level in blood is expressed as international unit (U/L). One international unit is catalytic activity, which is required to convert **one millimole of substrate to product in one minute, in one litre of serum**.

Decrease in the serum level of functional enzymes and increase in the serum level of non-functional enzyme has got significance in the diagnosis of diseases.

Certain enzymes or their isoforms are rich in certain tissues.

- Creatine kinase MB isoform (CK-MB) and LD1 (LDH) isoform are rich in cardiac muscle.
- Alanine transaminase (ALT) activity is rich in liver cells.
- Amylase activity is found in the pancreas and salivary gland.

When there is damage to the cells of the tissue, the enzyme or the isoform of the enzyme, which is specific for that tissue, leaks out of these cells and appears in the blood. Hence, their concentration in the circulation increases.

The enzymes of diagnostic importance and the tissues for which they are specific are listed in Table 5.2.

Table 5.2: Enzymes of diagnostic importance

Enzyme	*Tissue to which it is specific*
1. Alanine transaminase (ALT)	Liver cell damage (liver disease)
2. Aspartate transaminase (AST)	Heart and liver
3. Alkaline phosphatase (ALP)	Obstructive liver disease and bone disease
4. Acid phosphatase (ACP)	Prostatic cancer
5. Amylase	Pancreatic disease
6. Lipase	Pancreatic disease
7. Lactate dehydrogenase (LDH)	Heart disease

Contd.

8. Total creatine kinase (Total CK)	Skeletal muscle dystrophies
9. Creatine kinase MB isoform (CK-MB)	Damage to myocardium
10. Gammaglutamyltranspeptidase (GGT)	Alcoholic liver disease
11. 5′Nucleotidase	Obstructive liver disease
12. Serum cholinesterase	Organophosphorus poisoning

The group of enzymes, which helps to diagnose liver diseases, are ALT, AST, ALP, GGT, and 5′ nucleotidase. If there is damage to the liver cells, the level of ALT and AST is increased. If there is an obstruction in the flow of bile, the blood level of alkaline phosphatase, 5′nucleotidase, and GGT is increased. GGT, more specifically, indicates the alcohol liver disease.

The markers of myocardial damage are AST, LDH, CK-MB and troponin I or troponin T. Cardiac troponins are proteins present along with the actin filament. Their levels are increased if there is damage to the myocardium.

QUESTIONS

Long Essay

1. What are enzymes? Discuss in detail the factors affecting the enzyme activity.
2. Discuss the regulation of enzyme activity.
3. Discuss in detail the enzymes of diagnostic importance.

Short Essay

1. Classify the enzymes with suitable examples.
2. Describe competitive inhibition.
3. Discuss suicide inhibition.
4. What are coenzymes? Discuss their importance.
5. Write briefly about different models, which help to explain the enzyme-substrate complex.
6. What are isoenzymes? Add a note on isoenzymes of LDH and CK.

6

Chemistry of Carbohydrates

INTRODUCTION

Plants are the primary producers of carbohydrates. They produce it by using carbon dioxide and sunlight in a process called photosynthesis. Carbohydrate is the main source of energy for animals. Energy derived from carbohydrates by oxidation is used for different purposes like locomotion, transport of molecules across the membranes, etc.

Carbohydrates contain carbon, hydrogen, and oxygen. The ratio of hydrogen to oxygen is similar to that present in water. For example, the molecular formula of glucose is $C_6 H_{12} O_6$. If we write it as $C_6 (H_2O)_6$ or $C(H_2O)$, it appears as if carbon is hydrated, hence the name carbohydrates. Carbohydrates are carbon compounds that contain large number of hydroxyl groups.

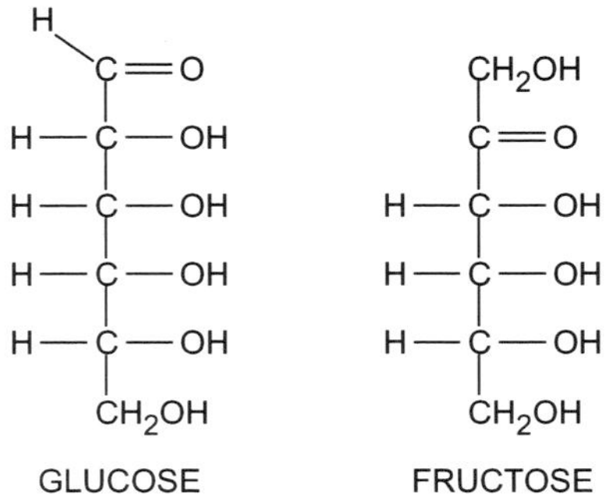

Fig. 6.1

Some carbohydrates contain an aldehyde group, while others have ketogroup. In the above example, glucose (Fig. 6.1) contains many hydroxyl groups, and the first carbon atom is an aldehyde group. In case of fructose (Fig. 6.1), many hydroxyl groups are present, and the second carbon atom is a keto group. **Carbohydrates can be defined as polyhydroxy aldehydes or polyhydroxy ketones, or the compounds from which they can be formed by hydrolysis**.

FUNCTIONS OF CARBOHYDRATES

1. Major source of energy. Glucose is a ready source of energy.
2. Glycogen is the storage form of energy.
3. Carbohydrates are structural components of animals and plants.
4. Ribose and deoxyribose are the constituents of nucleic acids.
5. Carbohydrates are required for cell-to-cell interactions.
6. Ascorbic acid, a derivative of carbohydrate, is a water-soluble vitamin.
7. Amino sugars (derivatives of carbohydrates) are components of antibiotics like erythromycin, streptomycin, etc.
8. Blood group substances are carbohydrate in nature.

CLASSIFICATION OF CARBOHYDRATES

Carbohydrates are classified based on the number of polyhydroxy aldehyde or polyhydroxy ketone units present in them.

1. Carbohydrates containing only one polyhydroxy aldehyde unit or polyhydroxy ketone unit are called **monosaccharides**.
2. Carbohydrates containing **two to ten monosaccharides are called oligosaccharides**.
3. Carbohydrates containing **more than ten monosaccharides are called polysaccharides**.

MONOSACCHARIDES

Monosaccharides are the simplest group of carbohydrates often referred to as **simple sugars**. They cannot be hydrolyzed further. They are sweet to taste and soluble in water (so they are called sugars). For example, glucose and fructose are monosaccharides.

The sugars containing aldehyde group as the functional group are called **aldoses**, and those containing keto group as the functional group are called **ketoses**. For example, glucose is an aldose, and fructose is a ketose. The most abundant monosaccharide in nature is the six-carbon containing aldose—glucose.

Monosaccharides are further divided depending upon the number of carbon atoms present in them. Monosaccharides with biological importance contain three to seven carbon atoms.

- Monosaccharides that contain three carbon atoms are called **trioses**. If triose contains an aldehyde group, it is called aldotriose. If triose contains keto group, it is called ketotriose.

  For example, glyceraldehyde is an aldotriose; dihydroxy acetone is ketotriose.
- Monosaccharides that contain four carbon atoms are called **tetroses**. If tetrose contains an aldehyde group, it is called aldotetrose. If tetrose contains keto group, it is called ketotetrose.

  For example, erythrose is an aldotetrose; erythrulose is a ketotetrose.
- Monosaccharides that contain five carbon atoms are called **pentoses**. The aldehyde group containing pentose is called aldopentose. The keto group containing pentose is called ketopentose.

  For example, ribose is an aldopentose, and ribulose is a ketopentose.

- Monosaccharides that contain six carbon atoms are called **hexoses**. If hexose contains an aldehyde group, it is called aldohexose. If hexose contains keto group, it is called ketohexose.
  For example, glucose is an aldohexose, and fructose is a ketohexose.
- Monosaccharides that contain seven carbon atoms are called **heptoses**.
  For example, sedoheptulose.

Properties of Monosaccharides

Compounds with same molecular formula but different structural formula are called **isomers**. Both glucose and fructose have same molecular formula but their structural formula differs. So, they can be called isomers. One is aldose; another is ketose, so they are aldose ketose isomers.

Stereoisomerism

Carbon atom has four valencies. The four valencies of the carbon should be satisfied. If four different groups or atoms occupy the four valencies, then the carbon is called an **asymmetric carbon atom (chiral centre)** (Fig 6.2).

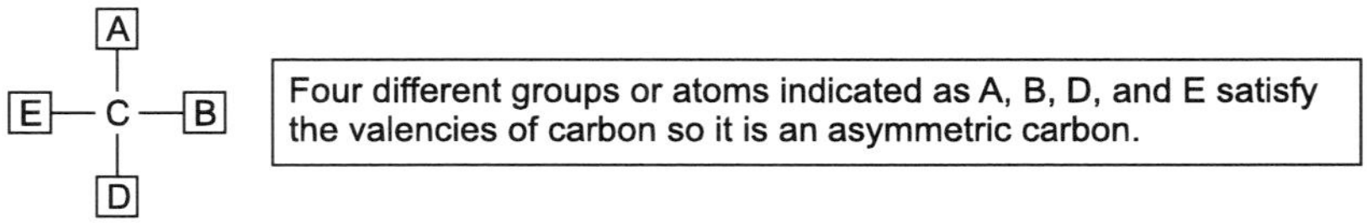

Fig. 6.2 Asymmetric carbon atom.

Monosaccharide containing the asymmetric carbon atom exhibits stereoisomerism. Stereoisomerism is the existence of the same compound in more than one form due to the different spatial arrangements of the groups or atoms attached to its asymmetric carbons. Number of stereoisomerism possible by a compound containing asymmetric carbon atoms depends upon the number of asymmetric carbon atoms present. It can be calculated by the formula 2n, where n is the number of asymmetric carbon atoms.

Glyceraldehyde, an aldotriose, contains three carbon atoms, where the middle carbon is an asymmetric carbon atom. So, glyceraldehyde exhibit stereoisomerism.

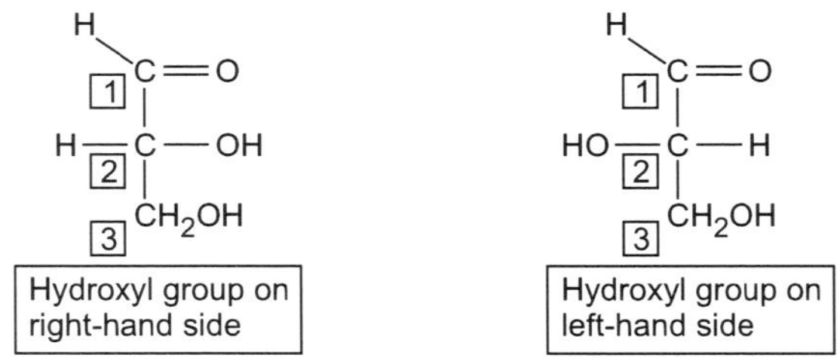

D-Glyceraldehyde L-Glyceraldehyde

Fig. 6.3 D and L glyceraldehyde.

In glyceraldehydes, the arrangement of OH and H around the second carbon atom is different. If OH is on the right hand side, it is called D-glyceraldehyde, and if it is on the left-hand side, it is called L-glyceraldehyde.

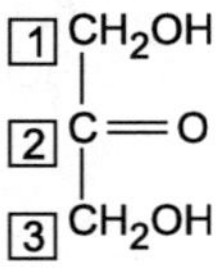

Dihydroxyacetone phosphate

D-glyceraldehyde looks like the mirror image of L-glyceraldehyde (Fig. 6.3). So, 'D' and 'L' forms of the sugars are mirror images. Naturally occurring monosaccharides present in the mammalian tissues are mostly of D-configuration. Dihydroxy acetone does not contain any asymmetric carbon atom.

In case of other sugars containing more than one asymmetric carbon atoms, depending upon the arrangement of OH and H around the penultimate carbon atom (last but one), they are named 'D' or 'L' sugar. D and L forms are mirror image forms. They are **enantiomers**.

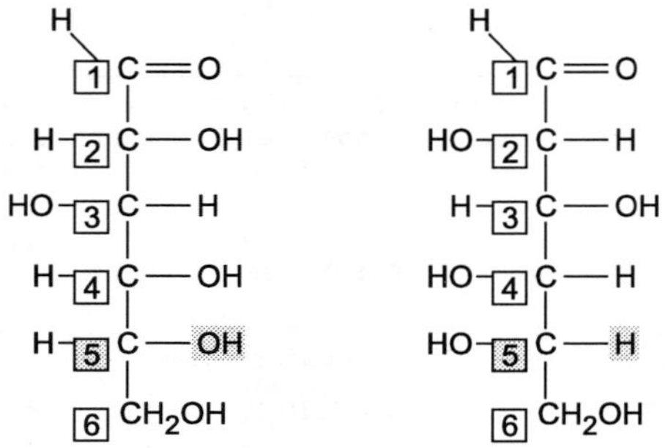

D-Glucose L-Glucose

In glucose 2nd, 3rd, 4th, 5th carbon atoms are asymmetric carbon atoms based upon the orientation of –OH and H around the 5th carbon atom i.e., the penultimate carbon atom glucose designated D or L. D-Glucose and L-Glucose are mirror image forms.

Fig. 6.4 D and L glucose.

Optical Isomerism

Optical activity is the ability of a compound to change the plane of vibration of polarized light when it is passed through it. Presence of asymmetric carbon atoms confers optical activity on the compounds. Monosaccharide containing asymmetric carbon atoms are **optically active**.

When the carbohydrate solution changes the plane of vibration of polarized light to right, it is called **dextrorotatory ('d' or +)**. If the plane of vibration of polarized light is changed to left, it is called **levorotatory ('l' or –)** (Fig. 6.5).

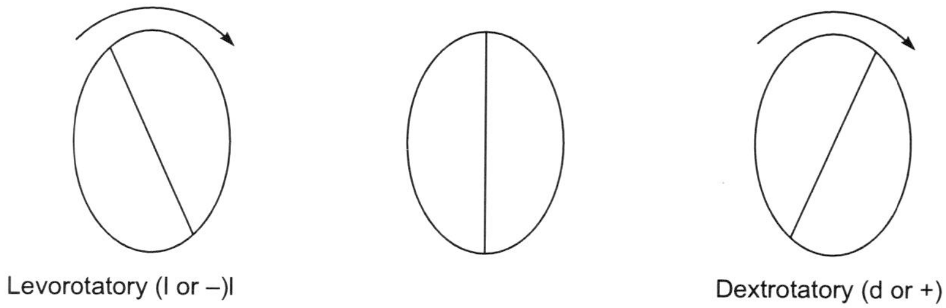

Fig. 6.5 Dextro and levorotation.

Depending upon the ability to change the plane of vibration of polarized light, there are two isomeric forms, 'd' or (+) and 'l' or (–) forms. For example, d-glyceraldehyde and l-glyceraldehyde are optical isomers.

Optical activity is measured with the help of **polariscope**, and it is expressed as specific rotation. **Specific rotation** is the rotation in degrees produced at a specified temperature (25°) when the polarized light of a standard wavelength passes through 1 cm long column of a solution containing 1 gm of the substance/mL.

For example, D-glucose is + 52.2° (glucose turns the plane of vibration of polarized light to right side by 52.5°). L-glucose is –52.2° (glucose turns the plane of vibration of polarized light to left side by 52.2°). Usually, only one of the optical isomers of a substance is biologically active.

Naturally occurring forms of glucose, mannose, ribose, sucrose, lactose, maltose, dextrin, starch, and glycogen are dextrorotatory; natural fructose is levorotatory.

Anomerism

The aldehyde and ketone group of the carbohydrates with five and six carbons will react spontaneously with alcohol groups present in neighbouring carbons to produce **hemiacetals** or **hemiketals**, respectively. Hemiacetal or hemiketal is a ring structure.

The aldehyde group of aldohexose can react with the hydroxyl group of 5th or 4th carbon atom. The keto group of ketohexose can react with the hydroxyl group of 5th or 6th carbon atom. These reactions result in the formation of five- or six-membered ring. The five-membered ring structure resembles the organic molecule, furan. The ring structure of the sugar, which resembles furan ring is termed furanose (Fig. 6.6). The sugars with six-member rings resemble the pyran and are called pyranose (Fig. 6.6).

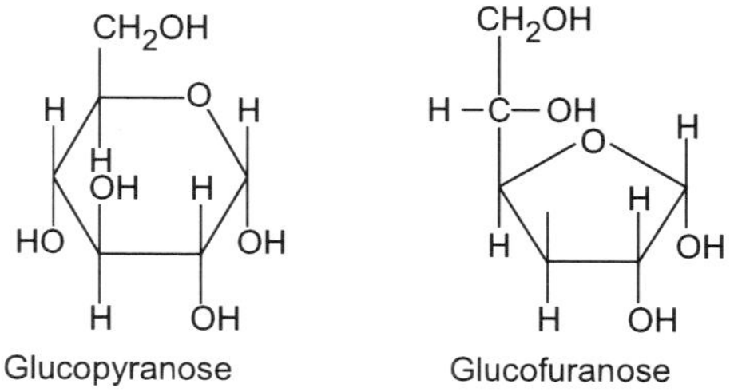

Fig. 6.6 Pyranose and furanose forms of glucose.

For example, glucose present in pyran ring form is called glucopyranose; furan ring form is called glucofuranose. The pyranose rings might exist either as a boat or chair conformation. The chair form is thermodynamically more stable than the boat form (Fig. 6.7).

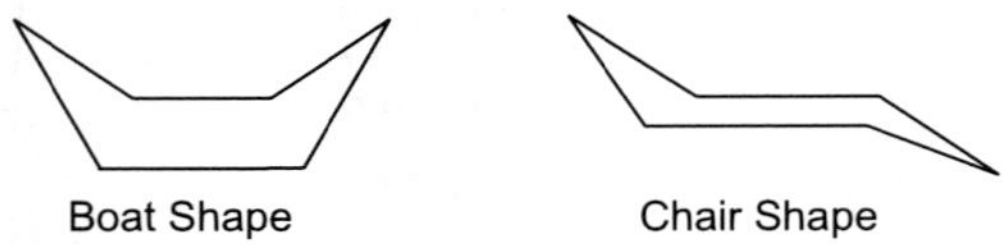

Fig. 6.7 Boat and chair conformation of pyranose.

The first carbon atom of aldose and the second carbon atom of ketose become asymmetric after the formation of hemiacetal and hemiketal ring structure. These carbon atoms are called **anomeric carbon atoms**. Depending upon the arrangement of the hydroxyl and hydrogen atom around these carbon atoms, sugars can present in α or β forms. These forms are called anomeric forms. Stereoisomerism is called **anomerism**.

In the Fischer projection (Fig. 6.8), the α configuration is indicated by placing the hydroxyl group attached to the anomeric carbon to the right side of the ring, and the β configuration is indicated by placing the hydroxyl group attached to the anomeric carbon to the left side of the ring.

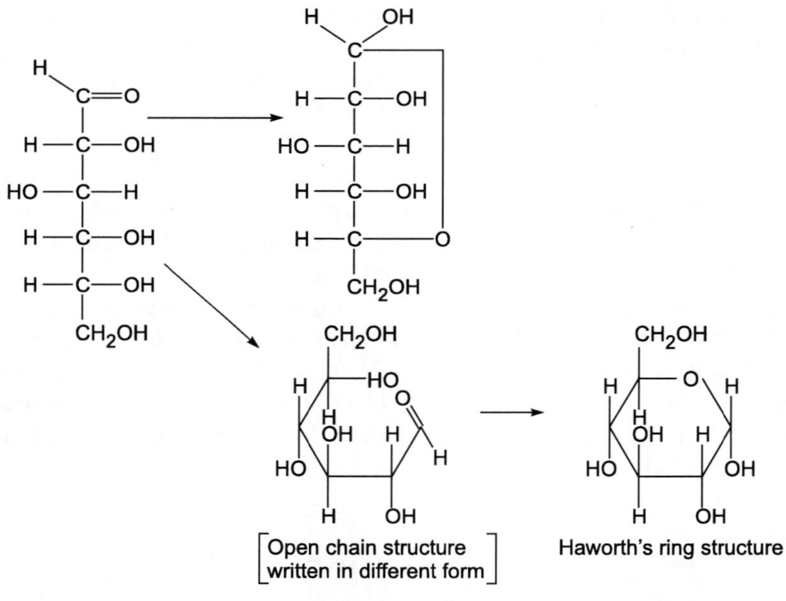

Fig. 6.8 Fischers projection formula and Haworth's ring structure.

When drawn in the Haworth projection (Fig. 6.8), the α configuration places the hydroxyl group downward, and the β configuration places the hydroxyl group above the plane of the ring.

Mutarotation

Sugars are crystallized in any one of the anomeric forms. When dissolved in the solvent, all the sugars present in the solution have the same anomeric form. If this freshly prepared sugar solution is allowed to stand over a period of time, some sugars change their anomeric form, which leads to the formation of an equilibrium mixture of α and β anomeric forms. Since the specific rotation of the α anomeric form and β anomeric form is different, change in the specific rotation of the freshly prepared sugar solution is observed when it is kept over a period of time.

For example, when alpha glucose is dissolved in water, the specific rotation of this solution is +112°. If the sugar solution is allowed to stand over a period of time, the specific rotation will change to +52°. This is because of the formation of an equilibrium mixture of α and β anomeric forms. In the equilibrium mixture, 2/3 of the glucose is present in the β anomeric form, and 1/3 of the glucose is present in the α anomeric form. The specific rotation of β anomeric form is +19°. So, the resultant specific rotation is +52°.

Mutarotation is the change in specific rotation of a freshly prepared sugar solution on standing due to the formation of equilibrium mixture of α and β anomeric forms.

Epimerism

It is a type of stereoisomerism, exhibited by compounds containing more than one asymmetric carbon atoms. Isomers differ from each other with respect to the arrangement of groups around a single asymmetric carbon atom other than the anomeric carbon atom. For example, glucose and mannose, and glucose and galactose are C_2 and C_4 epimers respectively (Fig. 6.9).

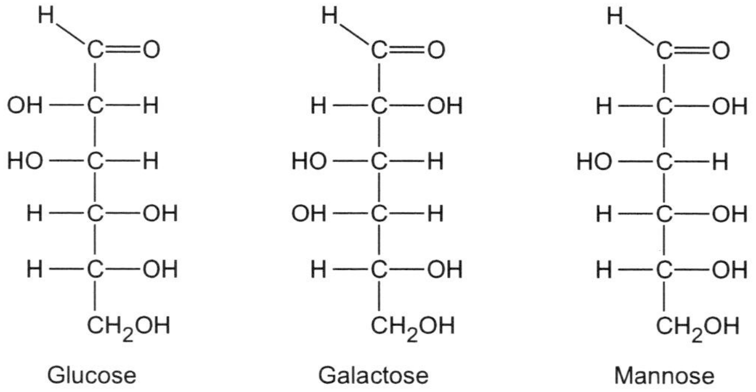

Fig. 6.9 Epimerism.

CHEMICAL PROPERTIES

Action of Concentrated Acid

When carbohydrates are treated with concentrated sulphuric acid, sugars get dehydrated to form furfural derivatives. Furfural derivatives can get condensed with phenolic compounds to give a colored complex. This is the basis for Molisch test and Seliwanoff's test.

Molisch test

To 1 mL of sugar solution, add one drop of freshly prepared alpha-naphthol. Mix well. Add concentrated sulphuric acid slowly along the side of the tube, so that the acid forms a separate layer below the sugar solution. At the junction of the two solutions, reddish-violet colour develops. This indicates the presence of carbohydrate in the given solution. It is a general test for carbohydrates.

Seliwanoff's test

To 3 mL of Seliwanoff's reagent, add 1 mL of sugar solution. Mix well. Boil for 30 seconds. Keto sugar gives a cherry red colour. Seliwanoff's reagent contains dilute HCl and resorcinol. When sugar is heated with the Seliwanoff 's reagent under the test conditions, keto sugar gets dehydrated first and forms furfural derivatives. It reacts with resorcinol and gives a cherry red colour.

Action of Mild Alkali

When sugars containing free aldehyde or keto group are treated with a mild alkali, they tautomerise to form enediol. Enediols are powerful reducing agents. They can reduce cupric ions to cuprous ions.

Benedict's test

It is a test for reducing sugars. Benedict's reagent consists of copper sulphate, sodium citrate, and sodium carbonate. Copper sulphate provides cupric ions; sodium citrate helps to keep cupric ions in solution; and sodium carbonate acts as a mild alkali.

Procedure: To 5 mL of Benedict's reagent, add 8 drops of sugar solution. Mix well and boil it for 2 minutes. If the blue colour of Benedict's reagent changes to green, yellow, orange, or brick-red colour, it indicates that the given sugar is a reducing sugar.

When sugars containing free aldehyde or keto groups are treated with Benedict's reagent, sugars are converted to enediols. Enediols are powerful reducing agents. They reduce cupric ions to cuprous ions, which is responsible for the change in the colour of Benedict's reagent (Table 6.1).

This test is used in clinical practice to detect the reducing substance present in the urine. It is a **semiquantitative test**. At different concentrations of sugar, different colours are developed. Benedict's test is not a specific test for glucose. The presence of reducing sugar (glucose, galactose, fructose, etc.) or any reducing substances (vitamin C, excess amount of uric acid) in urine can give positive Benedict's test.

Table 6.1: Results of Benedict's test for reducing sugar

Colour	*Approximate amount of sugar*	*Reported as*
Green	0.5%	+
Yellow	1%	++
Orange	1.5%	+++
Red	2.0%	++++

Glucose oxidase containing strips are now available to detect glucose present in the urine. Glucose oxidase acts specifically on glucose and produces hydrogen peroxide. Hydrogen peroxide is acted upon by an enzyme "peroxidase" to liberate oxygen. This oxygen oxidizes a colourless chemical substance to coloured product, indicating the presence of glucose. It is a specific test for glucose.

Oxidation

Mild oxidizing agents oxidize only the aldehyde group of the sugar to form **aldonic acids**. Glucose is oxidized to gluconic acid, and galactose is oxidized to galactonic acid. Strong oxidizing agents oxidize both aldehyde and primary alcohol group of sugar to form **saccharic acid**. For example, glucose is oxidized by a strong oxidizing agent to form glucosaccharic acid. Galactose is oxidized by a strong oxidizing agent to form galactosaccharic acid (mucic acid).

Under special conditions, only primary alcohol group of sugar is oxidized to form **uronic acid**. If the primary alcohol group of glucose is oxidized with its functional group intact, the product is called glucuronic acid. Glucuronic acid is produced in human body by uronic acid pathway. It is an enzyme catalyzed reaction, which converts glucose to uronic acid.

Glucuronic acid is required for detoxification reactions. For example, bilirubin is an endogenous toxic substance produced from heme. It is made less toxic by conjugating it with glucuronic acid in the liver. Glucuronic acid is a component of heteropolysaccharides.

Reduction

Sugars are reduced to form sugar alcohols, e.g., glucose is reduced to sorbitol. In case of uncontrolled diabetes mellitus, when the blood glucose level increases, glucose is converted to sorbitol in the lens of the eye. Since sorbitol is osmotically active, lens cells imbibe a lot of water. This damages the lens cells. Lens of the eye becomes opaque. This condition is called cataract.

In galactosemia, due to the defect in galactose metabolism, galactose gets accumulated in the cells of the lens and is converted to galcitol (dulcitol). It is the alcohol form of galactose, which is responsible for cataract in galactosemia. Mannitol is the alcohol form of mannose. It can be used to produce forced diuresis in case of renal failure and to reduce intracranial pressure.

GLYCOSIDE

Covalent bonds between the anomeric hydroxyl group of a cyclic sugar and the hydroxyl group of a second sugar (or another alcohol containing compound) are termed **glycosidic bonds** and the resultant molecules are glycosides. The linkage of two monosaccharides to form disaccharides involves a glycosidic bond. Several physiologically important disaccharides are sucrose, lactose, and maltose.

Glycosidic bonds can be of two types:

1. O-glycosidic bond.
2. N-glycosidic bond.

The glycosidic linkage found in the oligosaccharides and polysaccharides is O-glycosidic linkage. The glycosidic linkage between the ribose and the nitrogen containing base is N-glycosidic linkage. N-glycosidic linkage is seen in many biologically important glycosides, nucleic acids, and various glycoproteins.

Some biologically important glycosides are:

1. Digitoxine
2. Erythromycin
3. Streptomycin, etc.

DISACCHARIDES

Sucrose

It is **cane sugar**. It contains α-D-Glucose and β-D-Fructose linked to one another by α-β (1,2)-glycosidic bond (Fig. 6). Since both of the anomeric carbon atoms are involved in the glycosidic linkage, it is not able to reduce cupric ions to cuprous ion. So, sucrose is a non-reducing sugar.

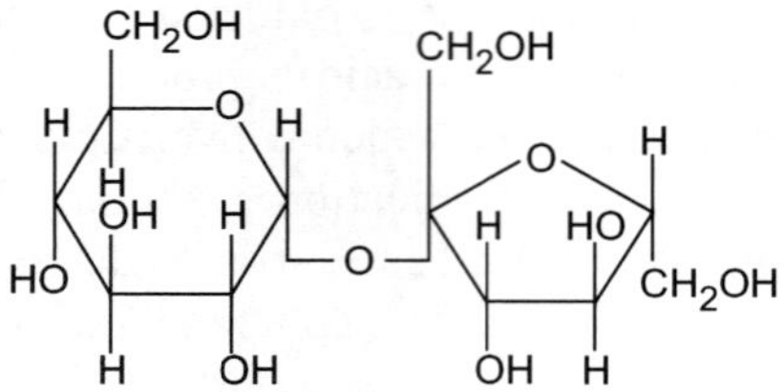

Sucrose

Sucrase is an enzyme involved in the digestion of sucrose. Sucrose can be hydrolyzed by strong mineral acids like concentrated HCl. The hydrolyzed mixture contains equimolar glucose and fructose. The hydrolyzed mixture of sucrose is initially dextrorotatory. On standing over a period of time, the optical activity of the hydrolyzed mixture is changed to levorotation. This process is called **inversion**. The hydrolyzed mixture of sucrose is called **invert sugar**. Sucrase is also called **invertase**.

Lactose

It is found exclusively in the milk of mammals, therefore it is also called **milk sugar**. Lactose consists of β-D-Galactose, joined by β-D-Glucose in a β-(1,4) glycosidic bond (Fig. 7). Since, the anomeric carbon atom of one of the monosaccharide units is free, it is capable of reducing cupric ions to cuprous ions. Lactose is a reducing sugar. It gives powderpuff-shaped or cotton ball-shaped osazone crystals.

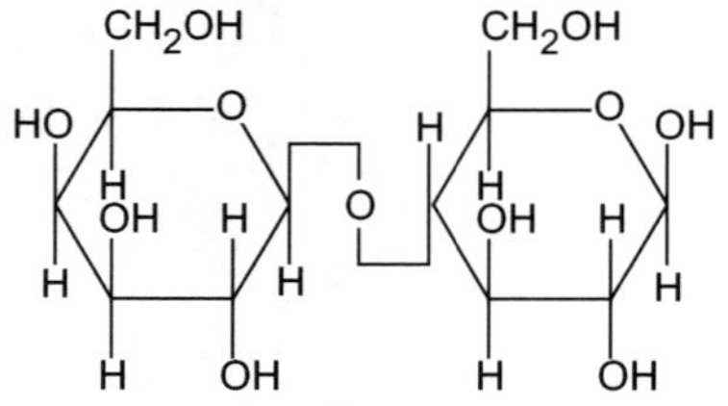

Lactose

Lactose is digested by lactase present in the intestine. **Deficiency of lactase** in the intestine gives rise to a condition known as **lactose intolerance**.

Maltose

It is a major degradation product of starch. Maltose is composed of two glucose units, which are joined by α-(1,4) glycosidic bond (Fig. 8). Maltose is a reducing sugar. It gives sunflower-shaped osazone crystals. Maltose is digested by the enzyme called **maltase**.

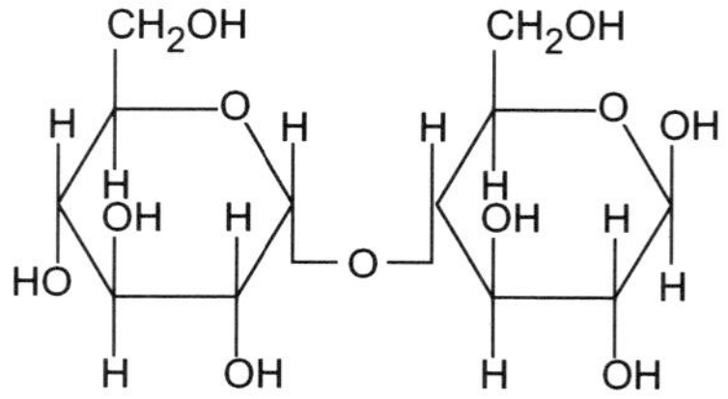

Maltose

POLYSACCHARIDES

Polysaccharides contain more than ten monosaccharides. They are high molecular weight substances. When polysaccharides are composed of a single type of monosaccharide building blocks, they are termed **homopolysaccharides**. For example, starch, glycogen, cellulose, inulin, chitin, dextrans, and agar. Polysaccharides composed of more than one type of monosaccharides are termed **heteropolysaccharides**, e.g., hyaluronic acid, chondroitin sulphate, and heparin, dermatan sulphate, keratan sulphate.

HOMOPOLYSACCHARIDES

Starch

Starch is the major form of stored carbohydrate in plant cells. Its structure is identical to glycogen, except for a much lower degree of branching (about every 20-30 residues). The unbranched portion of starch is called **amylose**; branched portion of starch is called **amylopectin**. Amylose is a water-soluble fraction of starch, where glucose units are linked to one another by α-1,4 glycosidic linkage.

Amylopectin portion of the starch is water insoluble. It has both α-1,4 glycosidic linkage and α-1,6 glycosidic linkage (Fig. 6.10).

Starch has both reducing ends and non-reducing ends. The number of non-reducing ends is more than the number of reducing ends. Starch is a non-reducing carbohydrate.

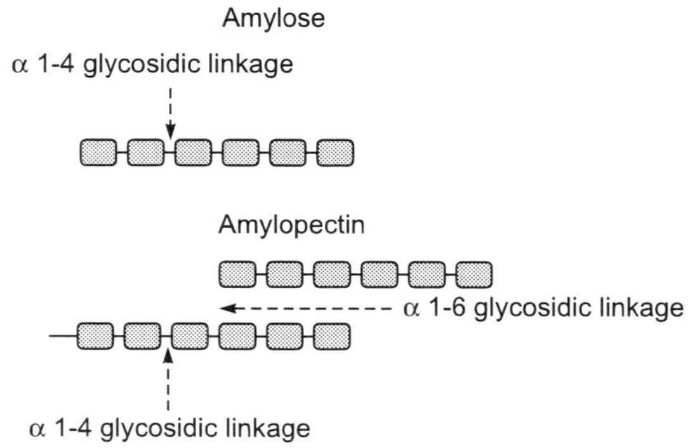

Fig. 6.10 Glycogen.

Iodine test

When starch solution is treated with iodine, it gives a **blue colour**. The blue colour develops because of the trapping of iodine in the helical coils of amylose. When this test solution is heated, the colour disappears due to the disruption of the helical coils of amylase. The colour reappears if the solution is cooled, due to the reformation of helical coils.

GLYCOGEN

Glycogen is the major storage form of carbohydrate in animals. It is a polymer of α-glucose. Glucose units are linked to each other by α-(1,4) linkage and α-(1,6) linkage at branch points. The branch point occurs at every 8-10 residues. The structures of starch and glycogen are similar, provided glycogen is more frequently branched compared to starch.

Glycogen is a compact structure, which results due to the coiling of the polymer chains. This compact arrangement of the polymer allows the storage of large amount of energy in a small volume, with little effect on cellular osmolarity.

During the breakdown of glycogen to release glucose, an enzyme called glycogen phosphorylase acts on the terminal glucose residue of the branches. Since glycogen is heavily branched, the number of sites on which this enzyme can act are more. So, at a time, large amount of glucose can be released.

Glycogen is stored mainly in the liver and muscle. The liver glycogen helps to maintain the blood glucose level, whereas the muscle glycogen provides energy during muscle contraction. Glycogen gives red colour with iodine.

HETEROPOLYSACCHARIDES

Mucopolysaccharides or Glycosaminoglycans (GAG'S)

Heteropolysaccharides are polysaccharides containing more than one type of monosaccharides. Glycosaminoglycans are long unbranched heteropolysaccharides. These are located primarily on the surface of cells or in the extracellular matrix (ECM) (Table 6.2).

Glycosaminoglycans have large number of negative charges. Negatively charged heteropolysaccharide chains tend to extend in solution because of repulsion between the like charges. A shell of water surrounds them. So, when they are brought together, they tend to slip past each other. This produces slippery consistency of mucous secretion and synovial fluid containing heteropolysaccharides.

Table 6.2: Localization of glycosaminoglycans

Glycosaminoglycans	*Localization*
(a) Hyaluronate	Synovial fluid, vitreous humour, ECM of loose connective tissue.
(b) Chondroitin sulphate	Cartilage, bone, heart valves, cornea.
(c) Heparan sulphate	Basement membranes, components of cell surfaces.
(d) Heparin	Mast cells, lungs, liver, and skin.
(e) Dermatan sulphate	Skin, blood vessels, heart valves.
(f) Keratan sulphate	Cornea, bone, cartilage aggregated with chondroitin sulphates.

GAGs are compressible. When they are compressed, water between the GAGs is squeezed out, so they are forced to occupy smaller volume. When pressure is released, they spring back to their original hydrated form. This property is responsible for the resilience of synovial fluid and vitreous humour of the eye.

It is composed of repeating disaccharide units. The disaccharide units contain one acidic sugar linked by glycosidic linkage to an amino sugar. Acidic sugar is either D-glucuronic acid or its epimer L-iduronic acid. Amino sugar is either [N-acetylgalactosamine (GalNAc) or N-acetylglucosamine (GlcNAc)]. In amino sugars, amino group is usually acetylated. These sugars may be sulphated on 4th or 6th carbon atoms.

The specific GAGs of physiological significance are hyaluronic acid, dermatan sulphate, chondroitin sulphate, heparin, heparan sulphate, and keratan sulphate (Fig. 6.11).

PROTEOGLYCANS

All the glycosaminoglycans except hyaluronic acid are covalently attached to protein, and this complex is called proteoglycans (Fig. 6.12). Proteoglycans have a core protein to which linear GAG's are covalently attached.

HYALURONIC ACID

Hyaluronic acid is composed of D-glucuronate (GlcVA) linked to N-acetylglucosamine (GlcNAc) by β (1,3) glycosidic linkage (Fig. 6.11). Hyaluronic acid is unique among the GAGs in that it **does not contain any sulphate** and is not found covalently attached to proteins as proteoglycan. It is attached to proteoglycans in the extracellular matrix by non-covalent bond.

Hyaluronic acid polymers are very large (with molecular weights of 100,000–10,000,000) and can displace a large volume of water. This property makes them excellent lubricators and shock absorbers.

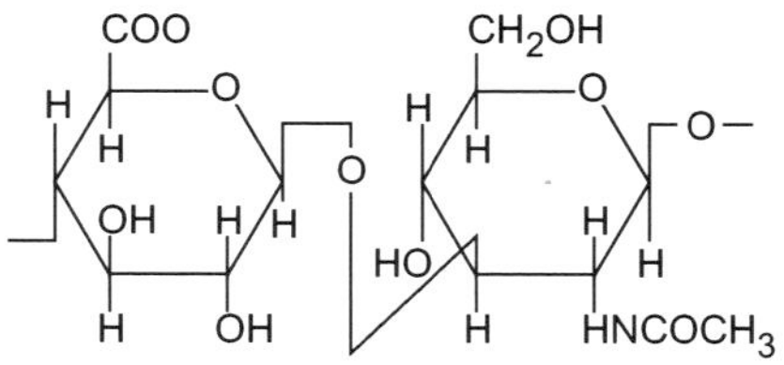

D-GlcUA GlcNAc

Hyaluronic Acid

D-glucuronate like to GlcNAc by β (1,3) glycosidic Linkage

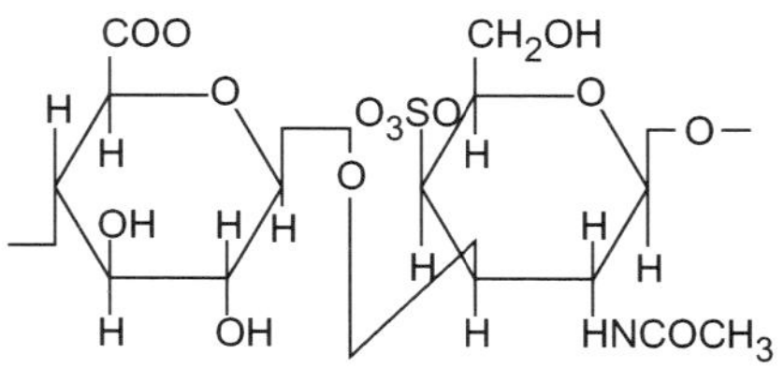

D-GlcUA GlcNAc 4-sulphate

Chondroitin sulphate

Chondroitin sulphated is composed of D-GlcUA + GalNAc-4- or 6-sulphate linked by β(1,3)glycosidic linkage

Contd.

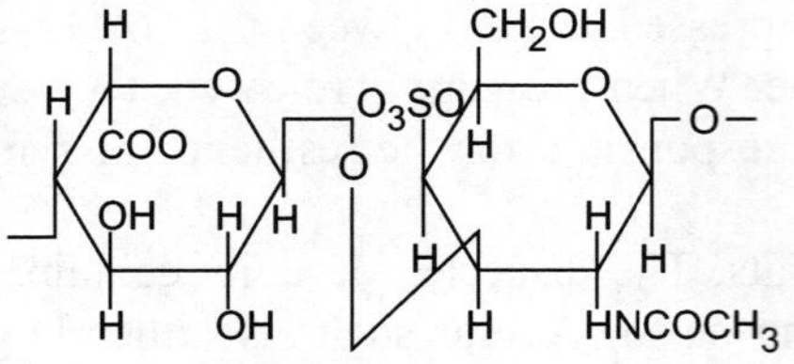

Dermatan sulphate

L-iduronic acid is linked to N-Acetylgalactoseamine 4-sulphate by β(1,3) glycosidic linkage

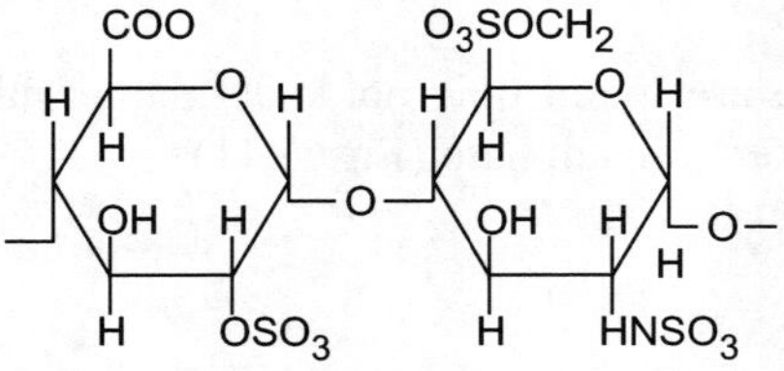

Heparin & heparan sulphate

Heparin and heparan sulphate are made up of glucosamine and any of the two uronic acids linked by α (1-4) glycosidic linkage. Heparin contains more of N sulphated glucosamine compared to heparin sulphate.

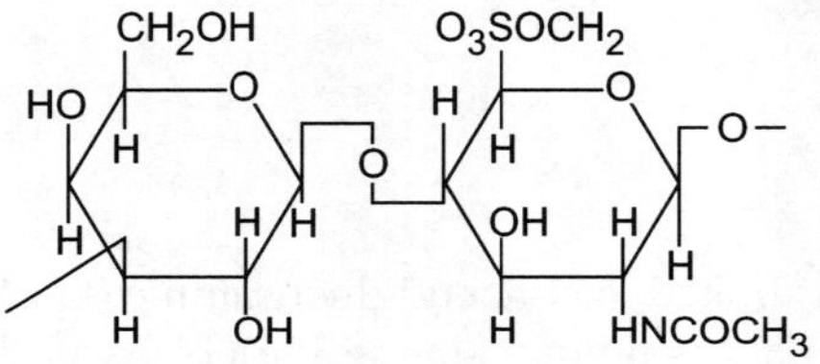

Ketatan sulphate

Composed of galactose and GlcNAc-6-sulphate linked by β (1,4) glycosidic linkage

Fig. 6.11 Heteropolysaccharides.

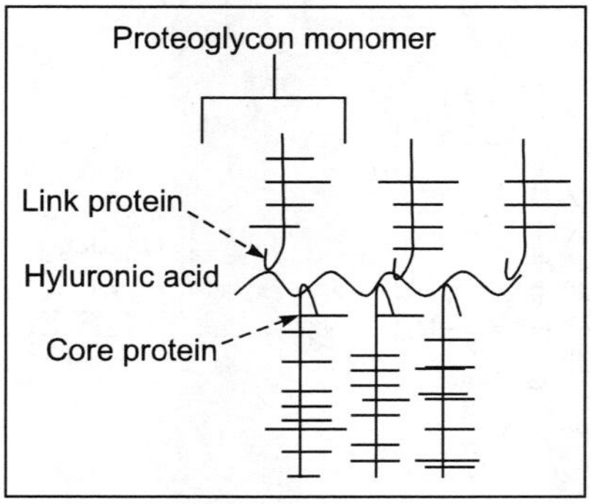

Fig. 6.12 Protoglycan.

QUESTIONS

Short Essay

1. Define carbohydrates. Classify them with examples.

2. Write short notes on:
   (a) Mutarotation
   (b) Stereoisomerism
   (c) Maltose
   (d) Lactose
   (e) Surcose
   (f) Inversion
3. What are polysaccharides? Classify them with suitable examples. Mention the structural components of starch.
4. What are heteropolysaccharides? Give any four examples and mention their functions.
5. Discuss briefly about the structure of starch. List the differences between the starch and glycogen.

Short Answers

1. What is glucuronic acid? Mention its function.
2. Name the test used to detect the reducing substance in the urine. Write its principle.
3. Mention any four carbohydrates having therapeutic importance.
4. Write a short note on epimerism.
5. Describe anomerism.
6. Why is surcose called non-reducing sugar?
7. Name the milk sugar and write its composition.
8. Name the cane sugar and write its composition.

7

Lipids

INTRODUCTION

Lipids form an important component of food. They are greasy materials, which can be extracted from animal and plant tissues. Lipids are a heterogeneous group of compounds, which are relatively insoluble in water and soluble in organic solvents like benzene, chloroform, ether, etc. Most of them contain or are derived from fatty acids (RCOOH). They have a great biochemical importance because they are a chief storage form of energy and have a role in cell structure and functions.

FUNCTIONS OF LIPIDS

1. Source of energy
   Triacylglycerol is the major storage form of energy in animals. Adipose tissue is a specialized tissue for lipid storage. It yields much more energy per gram when compared to carbohydrates and proteins (1 g of fat yields 9 kcals when compared to carbohydrates, which gives 4 kcal/gram).
2. Insulators
   Lipids are deposited under the skin and around the vital organs. They help to prevent loss of heat from the body, and protect the vital organs against mechanical shock and injury.
3. Important structural constituents of cell membranes
   Phospholipids and glycolipids are important constituents of the cell membrane.
4. Regulatory function
   Act as intra- and extracellular signals, e.g., steroid hormones.
5. Detergent and emulsifying action
   They are required for digestion and absorption of lipids and lipid-soluble vitamins.
6. Act as surfactants in lungs.

CLASSIFICATION OF LIPIDS

Lipids are classified as:

1. Simple lipids
2. Compound lipids
3. Derived lipids

Simple Lipids

These are esters of fatty acids with alcohol, e.g., fats and waxes. Fats are fatty acid esters of glycerol. Oils are fats having low melting points. They remain liquid at room temperature.

Waxes are esters of fatty acids with alcohol other than glycerol. The fatty acids present are generally long chain fatty acids, and alcohols may be aliphatic or alicyclic. For example, cetyl alcohol of the cetyl palmitate of sperm oil, and myricylalcohol of myricylpalmitate of beeswax.

Compound Lipids

These are esters of fatty acids with alcohol carrying other groups like carbohydrates, phosphates, nitrogenous base, or proteins.

1. ***Phospholipids:*** Phospholipid molecule is constructed from fatty acids, alcohol, phosphate and additional nitrogenous or non-nitrogenous substances.
2. ***Glycolipids:*** Glycolipids are fatty acid esters of sphingosine. But in addition to the fatty acids and sphingosine, they contain carbohydrates. Based on the carbohydrate part, glycolipids are classified into:
    - **Cerebrosides**: These contain glucose or galactose.
    - **Sulphatides** or sulpholipids: These contain sulphated galactose residues.
    - **Gangliosides**: These contain a branched oligosaccharide.
3. ***Lipoproteins:*** Lipoproteins are complexes of lipids and proteins. For example, high density lipoproteins (HDL), low-density lipoproteins (LDL), very-low-density lipoproteins (VLDL), and chylomicrons. Lipid part consists of phospholipids, cholesterol, cholesterol esters, fat, and free fatty acids.

Derived Lipids

Lipids derived from the hydrolysis of simple and compound lipids are called derived lipids e.g., fatty acids, prostaglandins, alcohols (like glycerol, sphingosine, and cetyl alcohol), steroids, polyisoprenoids (compounds containing two or more C5 hydrocarbon units are called isoprene units, for example, dolicol, tocopherols, carotenes, ubiqinones, and vitamin K) are derived lipids.

FATTY ACIDS

Fatty acids are one of the simpler molecular forms of lipids. They are **aliphatic carboxylic acids**. A fatty acid consists of a hydrophobic hydrocarbon chain with a terminal carboxyl group that has a pKa of 4.8. At physiological pH, the terminal carboxyl group –COOH ionizes to become $-COO^-$ (carboxylate).

$$R-COOH \rightleftharpoons R-COO^- + H^+$$

Fatty acids are depicted by the chemical formula RCOOH, where R stands for an alkyl chain.

Number or letters designates the carbon atoms of fatty acids. The **lettering system** describes the carbon atom next to the carboxyl group as alpha (α) carbon atom (Fig. 7.1), the next carbon as beta (β) carbon atom, and so on. The last carbon atom is called ω carbon in **numbering system**, the carboxyl group carbon is the first carbon atom with the next carbon as second carbon atom, and so on.

ω Carbon ↓ (on CH_3); β (on carbon 3); α (on carbon 2)

$$CH_3\ CH_2\ CH_2\ CH_2\ CH_2\ (CH_2)n\ \underset{5}{CH_2}\ \underset{4}{CH_2}\ \underset{3}{\overset{\beta}{CH_2}}\ \underset{2}{\overset{\alpha}{CH_2}}\ \underset{1}{COOH}$$

Fig. 7.1 Nomenclature of fatty acids.

Based on the number of carbon atoms present, fatty acids are grouped into

1. Short chain fatty acids—contain 2 to 4 carbon atoms.
2. Medium chain fatty acids—contain 6 to 10 carbon atoms.
3. Long chain fatty acids—contain 12 to 26 carbon atoms.

Based upon the presence or absence of double bonds, they are grouped into

1. **Saturated fatty acids:** These are fatty acids without double bonds. The general formula is $C_n H_{2n+1}$ COOH. For example, palmitic acid, stearic acid, etc.
2. **Unsaturated fatty acids:** Fatty acids having double bonds are called unsaturated fatty acids.
   (a) Fatty acids having one double bond are called **monounsaturated fatty acids**. For example, oleic acid and palmitolic acid are monounsaturated fatty acids.
   (b) Fatty acids with more than one double bond are called **polyunsaturated fatty acids**. For example, arachidonic acid, linoleic acid, etc.

Fatty acids containing double bonds may be present in different isomeric forms depending upon the arrangement of atoms or groups attached to the carbon atoms involved in the double bond.

Unsaturated fatty acids may be present in **cis** or **trans** form (Fig. 7.2). In the cis form, all the groups or atoms attached to the carbons involved in the double bond formation are on the same side, and in *trans* configuration, they are on the opposite side. **Fatty acids with *cis* configuration predominantly occur in the nature.**

Based upon the nutritional requirement, fatty acids are grouped into essential fatty acids and non-essential fatty acids. Some of the unsaturated fatty acids cannot be synthesized in the human body, so they should be supplemented in the daily diet. These fatty acids are called **essential fatty acids**. Linoleic acid, linolenic acid, and arachidonic acids are essential fatty acids. Corn oil, peanut oil, cottonseed oil, sunflower oil, etc. are rich in essential fatty acids.

Based upon the number of carbon atoms present, fatty acids are classified into two broad groups. Fatty acids having even number of carbon atoms are called **even chain fatty acids**, and fatty acids with odd number of carbon atoms are called **odd chain fatty acids**.

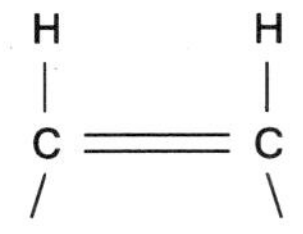

Cis configuration

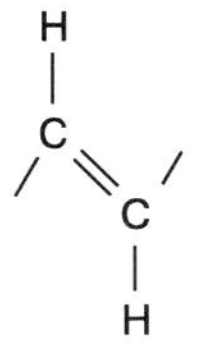

Trans configuration

Fig. 7.2 *Trans* and *cis* isomerism of fatty acids.

Based upon the presence or absence of branch point, fatty acids are classified into:

1. **Branched chain fatty acids:** These are fatty acids with the branched hydrocarbon chain. For example, isocaproic acid and isovelaric acid. Branched chain fatty acids are present in the waxes and sebaceous glands.
2. **Straight chain fatty acids:** The hydrocarbon chain doesn't have any branching. For example, palmitic acid and oleic acid.

Hydroxyl group containing fatty acids

Some of the fatty acids contain a hydroxyl group. They are called hydroxyl group containing fatty acids. For example, cerebronic acid and hydronervonic acid are predominantly present in the lipids of the nervous system.

Hydrocarbon chain of fatty acids may be linear or cyclic

The fatty acids containing the linear hydrocarbon chain are straight chain fatty acids, e.g., palmitic acid, stearic acid, etc. Fatty acids with cyclic hydrocarbon chains are cyclic fatty acids, e.g., eicosanoids (prostanoids and leukotriens).

Eicosanoids are derived from C20 polyunsaturated fatty acids. They are biologically very active. **Prostanoids** consist of prostaglandins, prostacyclins, and thromboxans. They are the products of cycloxygenase pathway. **Leukotriens** are the products of lipoxygenase pathway.

The prostaglandins were first isolated from the prostate. They are present almost everywhere in the human body, mainly in kidney, intestines, brain, thymus, etc. There are different types of prostaglandins like PGE1, PGF2a, PGE2, etc. depending upon the side chains and the number of double bonds present.

Functions of Eicosanoids

Prostaglandins

1. They have vasodilatory effect, so they decrease the blood pressure.
2. They decrease the gastric acid secretion and can be used in the treatment of peptic ulcer.
3. They increase the intestinal motility.

4. They produce uterine contraction, so they are used to induce labour.
5. PGE2 can be used in the treatment of bronchial asthma because of its effect on the tracheobronchial musculature.
6. They increase the renal perfusion and glomerular filtration rate (GFR) and produce diuresis.
7. They stimulate the renin secretion.

Prostacyclins

They are found in the arterial and capillary walls in normal healthy individuals. They prevent platelet aggregation.

Thromboxanes

They are produced mainly in the platelets, and are potent platelet aggregators.

Leukotrienes

They have a role in anaphylactic reaction.

TRIACYLGLYCEROL

Trihydric alcohol glycerol is esterified with three different fatty acids to form triacylglycerol (Fig. 7.3). It is also called triglyceride or simply as fat. Triacylglycerol is most abundant in nature. Triacylglyceride is the concentrated form of energy stored in the adipose tissue.

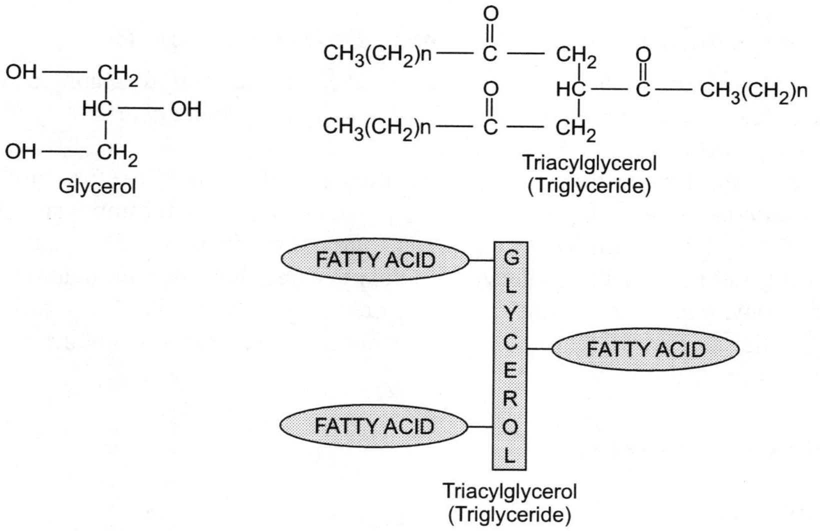

Fig. 7.3 Triacylglycerol.

The melting point of the fat depends upon the fatty acids present in it. Fats with saturated and long chain fatty acids have high melting point. Fats rich in unsaturated and medium chain fatty acids have low melting point. Fats with lower melting point are liquids at room temperature, and are called oils.

Triacylglycerol can be hydrolyzed to form fatty acids and glycerol when boiled with alkali or acid. In the gastrointestinal tract, it is hydrolyzed by an enzyme called **lipase**. When triacylglycerol is boiled with potassium hydroxide (KOH), it forms potassium salt of fatty acid called soap and glycerol. This is called **saponification**. Lipids which can undergo saponification are called saponifiable lipids. Triacylglycerol, phospholipids, glycolipids, and waxes can be saponified.

Saponification Number

The **number of grams of KOH required to saponify one gram of triacylglycerol is called saponification number**. Saponification number indicates the molecular weight of the fat. As the molecular weight increases, the saponification number decreases. Saponification number is **inversely related to the molecular weight of the fat.**

Halogenation

When fats containing unsaturated fatty acids are treated with halogens like iodine, bromine, etc., the halogens are absorbed by the unsaturated fatty acids and fatty acids are converted to halogenated saturated fatty acids.

Iodine Number

The **number of grams of iodine absorbed by 100 grams of fat is called iodine number**. The absorbed iodine is used only to saturate the double bonds. It is a measure of degree of unsaturation and **amount of unsaturated fatty acid present in the fat.**

Hydrogenation

Double bonds present in the unsaturated fatty acids of the fat can be saturated with hydrogen under a special test condition. This is called hydrogenation. Hydrogenation can convert the oils to fat by elevating their melting point.

Rancidity

When fat or oil is exposed to light or air for a long time, it becomes rancid. Rancidity may be due to hydrolysis of triacylglycerol and further decomposition of the fatty acids to form ketones, or it may be due to the oxidation of fatty acids.

Phospholipids

Phospholipids are fatty acid esters of alcohols esterified to phosphoric acid, and contain nitrogen containing or non-nitrogen containing bases (Fig. 7.4).

General structure

Based on the alcohol group present, phospholipids are classified as follows:

1. **Phosphoglycerides (Glycerophospholipids)**
   Glycerol is alcohol.
   - I. **Phosphoglycerides with nitrogenous compounds**
     - (a) Phosphatidylethanolamine (cephaline)
     - (b) Phosphatidylcholine (lecithine)
     - (c) Phosphatidylserine
   - II. **Phosphoglycerides containing non-nitrogenous compounds**
     - (a) Phosphatidylglycerol
     - (b) Diphosphatidylglycerol (cardiolipin)
     - (c) Phosphatidylinositol
   - III. **Ether phospholipids (plasmalogens)**
     - (a) Choline plasmalogen
     - (b) Ethanolamine plasmalogen
2. **Sphingophospholipids**
   Sphingosine is alcohol.

GLYCEROPHOSPHOLIPIDS

In glycerophospholipids, glycerol is esterified with the first and second carbon atom of glycerol. The fatty acid esterified to second carbon atom is unsaturated fatty acid. The third carbon atom of the glycerol is esterified with phosphoric acid.

Glycerol with fatty acids on first and second carbon atoms and phosphoric acid on the third carbon atom is called **phosphatidic acid** or **phosphatidate**. The phosphatidate is a phospholipid. Most of the glycerophospholipids contain nitrogenous or non-nitrogenous base attached to the phosphate of the phosphatidate (Fig. 7.4).

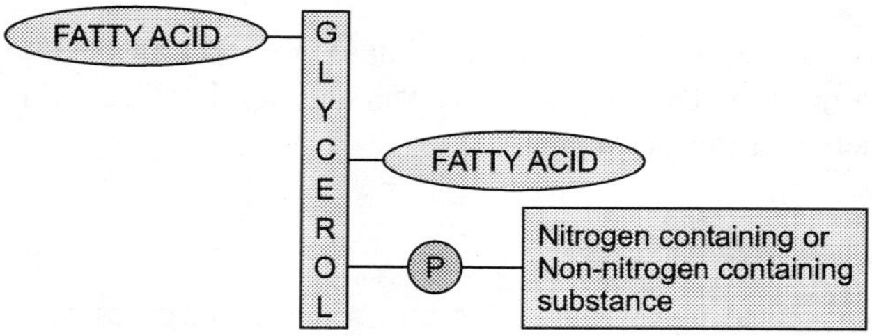

Fig. 7.4 Phospholipid.

In phosphatidylethanolamine (cephaline), the nitrogen containing base is ethanol amine. Similarly, phosphatidylcholine (lecithine) contains choline, and phosphatidylserine contains serine. In phosphatidylglycerol, phosphatidate is attached to glycerol, and in diphosphatidylglycerol, two phosphatidates are linked by glycerol.

(Phosphatidate + ethanolamine) ⟶ Phosphatidylethanolamine
(Phosphatidate + choline) ⟶ Phosphatidylcholine
(Phosphatidate + serine) ⟶ Phosphatidylserine
(Phosphatidate + glycerol) ⟶ Phosphatidylglycerol
(Phosphatidate + glycerol + phosphatidate) → Diphosphatidylglycerol

Phosphatidylcholine having two palmitates on the first and the second carbon atom of the glycerol is called dipalmitoyl phosphatidylcholine. It acts as a surfactant in the lungs and prevents the collapse of the alveoli during expiration. Deficiency of this phospholipid gives rise to respiratory distress.

Phospholipids have both hydrophilic and hydrophobic portions. So, they are amphipathic in nature. At a particular concentration in the aqueous medium, the amphipathic molecules aggregate to form stable droplets called **micelles**.

The polar head groups of the amphipathic molecules face the aqueous phase, and the hydrophobic portions of the molecules are aggregated to form hydrophobic core, which is devoid of water (Fig. 7.5). Phospholipids form micelles in the intestine along with other amphipathic molecules like bile salts and soaps. This helps in the digestion and absorption of lipids and fat-soluble vitamins.

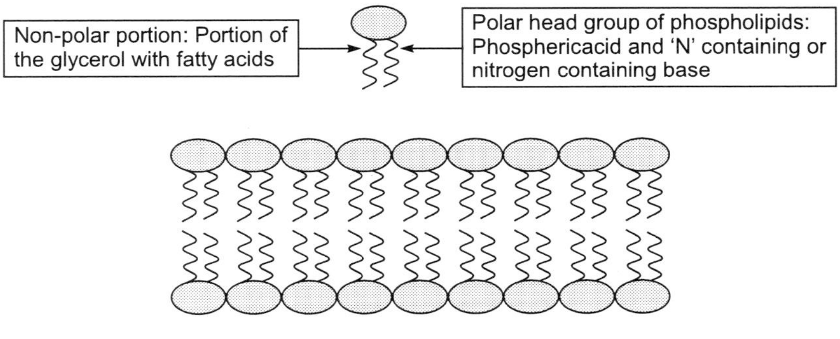

Fig. 7.5 Phospholipid bilayer.

Phospholipids at a higher concentration form a phospholipid bilayer (Fig.7.5). The phospholipid bilayer is made up of two phospholipid monolayers. In the phospholipid bilayer, the polar groups face the aqueous phase and the non-polar groups face each other as shown in Fig. 7.5. Cell membrane is a lipid bilayer in which many proteins are incorporated.

SPHINGOPHOSPHOLIPID

Sphingosine (Fig. 7.6) is alcohol. It is **synthesized from palmitoyl CoA and the serine** within the endoplasmic reticulum. It has amino group, a double bond, and a long hydrocarbon chain.

$$CH_3(CH_2)_{12}CH{=}CH{-}\underset{\large OH}{\underset{|}{CH}}{-}\underset{\large NH^{3+}}{\underset{|}{CH}}{-}CH_2{-}OH$$

Fig. 7.6 Sphingosine.

The amino group of sphingosine is linked to the carboxyl group of fatty acid to form **ceramide**. If the free hydroxyl group of ceramide is esterified to phosphocholine, it forms **sphingomyelin**. Sphingomyelin is a phospholipid.

GLYCOSPHINGOLIPIDS (GLYCOLIPIDS)

The hydroxyl group of ceramide, when linked to the sugar or oligosaccharide chain by the formation of glycosidic linkage, leads to the formation of glycosphingolipids (Fig. 7.7).

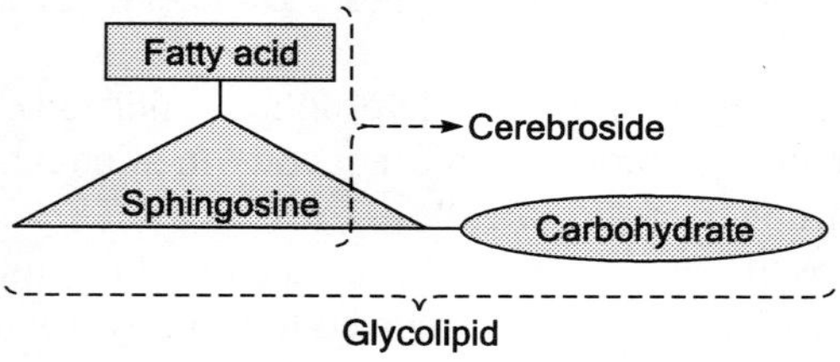

Fig. 7.7 Glycosphingolipid.

Cerebrosides

These are simple glycolipids. Ceramide is linked to **glucose** to form **glucocerebroside**, or to **galactose** to form **galactocerebrosides**. Galactocerebroside is a constituent of myelin sheath, and glucocerebroside is present in the extraneural tissues.

Sulphoglycolipids

If the galactose of galactocerebroside is sulphated, it is called sulphoglycosphingolipid.

Gangliosides

These are ceramide linked to the oligosaccharide chain containing sialic acid (NANA).

1. Glycolipids have a variety of functions.
2. They are present in the plasma membrane.
3. Involved in communication between the cells.
4. Antigenic in nature—blood group antigens are glycolipids.
5. Some of the glycosphingolipids act as receptors for certain bacterial toxins.

Steroids

Steroids are a group of compounds that contain **cyclopentanoperhydrophenanthrene ring** structure (Fig. 7.8). Cyclopentanoperhydrophenanthrene ring is present in steroid hormones, bile acids, cholesterol, etc. Steroid hormones are synthesized from cholesterol mainly in the adrenal cortex and gonads. They have regulatory effect on the metabolism and various target tissues. Cholesterol contains 27 carbon atoms.

Important compounds derived from cholesterol are text type:

1. Bile acids
2. Vitamin D
3. Steroid hormones

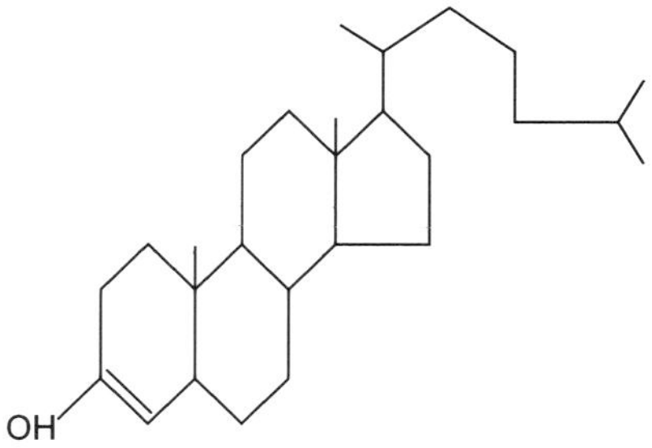

Fig. 7.8 Cyclopentanoperhydrophenanthrene ring.

Bile acids are synthesized in the liver from cholesterol, e.g., **cholic acid** and **chenodeoxycholic acid**. They are amphipathic in nature and therefore facilitate the digestion and absorption of lipids and lipid-soluble vitamins. Cholesterol is synthesized in our body as well as supplied in the daily diet.

There is no catabolic pathway in human body to destroy the cyclopentanoperhydrophenanthrene ring, so it is excreted as such after modification or in conjugation with hydrophilic substances. Cholesterol deposition in the extrahepatic tissues gives rise to **atherosclerosis**. Cholesterol deposition in the arteries leads to the narrowing of the lumen of the arteries and disturbance in the blood supply to the distal organs, which gives rise to ischemic damage. It may be a cause for various life-threatening conditions like myocardial infarction, cerebrovascular damage, etc.

Libermann-Burchard Reaction and Salkowski's Reaction

These tests were used as qualitative tests, but now the same principle is used to quantify cholesterol present in the serum. Both the reactions involve 3-hydroxy-5-ene part of cholesterol. The reagents are capable of dehydrating, oxidizing, and sulphonating properties.

Initially, cholesterol is dehydrated to form cholesta-3,5-diene. Later, it is oxidized and two molecules are linked to form bischolesta-3,5-diene. It is later sulphonated by sulphuric acid to form mono and di-sulphonic acids, which are coloured products. In Libermann-Burchard reaction, green colour is formed.

If excess of sulphuric acid is added and reaction mixture contains ferric ions, disulphonic acids are predominantly formed and give red colour. It is called Salkowski reaction.

QUESTIONS

Short Essay

1. What are lipids? Classify them with examples.

2. Write a note on the functions of lipids.
3. Write the composition of phospholipid. Mention any five phospholipids and their functions.
4. What are eicosanoids? Mention their functions.
5. Discuss briefly about triglycerides.
6. Discuss briefly about the functions of phospholipids.

Short Answer

1. Name the ring structure present in the cholesterol. Mention the normal serum total cholesterol level.
2. Name the biologically important products formed from cholesterol.
3. Describe Salkowski reaction.
4. Write a note on Leibermann-Burchard reaction.
5. Define bile acid. Mention any one of its functions.
6. Mention the composition and functions of sphingosine.
7. What are lipoproteins?
8. Write a note on fatty acids.

8

Digestion and Absorption

INTRODUCTION

The process of converting complex food particles into simple and easily digestible, water-soluble, and absorbable form is called digestion. The process of digestion involves mechanical processes like chewing and peristaltic movements along with various enzymatic reactions.

Chewing of food particles helps to convert large food particles to smaller ones. This increases the surface area available for the action of enzymes involved in digestion. The enzymes involved in digestion may be present in the lumen, on the mucosal surface, or within the epithelial cells lining the gastrointestinal tract.

Based upon the site where these enzymes carry out digestion, the process of digestion is divided into:

1. Luminal digestion
2. Surface digestion
3. Intracellular digestion

The enzymes involved in digestion **belong to the class hydrolase**, and therefore utilize water molecules to break covalent bonds.

DIGESTION AND ABSORPTION OF CARBOHYDRATES

Digestion of Carbohydrates

Glucose is mainly derived from food. The major dietary carbohydrate is starch. It constitutes more than 50% of the carbohydrate content of food. In addition to starch, diet also contains various oligosaccharides and monosaccharides.

The **digestion of starch begins in mouth**. Salivary juice contains an enzyme called **salivary amylase (ptyaline)**, which is involved in the digestion of starch. Salivary amylase is mixed with food in the mouth during the process of chewing, and digestion of starch begins.

When the food bolus reaches the stomach, amylase becomes inactive and digestion stops. Further digestion of starch takes place in the intestine with the help of pancreatic amylase.

POLYSACCHARIDES AND OLIGOSACCHARIDS

↓ DIGESTED IN THE GASTROINTESTINAL TRACT

MONOSACCHARIDES (SIMPLE SUGAR)

↓

ABSORBED BY THE INTESTINAL MUCOSA

Both salivary amylase and pancreatic amylase are **α-amylases**. They **require chloride ions** in their surrounding environment for their activity. Salivary and pancreatic α-amylases act on α-1,4 glycosidic linkage. They cannot act on α-1,6 glycosidic linkage and α-1,4 glycosidic linkage adjacent to α-1,6 glycosidic linkage.

The simplest carbohydrate formed from starch by the action of α-amylase is maltose. Amylase cannot act on α-1,4 glycosidic linkage present in maltose.

The products formed by the action of α-amylase on starch are maltose and α-limit dextrin (oligosaccharide having both α-1,4 and α-1,6 glycosidic linkage). These are further digested by the enzymes present on the surface of intestinal epithelial cells like **maltase**, **lactase**, **sucrase**, and **isomaltase**, leading to the formation of monosaccharides.

Absorption of glucose requires a transport protein called glucose transporter (Glut). Glucose is transported from the intestinal lumen to the intestinal cell along with Na^+ (Fig. 8.1). The absorption of glucose is by Na^+ Glucose Symport (co-transport) mechanism.

Na^+ is transported along the concentration gradient. Glucose is transported against the concentration gradient. Glucose, which enters into the intestinal cell, is transported to interstitial space with the help of another transporter, which is a uniport (carries only glucose out of the cell). Na^+ concentration is kept low within the cell with the help of Na^+ K^+ ATPase. The absorption of glucose is an example of secondary active transport mechanism.

Absorption of glucose requires sodium and vice versa. In oral rehydration salt (ORS), glucose is included along with sodium salt to facilitate the absorption of sodium.

LACTOSE INTOLERANCE (DEFICIENCY OF LACTASE)

Lactose intolerance is a condition, which occurs due to the **deficiency of an enzyme "lactase"**. Lactose is not digested, so it gets accumulated. Since lactose is osmotically active, it holds a lot of water, which is responsible for diarrhoea seen in lactose intolerance.

Lactose, which is accumulated in the intestine, is fermented by the bacteria present in the intestine, resulting in **abdominal pain and flatulence**. If lactose is absorbed it is not utilized, passes out in urine (lactosuria).

DIGESTION AND ABSORPTION OF PROTEINS

Digestion of Proteins

The digestion of proteins begins in the stomach, but mainly takes place and ends in the intestine. In the intestine, the alkaline pH is provided by the bicarbonate present in the pancreatic juice and bile.

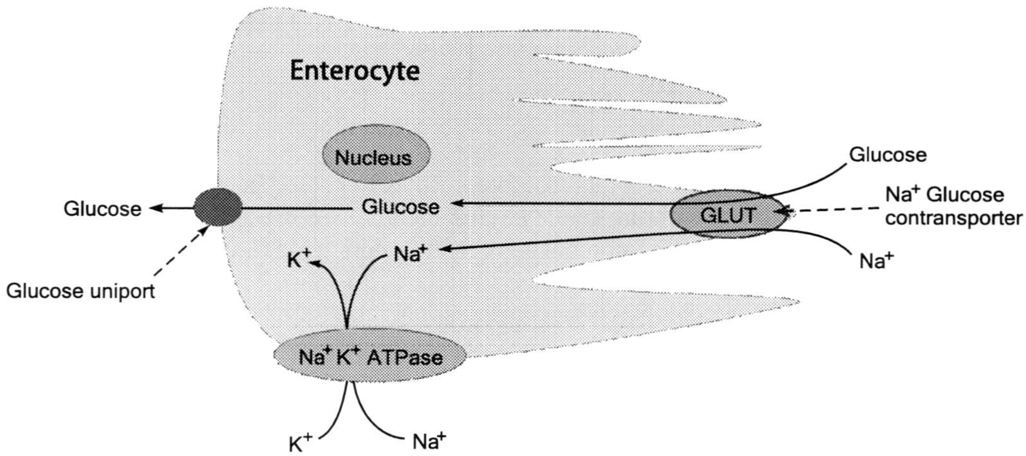

Fig. 8.1 Absorption of glucose.

In the stomach, acidic environment favours protein digestion, and is provided by strong mineral acid, **hydrochloric acid (HCl), secreted by the stomach**. Hydrochloric acid the gastric juice can **denature the proteins**. When proteins are denatured, the secondary, tertiary, and quaternary structures of the proteins are lost and peptide bonds situated deep inside the protein molecules are exposed. This increases the digestibility of the proteins. Hydrochloric acid of gastric juice helps to **kill any microorganism entering the stomach**.

The enzymes involved in the digestion of proteins are called **proteases (or peptidases)**. They hydrolyze specific peptide bonds. Peptidases, which act deep inside the core of the protein and convert proteins to small peptide fragments, are called **endopeptidases**. Peptidases, which act from the amino or carboxy-terminal ends of proteins or peptides are called **exopeptidases**.

Exopeptidases that act from the carboxy-terminal are called **carboxypeptidases**, and if they act from the amino-terminal, they are called **amino-peptidases**. Each time the exopeptidase acts on the protein or peptide, it releases one amino acid.

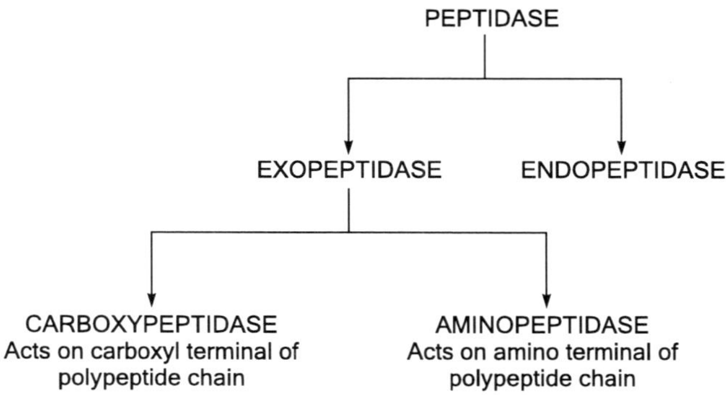

ENZYMES INVOLVED IN THE DIGESTION OF PROTEINS

Proteases of Gastric Origin

Pepsin

Pepsin is present in the gastric juice in its inactive form as **pepsinogen**. Pepsinogen is activated first when it comes in contact with the acidic environment where the pH is below 2. The acidic environment is created by the strong mineral acid, hydrochloric acid (HCl) present in the gastric juice. Pepsin has an optimum pH of 1.5 to 3.0.

$$\text{PEPSINOGEN} \xrightarrow{\text{pH} < 2} \text{PEPSIN}$$

$$\text{PEPSINOGEN} \xrightarrow[\text{Autocatalysis}]{} \text{PEPSIN}$$

The pepsin formed can act on another pepsinogen and convert it into pepsin. **Pepsin is an endopeptidase**. It acts on the peptide bonds formed by the carboxyl group of aromatic amino acids—methionine or leucine. The action of pepsin on proteins continues until the gastric contents reach the intestine.

Rennin

Rennin is present in the gastric juice in its inactive form as **prorennin**. The gastric pepsin activity converts prorenin to renin. The action of this enzyme has **significance in the early period of life**. It acts on the milk protein casein and coagulates it.

$$\text{PRORENNIN} \xrightarrow{\text{Pepsin}} \text{RENNIN}$$

It is an endopeptidase. It can act on the peptide bonds formed by the carboxyl groups of aromatic amino acids.

Proteases of Pancreatic Origin

Trypsin

Trypsin is synthesized and secreted in the inactive form as **trypsinogen** by the pancreatic gland. Trypsinogen is first activated to form trypsin by an intestinal enzyme **enteropeptidase**. (ENTEROKINASE)

$$\text{TRYPSINOGEN} \xrightarrow[\text{(Enteropeptidase)}]{\text{ENTEROKINASE}} \text{TRYPSIN}$$

$$\text{TRYPSINOGEN} \xrightarrow{\text{Trypsin}} \text{TRYPSIN}$$

Trypsin can further activate trypsinogen to trypsin. Trypsin can also activate other zymogens (inactive form of the enzymes) secreted by the pancreatic gland. **It is an endopeptidase**. Trypsin can act on the peptide bonds formed by the carboxyl group of basic amino acids.

Chymotrypsin

Chymotrypsin is secreted by the pancreatic gland in the inactive form as **chymotrypsinogen**. It is activated in the intestine by the action of trypsin to form chymotrypsin.

$$\text{CHYMOTRYPSINOGEN} \xrightarrow{\text{Trypsin}} \text{CHYMOTRPYSIN}$$

Chymotrypsin is an endopeptidase. It can hydrolyze the peptide bonds formed by the carboxyl group of the aromatic amino acid.

Elastase

Elastase is a pancreatic enzyme. It is secreted as **proelastase**. Proelastase is activated to form elastase by the action of trypsin.

$$\text{PROELASTASE} \xrightarrow{\text{Trypsin}} \text{ELASTASE}$$

Elastase is an endopeptidase. It acts mainly on the peptide bond in the proteins formed by the carboxyl groups of the neutral aliphatic amino acids like glycine, alanine, serine, etc. with small side chains.

Carboxypeptidase

Carboxypeptidase is present in the pancreatic juice as **procarboxypeptidase**. It is activated in the intestine by the action of trypsin.

$$\text{PROCARBOXYPEPTIDASE} \xrightarrow{\text{Trypsin}} \text{CARBOXYPEPTIDASE}$$

Carboxypeptidase is an exopeptidase. It cannot act on the dipeptides. So, carboxypeptidase acts on the polypeptide until a dipeptide is formed. It is a zinc-containing enzyme. There are two different types of carboxypeptidases:

1. **Carboxypeptidase A:** It acts on the last peptide bond on the carboxy terminal end if the last amino acid residue is an aromatic amino acid.
2. **Carboxypeptidase B:** It acts on the last peptide bond on the carboxy terminal end, provided the amino acid present on the carboxy terminal should be a basic amino acid.

Proteases of Intestinal Origin

Aminopeptidase

Aminopeptidase is present on the **surface of the intestine**. **It is an exopeptidase**. It acts on the amino-terminal of the peptide chain and releases one amino acid at a time. The action of this enzyme continues until a dipeptide is formed. Aminopeptidase cannot act on the dipeptides.

Dipeptidase

Dipeptidases are present on the surface of the intestine as well as the cytoplasm of the enterocytes. Dipeptidases act on dipeptides and hydrolyze the peptide bond forming free amino acids.

Absorption of Amino Acids

The free amino acids formed by the digestion of proteins are absorbed in the intestine by a Na^+ dependent carrier mechanism called sodium amino acid symport, and Na^+ independent mechanism called gamma glutamyl cycle or Meister's cycle.

Na^+ dependent absorption of amino acids

It is a Na^+ amino acid symport mechanism. Amino acids and Na^+ are transported together from the lumen of the intestine into the cytoplasm of the enterocytes as in glucose transport. The concentration of Na^+ in the enterocytes is kept low with the help of Na^+ K^+ ATPase.

There are different carrier systems present on the luminal surface of the intestine to absorb amino acids.

1. Transporter for acidic amino acids
2. Transporter for basic amino acids
3. Transporter for aromatic amino acids and methionine
4. Transporter for neutral amino acids
5. Transporter for imino acids

Na^+ independent absorption of amino acids

Na^+ independent mechanism of amino acid transport requires a tripeptide called glutathione.

DIGESTION AND ABSORPTION OF LIPIDS

Digestion of Lipids

Our daily diet contains about 60 to 150 g of lipids, the major portion of which is made up of fat (triacylglycerol). In addition to fat, dietary lipids may be phospholipids, cholesterols, cholestryl esters, and free fatty acids.

Lipids are hydrophobic in nature. The enzymes involved in the digestion of lipids are hydrophilic. Hence, these enzymes mainly act on the lipids present on the surface of the lipid droplet, as they cannot enter the core of the lipid aggregate. During the process of digestion, however, this obstacle is overcome by breaking the big lipid aggregates with the help of amphipathic molecules to form smaller lipid aggregates. This process is called **emulsification**.

Emulsification of lipids **helps to increase the surface area of the lipids** exposed to the lipid digestive enzymes present in the aqueous phase.

Emulsification

The digestion of lipids requires bile secreted from the liver. Bile contains bile acids and phospholipids. Both these molecules are amphipathic in nature. They have both hydrophilic as well as hydrophobic portions. Hydrophilic groups can interact with the water phase, and hydrophobic portions can interact with the lipid phase.

When the lipid content is constant, if the concentration of amphipathic molecules is increased (as happens in the intestine during the digestion of lipids), the lipid-water interphase increases. This causes the lipid phase to disperse as smaller droplets (Fig. 8.2).

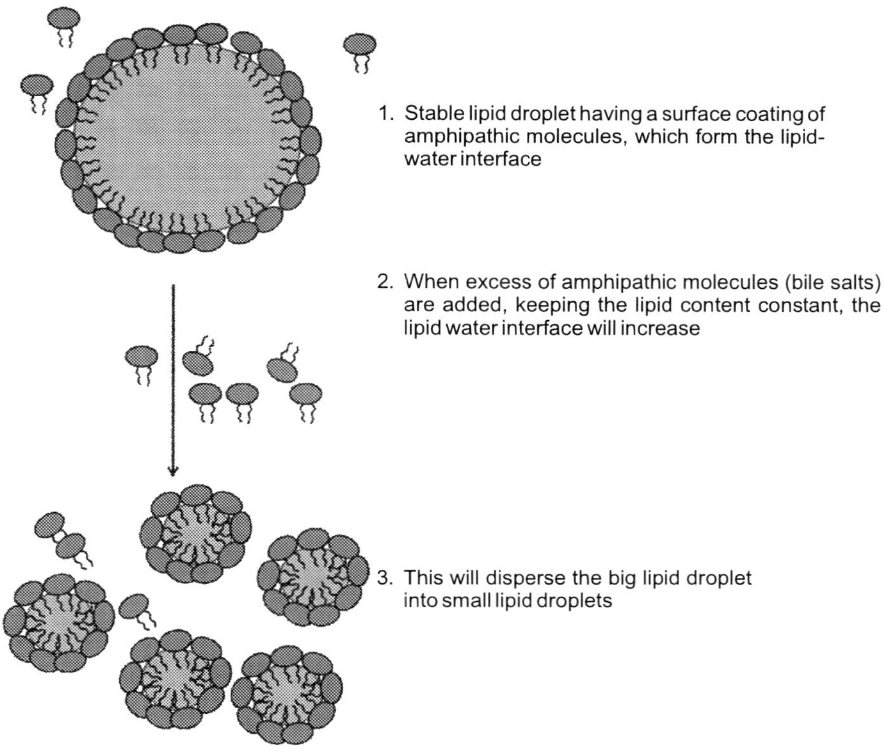

Fig. 8.2 Emulsification.

Once the lipids are emulsified with the help of bile salts, the surface area of the lipid droplets exposed to the digestive enzymes is increased.

ENZYMES INVOLVED IN THE DIGESTION OF LIPIDS

Digestion of Lipids in Stomach

Digestion of lipids starts in the stomach by an enzyme called **acid stable lipase**. In the stomach, this enzyme acts on a small portion of triacylglycerol and hydrolyzes it to form glycerol and fatty acids.

Digestion of Lipids in the Intestine

In the intestine, the digestion of lipids is mainly carried out by the pancreatic enzymes. The enzymes present in the pancreatic juice are as follows:

1. Pancreatic lipase
2. Pancreatic colipase
3. Phospholipase
4. Cholesterol esterase

Lipase

Lipase hydrolyzes the ester linkage between the fatty acids (preferably fatty acids having more than 10 carbon atoms) and the alcohol group of glycerol.

Pancreatic lipase is especially specific for the ester linkage present on the first carbon of glycerol. The next prefered ester linkage is present on the third carbon atom. Lipase has a limited effect on the ester linkage present on the second carbon atom.

$$\text{TRIGLYCEROL} \xrightarrow{\text{LIPASE}} \text{2,3 DIACYLGLYCEROL + FATTY ACID}$$

$$\text{2,3 DIACYLGLYCEROL} \xrightarrow{\text{LIPASE}} \text{2- MONOACYLGLYCEROL + FATTY ACID}$$

2 monoacylglycerol is stable in the intestinal pH. Some amount of 2 monoacylglycerol can be isomerized to 1 monoacylglycerol and acted upon by the lipase to form glycerol and fatty acid.

$$\text{2-MONOACYLGLYCEROL} \xrightarrow{\text{ISOMERIZED}} \text{1-MONOACYLGLYCEROL}$$

$$\text{1-MONOACYLGLYCEROL} \xrightarrow{\text{LIPASE}} \text{GLYCEROL + FATTY ACID}$$

The fatty acids formed can inhibit the activity of lipase. The inhibitory effect of fatty acid is prevented with the help of Ca^{2+}. Calcium reacts with the fatty acid to form calcium soaps and decreases the concentration of fatty acids.

Colipase

Colipase is present in the pancreatic juice in the inactive form as **procolipase**. It is activated in the intestine by the action of the enzyme trypsin to form colipase. Colipase is required for the action of lipase on triacylglycerol.

$$\text{PROCOLIPASE} \xrightarrow{\text{Trypsin}} \text{COLIPASE}$$

Lipase is inhibited by bile salts present in the surrounding media. It cannot penetrate the surface monolayer of bile salt present on the fat droplets. Colipase displaces the bile salts (emulcifier) and helps the lipase to get access to triacylglycerol.

Phospholipase

Phospholipase of the pancreatic juice is secreted in the inactive form as **prophospholipase**. It is activated by the action of the enzyme, trypsin.

$$\text{PROPHOSPHOLIPASE} \xrightarrow{\text{Trypsin}} \text{PHOSPHOLIPASE}$$

Phospholipase acts on the phospholipids and hydrolyzes the ester bond present on the second carbon atom of glycerol and converts phospholipids into lysophospholipids.

Cholesterol Esterase

Cholesterol esterase of the pancreatic juice acts on cholesterol ester and forms free cholesterol and fatty acid.

Absorption of Lipids

The digestive products of lipids are monoacylglycerol, fatty acids, glycerol, lysophospholipids, and cholesterol. The short chain fatty acids and medium chain fatty acids are absorbed directly and enter the portal circulation.

The theory of the absorption of lipids was given by **Bergstrom**. Long chain fatty acids and other lipid digestion products are predominantly hydrophobic in nature. They form micelles along with bile salts, which are water-soluble. From these micelles **(mixed micelles)**, lipid digestive products are absorbed.

Once inside the enterocytes, these digestion products are converted back to their original form (Fig. 8.3). Monoacylglycerol is esterified to form triacylglycerol; lysophospholipid is converted to form phospholipids; and cholesterol is converted to cholesterol ester. These lipids are packed into a **lipid-protein complex (lipoprotein)** and secreted into the lymphatic circulation. Lipoprotein is one micron in diameter and is **called chylomicron**. Later, these chylomicrons enter the systemic circulation and deliver lipids to those tissues, which are capable of metabolizing it.

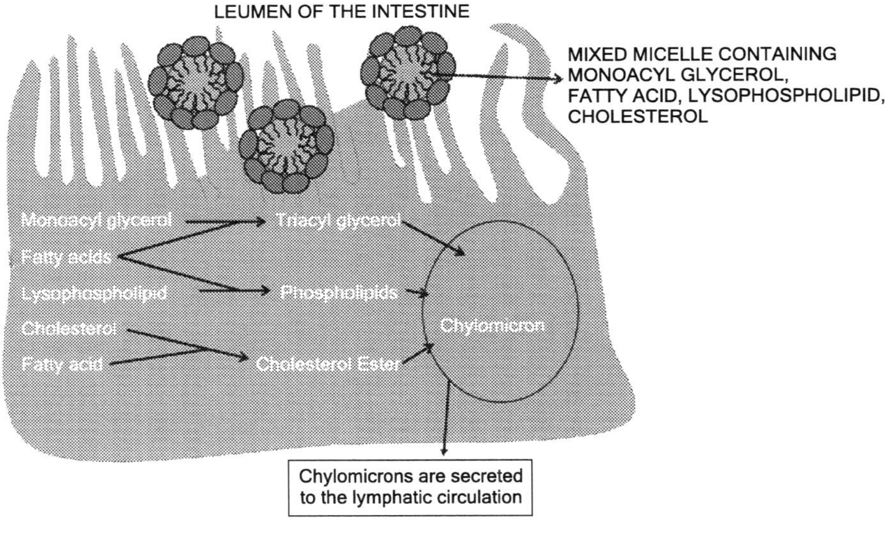

Fig. 8.3 Absorption of lipids

Steatorrhoea

The excretion of **excess of fat in the faeces is called steatorrhoea**. This may be due to malabsorption (defect in absorption) or maldigestion (defect in digestion). Steatorrhoea due to malabsorption leads to **split fat excretion** in faeces, whereas steatorrhoea due to maldigestion contains **unsplit fat in the faeces**.

Causes

1. Defect in the synthesis and secretion of pancreatic juice, as in various diseases of pancreas.
2. Defective secretion of bile, which is seen in case of obstructive jaundice or diseases of the liver, where bile synthesis may be decreased.
3. Diseases where absorptive surface of the intestine may be abnormal, which leads to defective absorption of fat.

Faecal fat content can be measured and the presence of split and unsplit fat content can be estimated. In steatorrhoea, fat-soluble vitamin absorption may also be defective, and deficiency manifestations may be seen in long standing cases.

QUESTIONS

Long Essay

1. Discuss in detail about the digestion and absorption of lipids.
2. Discuss in detail about the digestion and absorption of carbohydrates.
3. Discuss in detail about the digestion and absorption of proteins.

Short Essay

Write short notes on the following:

1. Absorption of carbohydrates.
2. Digestion of lipids.
3. Absorption of proteins.
4. Absorption of lipids.

Short Answer

1. What is the defect in lactose intolerance? Write the signs and symptoms of lactose intolerance.
2. Describe Hartnup's disease.
3. What is steatorrhoea? Mention any two causes for the same.
4. Discuss about the digestion of lipids.
5. How is the glucose absorbed in the intestine?
6. Write a note on the digestion of proteins.
7. Mention the functions of hydrochloric acid secreted by the stomach.

9

Intermediary Metabolism

INTRODUCTION

All living organisms require energy to sustain life. Living organisms may be **autotrophs** or **heterotrophs**. Autotrophs are able to synthesize their own food. For example, green plants are capable of synthesizing their own food using carbon dioxide and sunlight by a process called photosynthesis. Heterotrophs use organic molecules produced by autotrophs and oxidize it to derive energy. During the oxidation of nutrients, carbon dioxide is released and returned to the atmosphere.

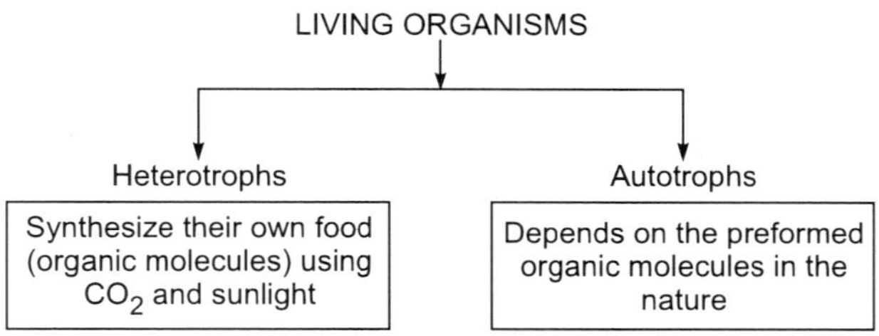

Fig. 9.1 Energy source.

Cells of higher animals are heterotrophs. They need preformed nutrients to derive energy. Food produced by plants is consumed by animals. Dietary components are digested and absorbed. After absorption, the small biomolecules reach various parts of the body through circulation, where different biochemical reactions take place.

The fate of the dietary components after digestion and absorption constitute intermediary metabolism. Metabolism is a highly regulated and coordinated cellular activity. It helps to convert nutrient molecules into cell's own characteristic biomolecules and to obtain energy. Series of biochemical reactions, which constitute metabolism, are called metabolic pathways (Fig. 9.2).

$$A \xrightarrow{E1} B \xrightarrow{E2} C \xrightarrow{E3} D$$

Fig. 9.2 Metabolic pathway.

In metabolic pathways consisting of more than one biochemical reaction, the product of one reaction is substrate for the next enzyme in the reaction sequence. The intermediates formed are called metabolites. Each consecutive step in such a pathway brings about small chemical changes like removal, addition, and transfer of specific groups, molecules, or atoms.

Metabolic Pathways

Metabolic pathways fall into three different categories:

1. Anabolic pathways
2. Catabolic pathways
3. Amphibolic pathways

Anabolic Pathway

It is the metabolic pathway involved in the synthesis of compounds constituting the structural and functional components of the cells. Here simple precursors are used to build large complex molecules. Anabolic pathways require input of energy.

Amino acids ⟶ Proteins

Glucose ⟶ Glycogen

Catabolic Pathway

It is an **oxidative pathway** in which energy is released in the form of ATP or reducing equivalents. Energy can be produced in the form of ATP at the metabolic pathway level. Then it is called **substrate level phosphorylation** (Fig. 9.3). During oxidative pathway, reducing equivalents are removed from the intermediates of metabolic pathway. These are then donated to the electron transport chain in the mitochondria.

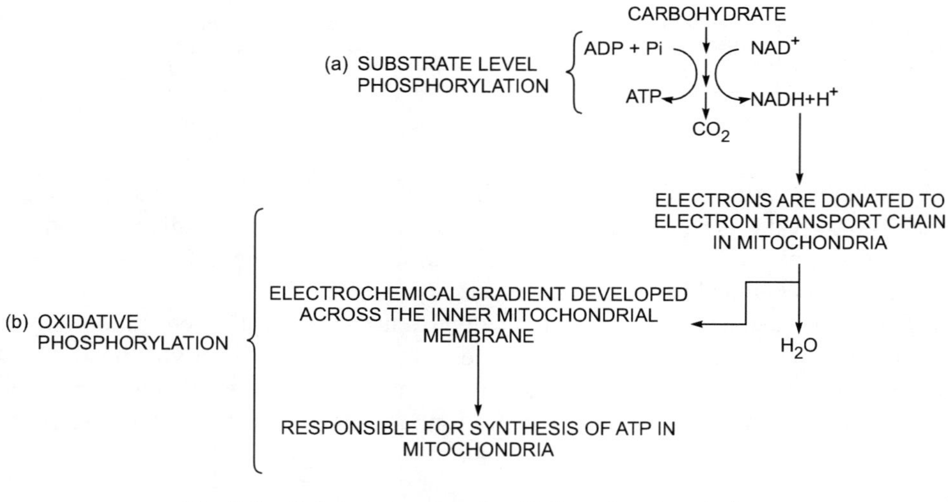

Fig. 9.3 Substrate level and oxidative phosphorylation.

When these electrons flow through the electron transport chain, electrochemical gradient is developed across the inner membrane of mitochondria. This is responsible for the formation of ATP in the mitochondria (chemiosmotic theory). This process is called **oxidative phosphorylation** (Fig. 9.3).

ELECTRON TRANSPORT CHAIN

Electron transport chain is situated in the inner membrane of mitochondria. It consists of a series of electron carriers. These carriers may be proteins or non-protein carriers. They are present as complexes. Electron transport chain is made up of different complexes. They are:

1. Complex I
2. Complex II
3. Complex III
4. Complex IV
5. Complex V (it is the ATP synthase complex).

Electrons removed from the metabolic intermediates during the metabolic pathway are transferred to the electron transport chain. These complexes are arranged in the **increasing order of their redox potential**. The component of the electron transport chain (ETC) at the beginning has negative redox potential, and the last component of the ETC has a positive redox potential.

The first complex of electron transport chain is called **NADH dehydrogenase** (Fig. 9.4). The reducing equivalents received by the coenzyme. NAD^+ from the intermediates of catabolic pathway are given to the complex I of the electron transport chain. From complex I, electrons are transferred to complex III through coenzyme Q.

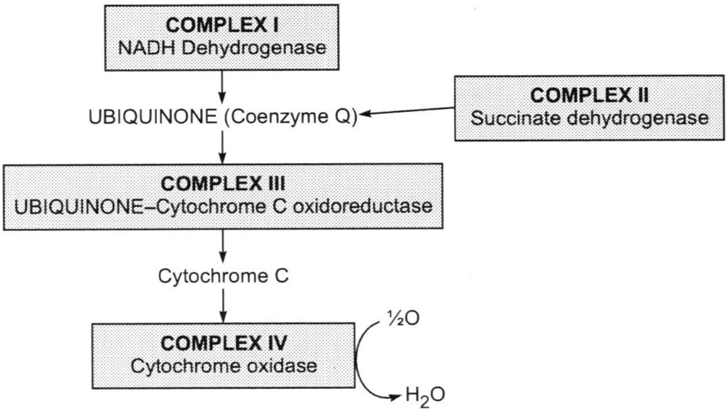

Fig. 9.4 Substrate level and oxidative phosphorylation.

Complex II of electron transport chain receives electrons from $FADH_2$ generated during the reaction catalyzed by succinate dehydrogenase of TCA cycle and glycerol-3-P dehydrogenase. From complex II, electrons are transferred to complex III via coenzyme Q.

The electrons later flow from complex III to complex IV via cytochrome C. Complex IV at the end reduces molecular oxygen by using these electrons to form water.

OXIDATIVE PHOSPHORYLATION

Oxidation is coupled to the phosphorylation of ADP to form ATP with the help of electron transport chain. It is explained by the **chemiosmotic theory of Peter Mitchell**. When electrons flow through the electron transport chain, protons are pumped at three different sites from the mitochondrial matrix to the intermembranous space. These three sites are:

1. Complex I
2. Complex III
3. Complex IV

This creates an **electrochemical gradient across the inner mitochondrial membrane**. Inner mitochondrial membrane contains another protein called **ATP synthase complex** (Fig. 9.5). When the protons present in the intermembranous space are carried along this protein back to the mitochondrial matrix, ADP is phosphorylated to form ATP.

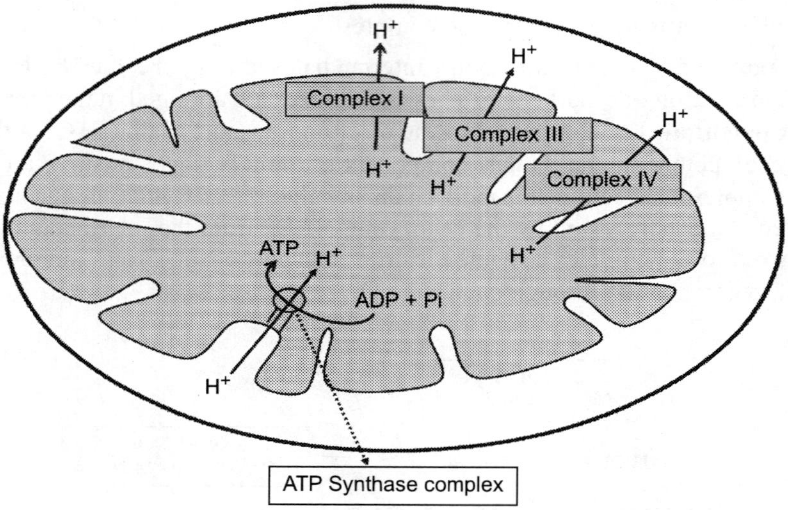

Fig. 9.5 Oxidative phosphorylation.

P:O RATIO

The relationship between the ATP generated and the oxygen consumed is given by the P:O ratio. When electrons are transported along the electron transport chain, the number of ATPs generated per molecule of oxygen consumed gives the P:O ratio.

When NADH+H^+ donates its electrons to the electron transport chain, three ATPs are synthesized and one oxygen atom of molecular oxygen is reduced to water. So, the P:O ratio is 3. The electrons are directly donated to the complex II of electron transport chain in case of dehydrogenases containing FAD as the coenzyme. The number of ATPs produced is two, and the number of oxygen atoms consumed is 1. So, the P:O ratio is 2.

Inhibitors of electron transport chain		
Complex I	*Complex III*	*Complex IV*
Amytal Piericidin Rotenone, etc.	Antimycin British antilewisite (BAL)	Cyanide Hydrogen sulphide Carbon monoxide

Uncouplers: Uncouplers are substances which uncouple (dissociate) oxidation from phosphorylation, e.g., (a) 2,4 dinitrophenol, (b) Pentachlorophenol, (c) Dinitrocresol, (d) Bilirubin.

QUESTIONS

Short Essay

1. Discuss briefly the components of electron transport chain and its inhibitors.
2. Describe chemiosmotic theory.

Short Answer

1. What are uncouplers? Give examples.
2. Mention any four inhibitors of oxidative phosphorylation.

10

Carbohydrate Metabolism

INTRODUCTION

Glucose is the major source of energy. There are certain tissues like erythrocytes, brain, and renal medulla, which mainly depend upon glucose as the source of energy. A continuous supply of glucose is required for the proper functioning of human body. Fasting plasma glucose level in normal individuals is 75 to 110 mg/dL. A normal individual has the capacity to maintain the fasting glucose level with the help of various hormones.

Carbohydrates are immediate source of energy. Glucose is oxidized in the body to form carbon dioxide and water. During this process, energy is extracted from it. Hence, most of the tissues contain enzymes required for carbohydrate metabolism. The major metabolic pathways by which glucose is metabolized are:

1. Glycolysis
2. Glycogenolysis
3. Glycogenesis
4. Gluconeogenesis
5. Hexose monophosphate shunt (HMP)
6. Uronic acid pathway
7. Galactose and fructose are also metabolized in our body by specific metabolic pathways.

GLYCOLYSIS

Glycolysis is a pathway by which glucose is converted into pyruvate under aerobic condition, and lactate under anaerobic condition.

Tissues where glycolysis takes place

- RBCs depend entirely upon glycolysis for energy production
- When the muscle is contracting, energy is provided by glycolysis
- Glycolytic pathway takes place in almost all the tissues.

Subcellular Site

- Reactions take place in the cytosol.

End products formed: Under aerobic condition (presence of oxygen) pyruvate, and anaerobic condition (absence of oxygen) lactate.

The glycolytic pathway can be divided in following three phases:

1. Energy investment phase

Energy investment phase extends up to the formation of fructose 1,6-bisphosphate.

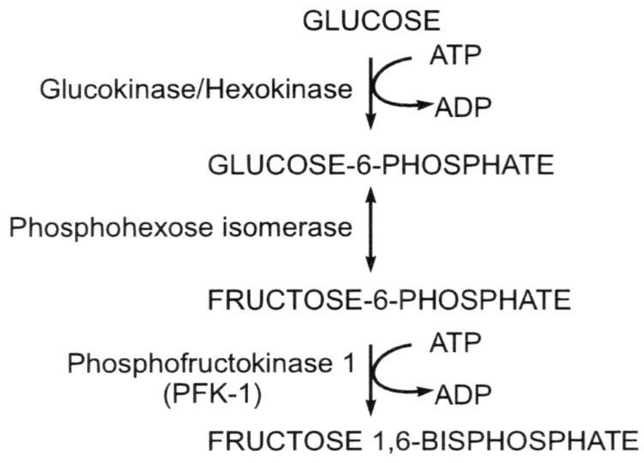

2. Splitting phase

Fructose 1,6-bisphosphate formed is split into two trioses, glyceraldehyde-3-phosphate and dihydroxy acetone phosphate by an enzyme called **aldolase B**.

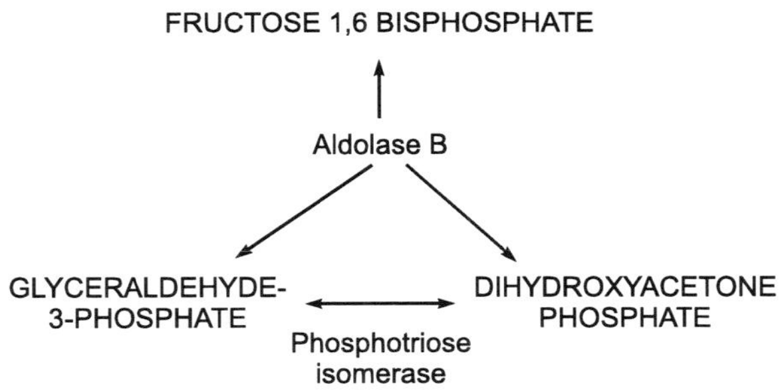

Dihydroxyacetone phosphate and glyceraldehyde-3-phosphate can be interconverted by an enzyme known as **phosphotriose isomerase**. Glyceraldehyde-3-phosphate enters into further glycolytic reactions. Hence, its concentration is kept low. This favours the conversion of dihydroxyacetone phosphate to glyceraldehyde-3-phosphate.

Under aerobic conditions, pyruvate formed at the end of glycolysis enters into mitochondria. Inside the mitochondria, pyruvate is converted into acetyl CoA, which is further metabolized.

Under anaerobic conditions, pyruvate is converted to lactate by an enzyme called **lactate dehydrogenase**. Lactate dehydrogenase utilizes the reducing equivalents from the NADH + H^+ and reduces pyruvate to lactate.

3. Energy extraction phase

Further reactions of glycolysis will convert glyceraldehyde-3-phosphate into pyruvate as shown below. During this phase, phosphoglycerate kinase and pyruvate kinase are the two enzymes that are capable of phosphorylating ADP to form ATP. So, energy is produced at the metabolic pathway level. This mechanism of production of ATP is called **substrate level phosphorylation**.

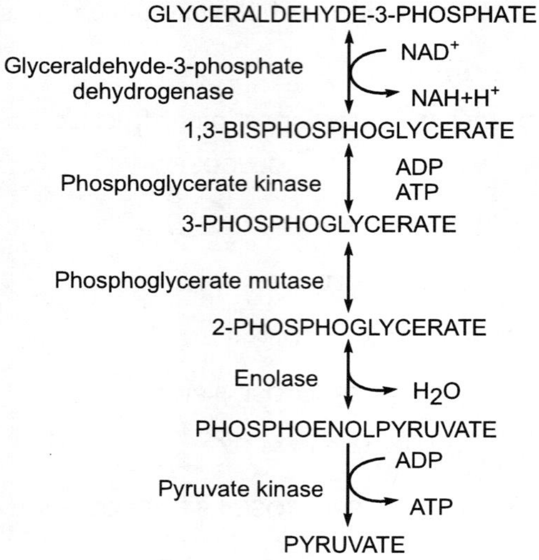

Energetics of Glycolysis

Initially, glucose is activated to form glucose-6-phosphate, and later, fructose-6-phosphate is converted to fructose 1,6-bisphosphate. Both these reactions utilize one ATP each. So, two ATPs are utilized for the activation of glucose. Activated glucose is split into two trioses:

1. One glyceraldehyde-3-phosphate, which enters into glycolysis, and produces 1 NADH + H^+, which is equivalent to **3 ATPs**.
2. **Two ATPs** are produced by the substrate level phosphorylation.

Since two glyceraldehyde-3-phosphates are formed from single glucose, **10 ATPs** are formed. Out of 10 ATPs, 2 are utilized during the energy investment phase.

1. The net ATPs produced by aerobic glycolysis are 8.
2. Under anaerobic conditions, NADH + H^+ is utilized for the conversion of pyruvate to lactate. So, the net ATPs formed are 2.

Regulation

- Phosphofructokinase-1 is the regulated enzyme of glycolysis.
- Fructose 2,6-bisphosphate and ADP are allostearic activators of the enzyme. ATP and citrate are allostearic inhibitors.
- Pyruvate kinase activity can also be regulated.

Rapoport Leubering Shunt

One of the intermediate of glycolysis, 1,3-bisphosphoglycerate can be converted into 2,3-bisphosphoglycerate. 2,3-bisphosphoglycerate has got a biological significance. It binds to haemoglobin and decreases its affinity for oxygen and increases the delivery of oxygen to the tissues.

2,3-bisphosphoglycerate is later converted to 3-phosphoglycerate, which enters glycolytic pathway. Since 1,3-bisphosphoglycerate is shunted through this pathway back to glycolysis, it is called 2, 3-bisphosphoglycerate shunt (2,3-BPG Shunt).

Citric Acid Cycle

(Kreb's Cycle, Tricarboxylic Acid Cycle)

Citric acid cycle (Fig. 10.1) **takes place inside the mitochondria**. Pyruvate, which is the end product of glycolysis, is transported into the mitochondria with the help of a transporter. Inside the mitochondria, it is converted into acetyl CoA by the action of **pyruvate dehydrogenase complex**. Pyruvate dehydrogenase complex is a multienzyme complex, which requires thiamine pyrophosphate (TPP), NAD^+, FAD, and lipoic acid.

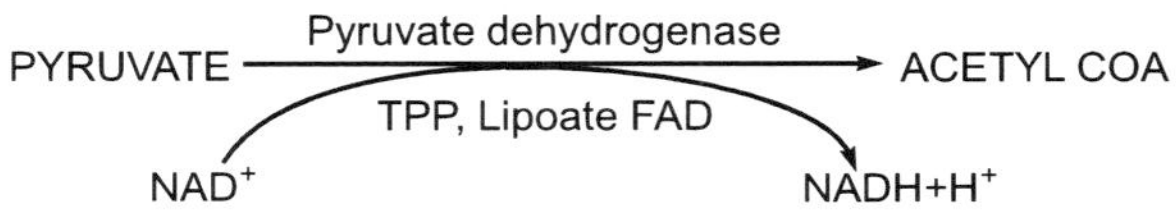

Acetyl CoA combines with oxaloacetate to form citrate. This is the first reaction of citric acid cycle. Further reactions proceed as shown in the reaction sequence, and at the end, oxaloacetate is regenerated. Reducing equivalents removed from the intermediates of TCA cycle are accepted by the coenzymes like NAD^+ and FAD, leading to the formation of NADH + H^+ and $FADH_2$. These reduced forms of coenzymes donate the electrons to the electron transport chain to generate three ATPs and two ATPs respectively.

Energetics

One acetyl CoA, which enters TCA cycle, is oxidized to produce **12 ATPs:**

1. Isocitrate dehydrogenase → 1 NADH + H^+ = 3 ATPs
2. α-ketoglutarate dehydrogenase → 1 NADH + H^+ = 3 ATPs
3. Succinate thiokinase (Substrate level phosphorylation) = 1 ATP
4. Succinate dehydrogenase → $FADH_2$ = 2 ATPs
5. Malate dehydrogenase → 1 NADH + H^+ = 3 ATPs

Total ATPs produced = 12 ATPs

From one glucose molecule, 2 pyruvate molecules and **8 ATPs** are formed from glycolytic pathway. Both pyruvate molecules enter mitochondria, where they are converted to acetyl CoA by pyruvate dehydrogenase complex. During this reaction, two NADH + H^+ is produced, which is equivalent to **6 ATPs**. Acetyl-CoA enters into TCA cycle and gets oxidized to CO_2 and form **24 ATPs**. So one mole of glucose when completely oxidized by glycolysis, followed by TCA cycle, it leads to the formation of **38 ATPs**.

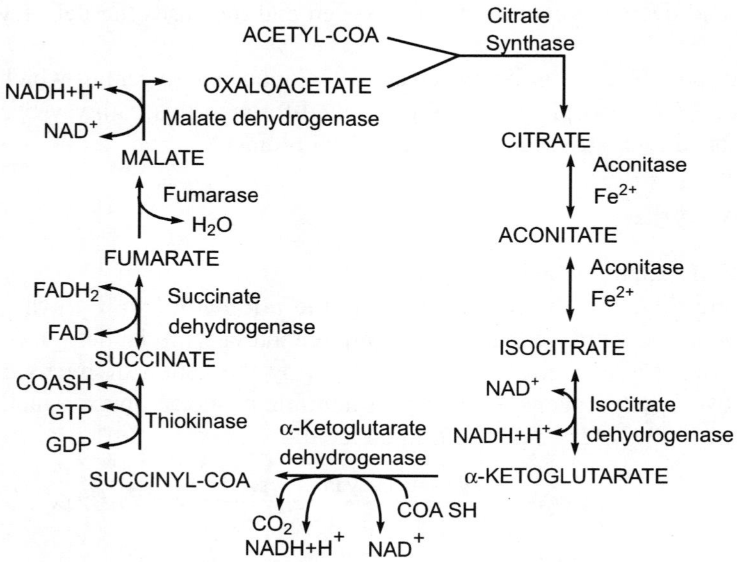

Fig. 10.1 Citric acid cycle.

Regulation

Increase in the NADH+H$^+$ will inhibit isocitrate dehydrogenase and alpha ketoglutarate dehydrogenase.

Anaplerosis

Reactions which replenish the intermediates of TCA cycle are called anaplerotic reaction and the process is called anaplerosis.

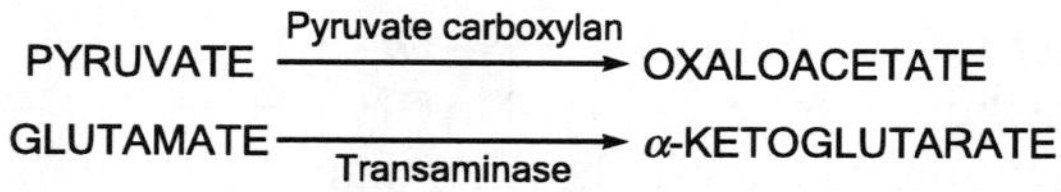

TCA Cycle is an Amphibolic Pathway

TCA cycle is an amphibolic pathway. Metabolism of amino acids and fatty acids produces intermediates of TCA cycle. They are catabolized through TCA cycle to produce CO_2 and energy. For example, propionic acid is produced during the β-oxidation of the odd chain fatty acids. Later, it is converted to succinyl CoA and enters TCA cycle. Alanine and glutamate are converted to pyruvate and α-ketoglutarate respectively. They are further oxidized by TCA cycle to yield energy.

Pyruvate can be converted to acetyl CoA and oxidized by TCA cycle. α-ketoglutarate is an intermediate of TCA cycle, which can be oxidized by TCA cycle to yield energy.

Intermediates of TCA cycle can be utilized for the synthesis of various biologically important substances (anabolism).

For example:

1. Succinyl CoA is used for the synthesis of heme.
2. α-ketoglutarate is used for the synthesis of non-essential amino acid, glutamate.
3. Oxaloacetate is used for the synthesis of aspartate. It is required for protein synthesis.

Since TCA cycle has got role both in anabolism as well as catabolism, it is an **amphibolic pathway**.

Gluconeogenesis

Gluconeogenesis or neoglucogenesis is the synthesis of new glucose molecules from noncarbohydrate substances. When there is a deficiency of glucose in our body, as in the case of starvation, gluconeogenesis provides glucose to those tissues (neurons), which entirely depend upon glucose (RBCs) or are mainly dependent on glucose as a source of energy.

Gluconeogenic Substrates

Non-carbohydrate substances that can be converted to glucose are called gluconeogenic substrates. **Lactate**, **glycerol**, **propionic acids**, and **various amino acids** called glucogenic amino acids are gluconeogenic substrates.

Lactate is formed during anaerobic glycolysis. Glycerol is produced when the fat in the adipose tissue is mobilized. The last three carbons of the odd chain fatty acids are released as propionic acid. Proteolysis leads to the release of amino acids.

Gluconeogenesis involves most of the reactions of glycolysis, but in the reverse direction, and a few reactions of TCA cycle. Simple reversal of glycolysis is impossible because of the irreversible nature of few of the reactions of glycolysis. These reactions are catalyzed by the following enzymes:

1. **Pyruvate kinase:** Catalyzes the conversion of phosphoenolpyruvate to pyruvate.
2. **Phosphofructokinase 1:** Catalyzes the phosphorylation of fructose-6-phosphate to form fructose 1,6-bisphosphate.
3. **Glucokinase:** It converts glucose to glucose-6-phosphate.

Key Enzymes of Gluconeogenesis

The irreversible reactions of glycolysis are reversed with the help of special enzymes called key enzymes of gluconeogenesis. These enzymes are:

1. Pyruvate carboxylase
2. Phosphoenolpyruvate carboxykinase
3. Fructose 1,6-bisphosphatase
4. Glucose-6-phosphatase

Reactions Catalyzed by the Key Enzymes of Gluconeogenesis

The reaction catalyzed by pyruvate kinase in glycolysis is irreversible. To reverse this reaction, the sequential action of two different enzymes is required. First, pyruvate carboxylase converts pyruvate to oxaloacetate. It is a carboxylation reaction requiring carbon dioxide, water-soluble vitamin, biotin, magnesium, and energy (Fig. 10.2).

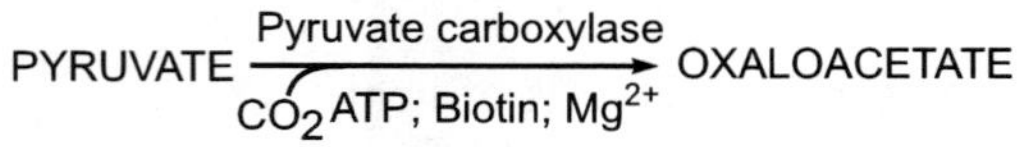

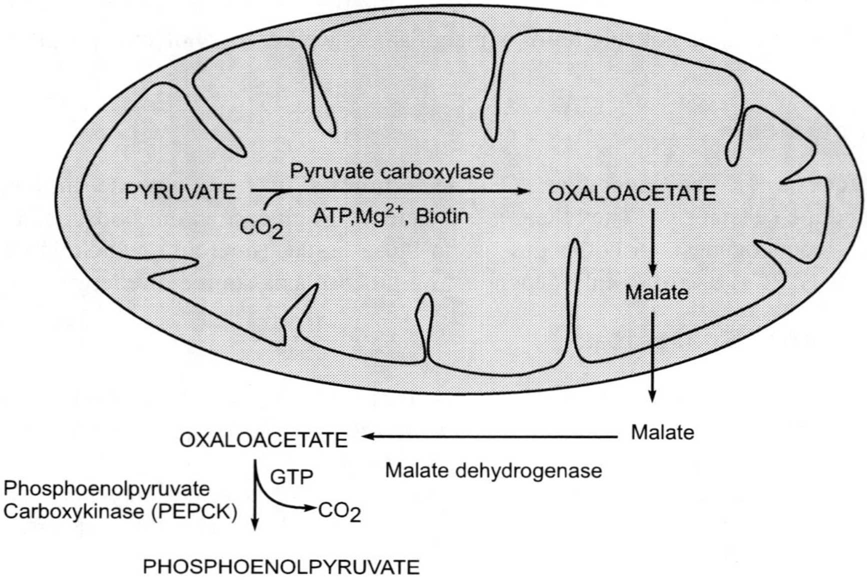

Fig. 10.2 Conversion of pyruvate to phosphoenolpyruvate.

Later, phosphoenolpyruvate carboxykinase (PEPCK) acts on the oxaloacetate to form phosphoenolpyruvate.

Oxaloacetate is impermeable through the mitochondrial membrane. To transport oxaloacetate across the mitochondrial membrane, it is first converted to malate by malate dehydrogenase. Malate is carried out of the mitochondria, and in the cytosol it is converted back to oxaloacetate by malate dehydrogenase.

PEPCK is present both in cytoplasm and mitochondria of human beings. PEPCK present inside the mitochondria can convert some amount of oxaloacetate to phosphoenolpyruvate. It is later transported to cytosol and enters further reactions of gluconeogenesis.

Phosphoenolpyruvate is converted to fructose 1,6-bisphosphate by the sequential action of various enzymes of glycolysis in the reverse direction. Conversion of fructose 1,6-bisphosphate to fructose-6-phosphate is an irreversible step. It requires another key enzyme of gluconeogenesis called fructose 1,6-bisphosphatase.

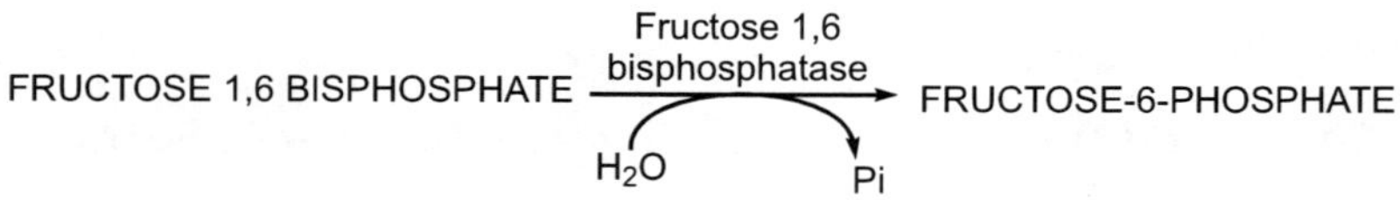

Fructose-6-phosphate formed is isomerized to form glucose-6-phosphate. Glucose-6 phosphate can be converted to glucose by the action of the enzyme glucose-6-phopshatase.

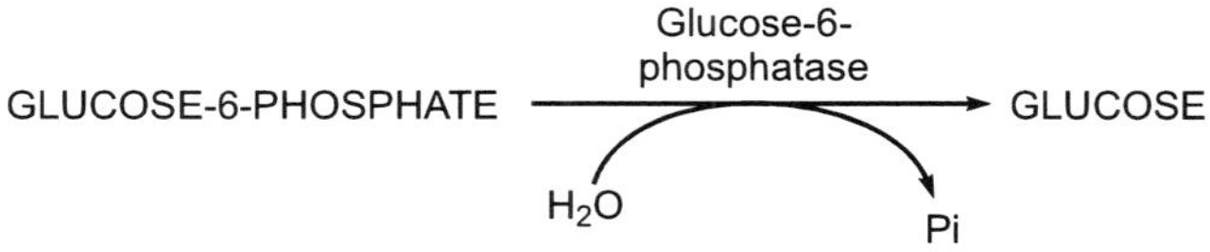

Glucose-6-phosphatase is present in the liver and medulla of the kidney. Substances that can form pyruvate or any intermediates of glycolysis and TCA cycle can be converted to glucose by the gluconeogenic pathway.

CORI'S CYCLE

During the muscle contraction, under anaerobic conditions, glucose is converted to lactate. Lactate is transported to liver through circulation. In the liver, lactate is converted back to pyruvate. By gluconeogenesis, pyruvate is converted to glucose. Glucose is transported back to the muscle. It is called Cori's cycle (Fig. 10.3).

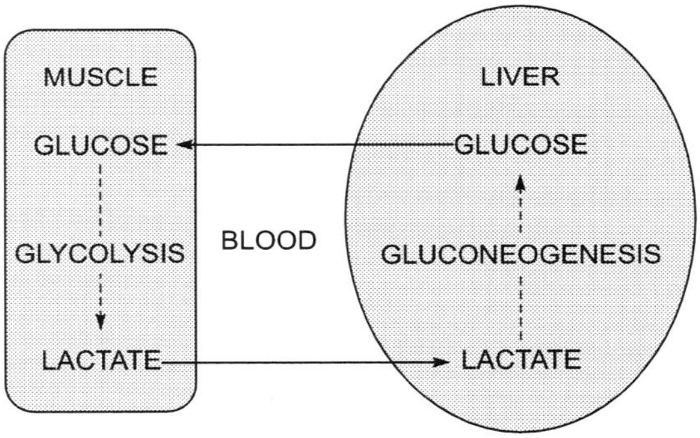

Fig. 10.3 Cori's cycle.

1. **Lactate** is converted to pyruvate by the action of lactate dehydrogenase. Later, pyruvate is converted to glucose by gluconeogenic pathway (Fig. 10.4).
2. **Glycerol** is converted to the intermediate of glycolysis, dihydroxyacetone phosphate. Later, it can be converted to glucose (Fig. 10.4).
3. **Various amino acids** can form pyruvate or intermediates of TCA cycle. They can be converted to glucose. These amino acids are known as glucogenic amino acids. Alanine, serine, cysteine, etc. form pyruvate during their metabolism. This can be later converted to glucose. Glutamate forms α-ketoglutarate and aspartate forms oxaloacetate. Both are later converted to glucose.
4. **Propionic acid** forms succinyl CoA, which is an intermediate of TCA cycle. Later it gets converted to glucose (Fig. 10.5). If the enzyme isomerase is defective or if there is a deficiency of vitamin B12, methyl malonic acid gets accumulated and is excreted in the urine. This condition is called **methylmelonic aciduria**.

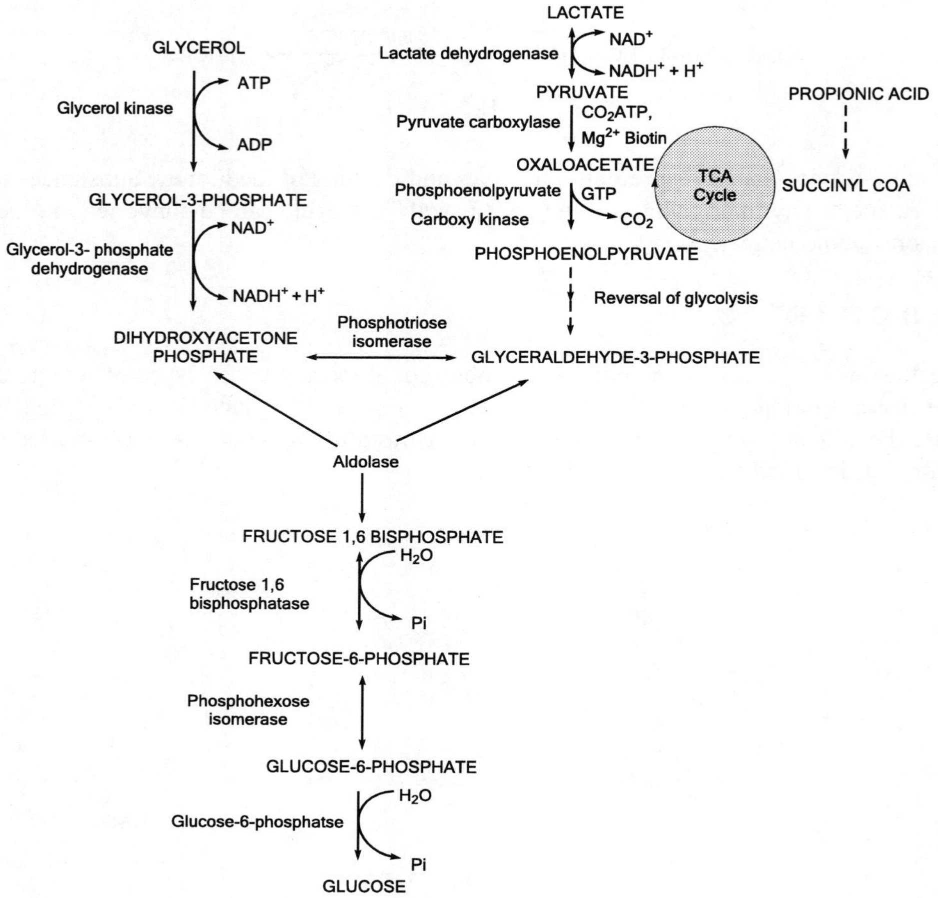

Fig. 10.4 Reactions of gluconeogenesis.

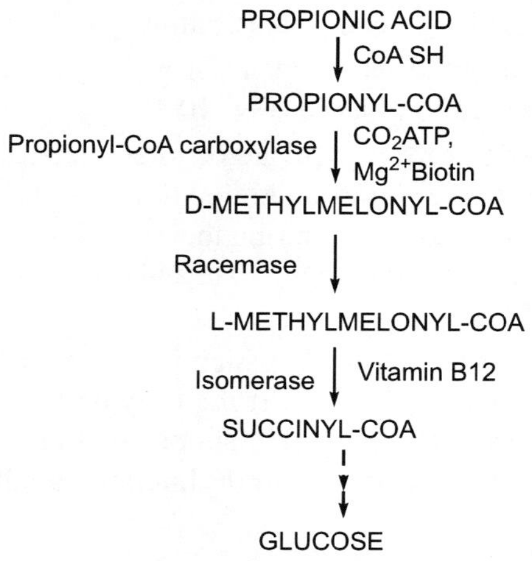

Fig. 10.5 Conversion of propionic acid to glucose.

Regulation of Gluconeogenesis

Insulin and **glucagon** are mainly involved in the regulation of gluconeogenesis. Insulin decreases the synthesis of the key enzymes of gluconeogenesis. Glucagon increases the synthesis of the key enzymes of gluconeogenesis.

Insulin increases the rate of glycolysis, whereas glucagon increases the rate of gluconeogenesis and decreases the rate of glycolysis. Most of the reactions of gluconeogenesis are reversal of glycolysis. So, both these pathways cannot operate simultaneously. When gluconeogenesis is predominant, the rate of glycolysis is decreased and vice versa.

Fructose 2,6-bisphosphate is formed in excess when the concentration of insulin is increased by the action of an enzyme called phosphofructokinase-2.

Fructose 2,6-bisphosphate is an inhibitor of fructose 1,6-bisphosphatase and an activator of phosphofructokinase-1 (Fig. 10.6).

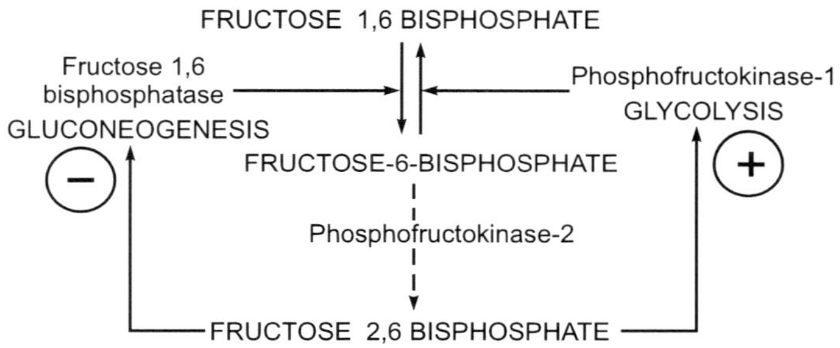

Fig. 10.6 Regulation by fructose 2, 6 bisphosphate.

In a well-fed state, when the blood glucose level is increased, insulin is secreted. It decreases the synthesis of the enzymes of gluconeogenesis and keeps phosphofructokinase-1 in an active state.

During fasting or starvation, blood glucose level is decreased. So, insulin level is decreased in the blood and there is a relative increase in glucagon level, which increases the rate of gluconeogenesis and decreases glycolysis.

GLYCOGEN METABOLISM

Glycogen is the storage form of carbohydrate in animals. It is a homopolysaccharide made up of glucose units. It is highly branched. Glycogen is stored in liver and muscles. When surplus amount of carbohydrate is present in the body, it is stored as glycogen. Synthesis of glycogen is called **glycogenesis**. If blood glucose level is decreased, the stored glycogen is broken down and glucose is released to maintain the blood glucose level. The breakdown of glycogen is called **glycogenolysis**.

Glycogenesis and glycogenolysis constitute glycogen metabolism.

GLYCOGENESIS

Glycogenesis is the set of biochemical reactions leading to the formation of glycogen. During glycogenesis, glucose is activated to form its nucleotide derivative called UDP glucose.

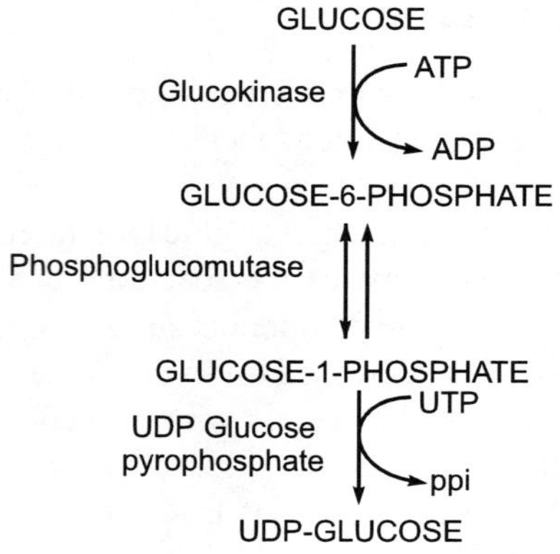

Synthesis of glycogen requires a pre-existing glycogen or a primer molecule. Glycogen primer can be synthesized on the backbone of a protein called glycogenin. This protein contains a tyrosine residue. The hydroxyl group of the side chain of tyrosine residue is linked to glucose by glycosidic linkage. Later, glucose residues are added to it, leading to the formation of glycogen primer, which is used for the synthesis of glycogen as shown in Fig. 10.7.

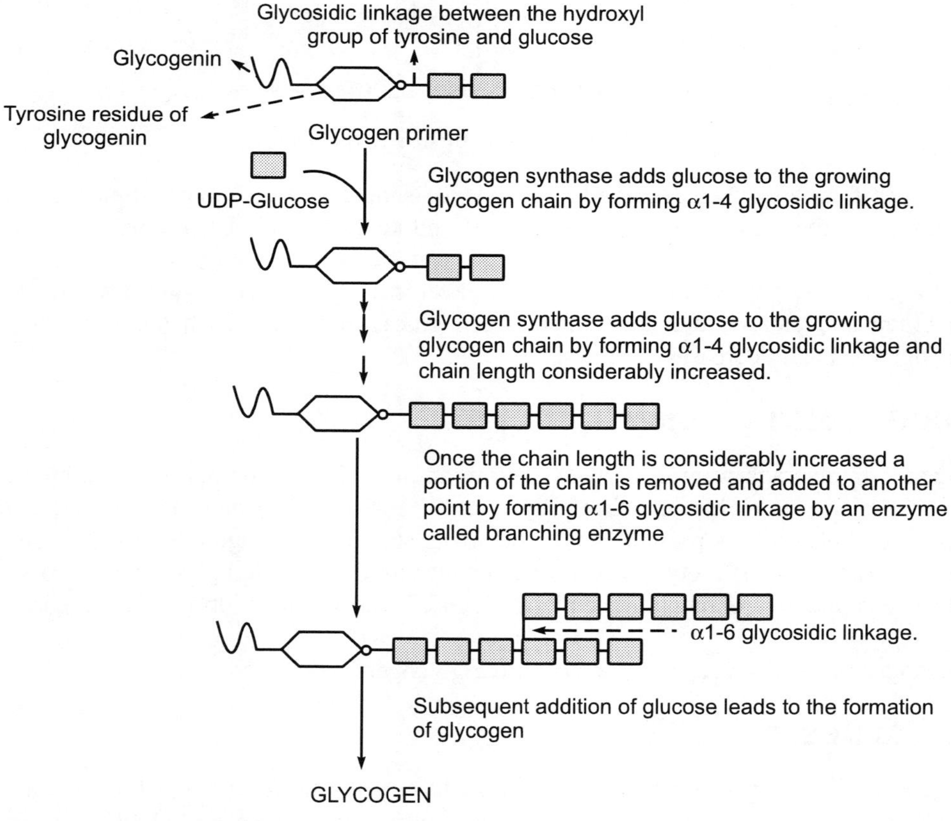

Fig. 10.7 Glycogenesis.

GLYCOGENOLYSIS

Free glucose can be formed from glycogen during glycogenolysis only in the tissues having **glucose-6-phosphatase** activities. Since the muscles don't have glucose-6-phosphatase activity, they cannot form free glucose, whereas the liver containing the glucose-6-phosphatase activity can convert glucose-6-phosphate to free glucose (Fig. 10.8).

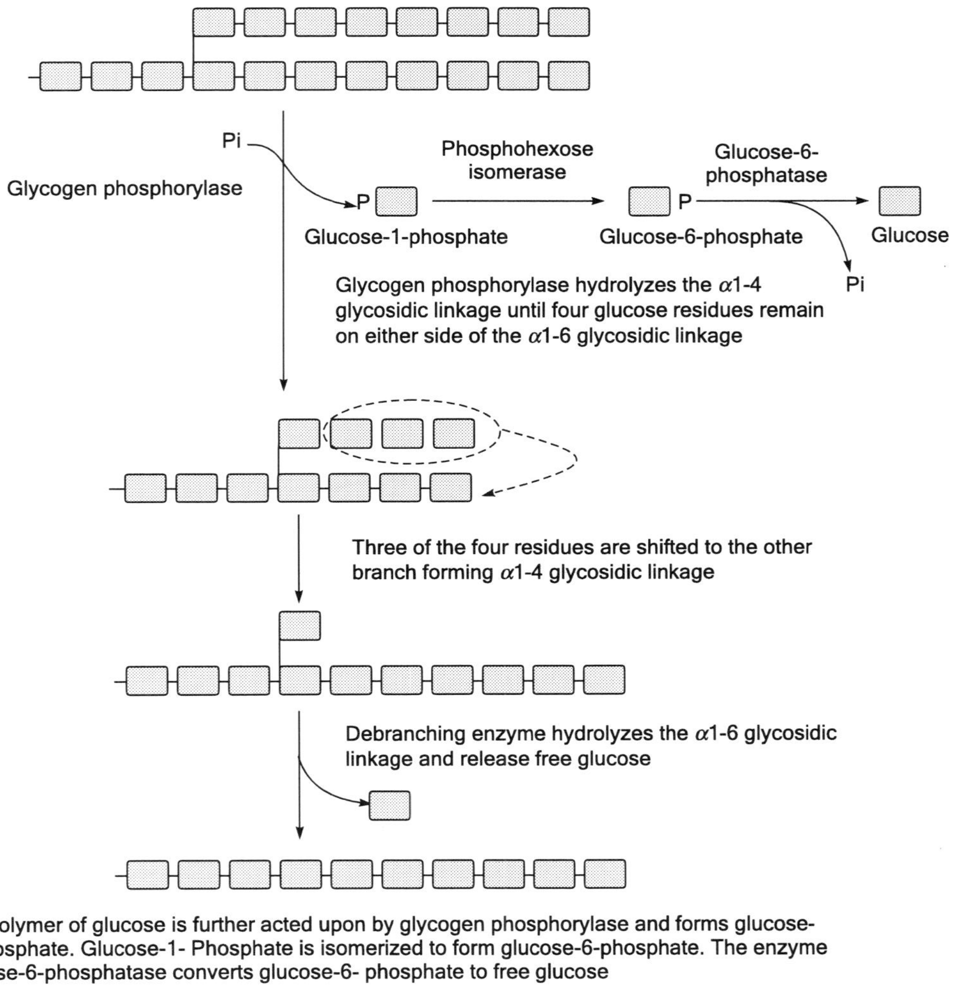

Fig. 10.8 Glycogenolysis.

Regulation of Glycogen Metabolism

Glycogen synthesis takes place in a well-fed state, and glycogen breaks down in fasting condition. Insulin favours glycogenesis, and hormones like glucagon and epinephrine favour glycogenolysis.

Epinephrine or glucagon binds to their respective receptors on the cell membrane and activate **adenylate cyclase**. Adenylate cyclase in the active form converts ATP to cAMP. Increase in the cAMP concentration activates cAMP dependent protein kinase.

Active form of the cAMP dependent protein kinase phosphorylates glycogen phosphorylase and glycogen synthase (Fig. 10.9). In the phosphorylated form, glycogen synthase is inactive, and glycogen phosphorylase is active.

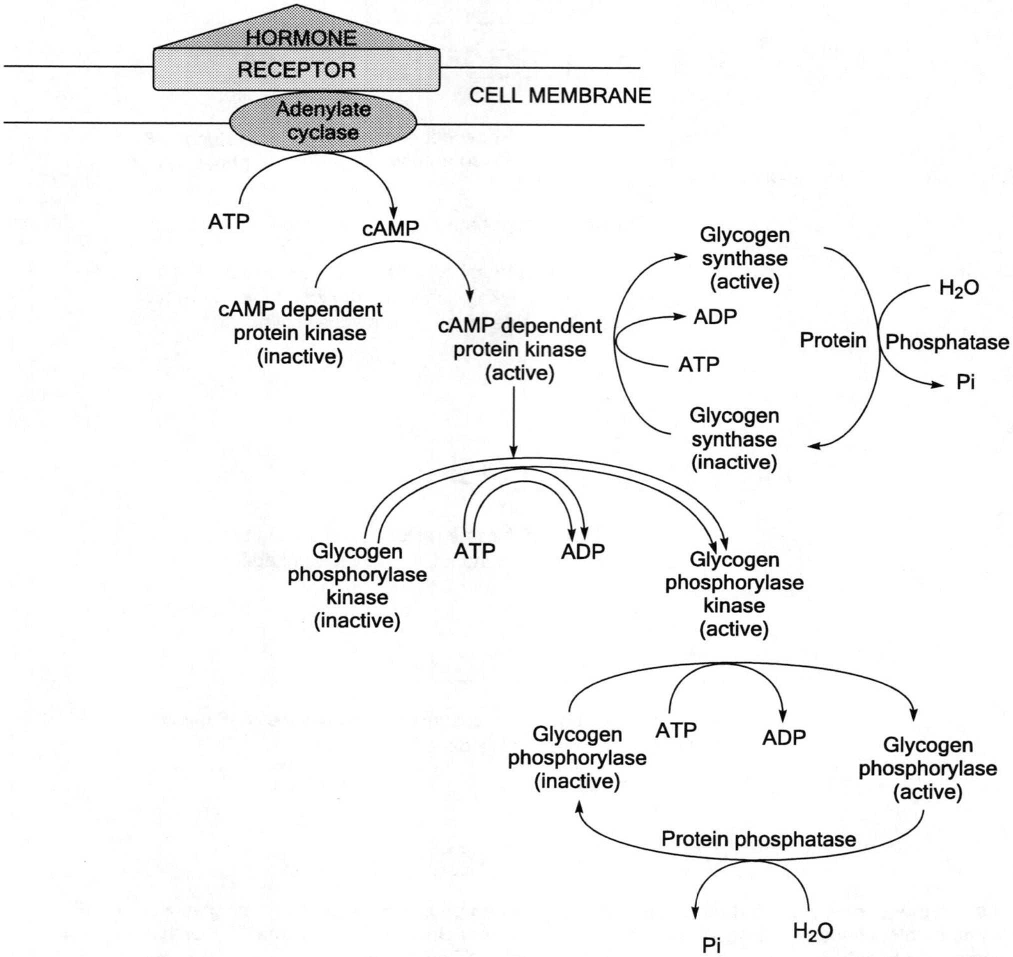

Fig. 10.9 Regulation of glycogen metabolism.

Phosphodiesterase is an enzyme, which converts cAMP to 5′AMP. Insulin activates this enzyme and decreases the cAMP content of the cell. It facilitates dephosphorylation of glycogen synthase and glycogen phosphorylase. In the dephosphorylated form, glycogen synthase is active whereas glycogen phosphorylase is inactive.

Glycogenstorage Disorders

Defect or deficiency in the enzymes involved in the metabolism of glycogen gives rise to glycogen storage disorders. There are eight different types of glycogen storage disorders (Table 10.1).

Table 10.1: Types of glycogen storage disorders

Type	*Deficiency*	*Name of the disease*
Type I	Glucose-6-phosphatase	Von Gierke's disease
Type II	α-1,4-glucosidase (Lysosomal)	Pompe's disease
Type III	Debranching enzyme	Cori's disease
Type IV	Branching enzyme	Andersen's disease
Type V	Glycogen phosphorylase (muscle)	McArdle's disease
Type VI	Glycogen phosphorylase (liver)	Hers's disease
Type VII	Phosphofructokinase	Tauri's disease
Type VIII	Phosphorylase kinase	

Von Gierke's Disease (Glucose-6-Phosphatase Deficiency)

In this disease, the enzyme glucose-6-phosphatase is defective. So, when glycogen is broken down, free glucose cannot be formed. Glucose gets accumulated as glucose-6-phosphate. Since free glucose is not formed during glycogenolysis, the affected individual suffers from fasting hypoglycemia. As the formation of glucose from glucose-6-phosphate is blocked, injection of glucagon or epinephrine is not able to increase the blood glucose level. Glucose-6-phosphate, which is accumulated, is diverted to HMP shunt pathway and glycolysis. This is responsible for lactic acidosis, hypertriglyceridemia and hyperuricemia seen in the type I glycogen storage disorder.

HEXOSE MONOPHOSPHATE SHUNT (HMP SHUNT)

[Pentose Phosphate Pathway (PPP)]

In HMP shunt, glucose is directly oxidized to produce NADPH + H^+ and CO_2. This pathway **takes place in the cytosol** of various tissues like the liver, adipose tissue, adrenal cortex, lactating mammary gland, erythrocyte, neutrophils, testis, etc.

Glucose is converted to glucose-6-phosphate by the action of glucokinase or hexokinase by utilizing phosphate group from the ATP. Glucose-6-phosphate is oxidized by the action of the enzyme glucose-6-phosphate dehydrogenase to form 6-phosphogluconolactone. The enzyme glucose-6-phosphate dehydrogenase requires $NADP^+$ as the coenzyme, and during the reaction the coenzyme is reduced (Fig. 10.10).

In the next reaction, an enzyme called 6-phosphogluconolactone hydrolase hydrolyses phosphogluconolactone to 6-phosphogluconate. This product is later oxidized and decarboxylated by the action of a dehydrogenase to form ribulose-5-phosphate. During this oxidative reaction, $NADP^+$, the coenzyme, is reduced to NADPH + H^+.

Some of the ribulose-5-phosphate isomerize to form ribose-5-phosphate and some are epimerized to form xylulose-5-phosphate. Ribose-5-phosphate and xylulose-5-phosphate react in the presence of an enzyme called transketolase to form sedoheptulose-7-phosphate and glyceraldehyde-3-phosphate. The enzyme transketolase requires thiamine pyrophosphate (TPP) as the coenzyme.

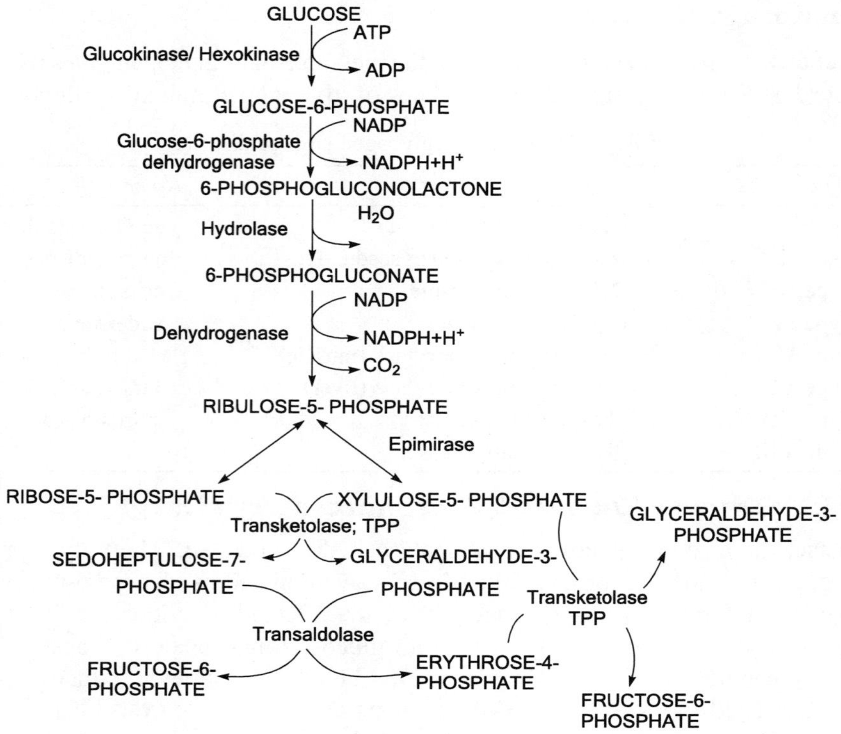

Fig. 10.10 Hexose monophosphate shunt (HMP)

Later, glyceraldehyde-3-phosphate and sedoheptulose-7-phosphate react in the presence of transaldolase to form erythrose-4-phosphate and fructose-6-phosphate.

Erythrose-4-phosphate and an another molecule of xylulose-5-phosphate react in the presence of transketolase to form fructose-6-phosphate and glyceraldehyde-3-phosphate. Glyceraldehyde-3-phosphate and fructose-6-phosphate enter into glycolytic pathway.

Glucose-6-phosphate is the substrate for glycolysis. It can enter into an alternative pathway, where it gets converted to fructose-6-phosphate that enters back into glycolysis. Hence, this pathway is called hexose monophosphate shunt.

The initial reactions of the pathway where oxidation takes place is called oxidative phase. At the end of the **oxidative phase**, ribulose-5-phosphate is formed. It is later isomerized to ribose-5-phosphate and xylulose-5-phosphate. This phase of the metabolic pathway is called **nonoxidative phase**.

Importance of Hexose Monophosphate Shunt

During HMP shunt, a large amount of reducing equivalents in the form of NADPH+ H^+ and pentose phosphate are produced.

I. The reducing equivalents can be used for the following purposes

1. Reductive synthesis of fatty acids, cholesterol, steroid hormone, glutamate, etc.

2. They can be utilized by cytochrome P450 enzyme system.
3. To reduce the oxidized form of glutathione by the enzyme glutathione reductase.
4. Help to keep haemoglobin in the reduced state.
5. In neutrophil, NADPH oxidase utilizes NADPH and produces superoxide, which helps to kill bacteria.

II. The pentose sugar, ribose-5-phosphate is utilized for the formation of nucleotides

1. The pentose sugars of dietary origin or produced inside the body can be converted to the intermediates of glycolysis.
2. Provide pentose sugars for the synthesis of oligosaccharide chain of glycoproteins.

Glucose-6-Phosphate Dehydrogenase Deficiency

In glucose-6-phosphate dehydrogenase deficiency, production of NADPH is reduced. Due to the decreased cell content of NADPH, glutathione remains in the oxidized state. This leads to the accumulation of free radicals in the cells, especially RBCs. It damages the cell membrane and gives rise to haemolysis. Usually, individuals with glucose-6-phosphate dehydrogenase deficiency are symptomless. When they are exposed to certain drugs like sulphonamides, primaquine, etc. or consumption of fava beans (Favism), the enzyme deficiency manifests and severe haemolysis can be seen.

METABOLISM OF FRUCTOSE

Fructose is a ketohexose. It is phosphorylated by fructokinase to form fructose-1-phosphate. Fructose-1-phosphate forms glyceraldehyde and dihydroxyacetone phosphate in the presence of an enzyme called aldolase.

Glyceraldehyde is converted to glyceraldehyde-3-phosphate by the action of a kinase, which utilises phosphate group from ATP. Glyceraldehyde-3-phosphate can enter into glycolytic pathway, or along with dihydroxyacetone phosphate it is converted to fructose 1,6-bisphosphate and enters gluconeogenic pathway (Fig. 10.11).

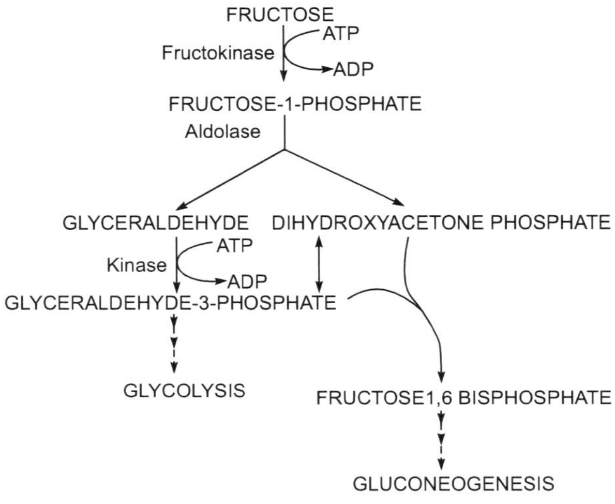

Fig. 10.11 Metabolism of fructose.

Fructose is mainly derived from the digestion of dietary sucrose. During the metabolism of fructose, it is converted to the intermediate of glycolysis and later to pyruvate, bypassing the regulated step of glycolysis. The fructose metabolism through the glycolytic pathway is unregulated. All the fructose gets converted to acetyl CoA. Acetyl CoA pool is increased. It is the substrate for fatty acid synthesis.

When the fructose content of the blood increases, blood glucose level also increases. This is responsible for the increase in insulin content of the blood. Insulin favours fat synthesis. So, individuals who eat more sweets may become obese due to the deposition of fat in the adipose tissues.

Essential Fructosuria

Defect in the enzyme fructokinase gives rise to essential fructosuria, in which fructose gets accumulated and is excreted in the urine.

Fructose Intolerance

Defect in the enzyme aldolase B is responsible for fructose intolerance.

METABOLISM OF GALACTOSE

Galactose is one of the constituents of milk sugar. It is converted to glucose in the liver (Fig. 10.12).

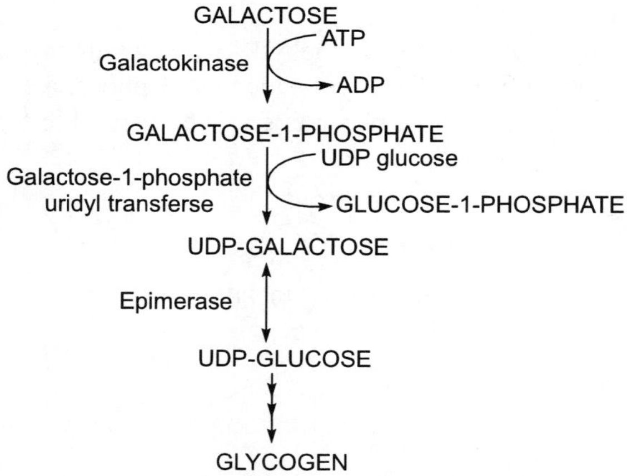

Fig. 10.12 Metabolism of galactose.

Galactosemia

In this disease, there is a **defect in the enzyme galactose-1-phosphate uridyl tranferase.** Galactose gets accumulated in the blood, so the condition is called galactosemia. Cataract is seen in children suffering from galactosemia.

BLOOD GLUCOSE HOMEOSTASIS

Normal fasting plasma glucose level is 75 to 110 mg/dL. After a carbohydrate rich diet, glucose level increases slowly and is below 140 mg/dL in normal individuals.

When the blood glucose level decreases, free glucose is poured into the circulation by different metabolic pathways in order to maintain the normal blood glucose level. When the blood glucose level increases, various metabolic pathways come into action to utilize the glucose or store the excess amount of glucose.

Metabolic pathways responsible for utilization and storage of glucose

1. Glycolysis and citric acid cycle
2. HMP shunt
3. Glycogenesis
4. Uronic acid pathway

Metabolic pathways responsible for the formation of glucose

1. Glycogenolysis
2. Gluconeogenesis

A perfect balance between these two groups of pathways of carbohydrate metabolism is responsible for maintaining the blood glucose level. This fine regulation is mainly done by the hormonal regulation of blood glucose hormones.

Insulin is the major hypoglycemic hormone, and glucagon is the major hyperglycemic hormone. In addition to glucagon, many other hormones are capable of increasing the blood glucose level by altering the rate of various metabolic pathways responsible for carbohydrate metabolism. These hormones are epinephrine, glucocorticoids, thyroid hormone, growth hormone, etc.

Increase in the Level of Blood Glucose

Insulin is secreted when the blood glucose level increases. Insulin increases the number of glucose transporters on the cell membrane of the peripheral tissues and increases the rate of transport of the glucose inside the cell. Insulin increases the rate of glycolysis, glycogenesis, and HMP shunt pathway and simultaneously suppresses the rate of glycogenolysis and gluconeogenesis.

Decrease in the Level of Blood Glucose

Glucagon level is increased when the blood glucose level decreases, as in case of fasting and starvation. Glucagon increases the rate of glycogenolysis and gluconeogenesis by keeping the regulated enzymes of these pathways in the active state and by increasing their synthesis. Glucagon decreases the rate of glycogenesis, glycolysis and HMP shunt pathway. Glucagon favours those pathways, which produce glucose, so it has got an **anti-insulin action**.

Epinephrine is synthesized in the adrenal medulla. It helps to increase the blood glucose level. It increases the rate of glycogenolysis and gluconeogenesis. Unlike glucagon, epinephrine increases the rate of glycogenolysis, both in liver and muscle. It can stimulate the release of glucagon and can inhibit the secretion of insulin. This action of epinephrine on the release of insulin and glucagon is responsible for increasing the blood glucose level.

Growth hormone is secreted in response to hypoglycemia. Growth hormone has hyperglycemic effect. Glucocorticoids and thyroxine have an effect on carbohydrate metabolism and tend to increase the blood glucose level.

DIABETES MELLITUS

Diabetes mellitus is a disorder, which results due to the deficiency of insulin secreted by the β-cells of pancreas, or defect in the action of insulin. In diabetes mellitus, the metabolism of carbohydrates, proteins, and lipids is altered.

Signs and Symptoms

Signs and symptoms of diabetes mellitus are:

1. Polyuria (excess urine excretion)
2. Polydypsia (increased thirst)
3. Polyphagia (excess hunger)

Diabetes mellitus may be Type 1 diabetes mellitus or Type 2 diabetes mellitus. Type 1 diabetes mellitus is insulin dependent diabetes mellitus (IDDM), and Type 2 diabetes mellitus is non-insulin diabetes mellitus (NIDDM).

Type 1 diabetes mellitus	*Type 2 diabetes mellitus*
1. Juvenile onset	1. Adult onset
2. Absolute deficiency of insulin	2. Resistance to the action of insulin
3. Ketoacidosis is commonly seen	3. Hyperosmolar coma is commonly seen

ALTERATION IN THE METABOLISM OF CARBOHYDRATE, PROTEIN AND LIPID METABOLISM

Carbohydrate Metabolism

Insulin is required for the uptake of glucose by the peripheral tissues. Insulin favours the metabolic pathways that are involved in the utilization and storage of carbohydrates. In case of diabetes mellitus, glucose is not taken up by the peripheral tissues and its utilization is decreased. This increases the blood glucose level, and the condition is called hyperglycemia.

When the blood glucose level is increased, it is excreted in urine and the condition is called **glycosuria**. Presence of the reducing sugar in urine can be detected by Benedict's test.

Lipid Metabolism

As a result of insulin deficiency, cells will be starved of glucose (source of energy). An alternate fuel, fatty acid, is released from the adipose tissue. When the blood level of fatty acid increases, its oxidation at the liver is also increased, leading to the accumulation of acetyl CoA. Acetyl CoA is later converted into ketone bodies.

If the level of ketone bodies increases in the blood, the condition is called **ketonemia**.

Excretion of excess of ketone bodies in the urine is called **ketonuria**.

Ketonemia and ketaonuria together is **ketosis**.

Since ketone bodies are acidic in nature, it might decrease the pH of the blood. If the blood pH is decreased due to the accumulation of ketone bodies, it is called **diabetic ketoacidosis**.

Protein Metabolism

In diabetes mellitus, protein synthesis decreases and protein breakdown increases. This leads to **negative nitrogen balance**.

Diagnosis of Diabetes Mellitus

1. Analysis of urine to detect the presence of glucose

Urine collected from the individual can be tested for the reducing substance. **Benedict's test** is done for the detection of reducing substance in the urine. Alternatively, glucose strip containing enzyme glucose oxidase and other reagents are available to detect the presence of glucose specifically.

2. Estimation of plasma glucose

Blood is collected from the patients by aseptic precautions. It is stored in a container containing **sodium fluoride and potassium oxalate (sugar tubes)**. Potassium oxalate acts as an anticoagulant, and **sodium fluoride is an inhibitor of glycolysis**.

If the fasting plasma glucose level is above 126 mg/dL on two separate occasions, it is diagnostic of diabetes mellitus.

3. Oral glucose tolerance test

Oral glucose tolerance test is done to assess the ability of the body to handle an excess amount of glucose. Plasma glucose level is measured before and after giving a load of glucose. Before performing the test, patient should be properly prepared.

A. Preparation of the patient

1. Patient should do his/her routine work.
2. If possible, drugs that might interfere with the test should be stopped.
3. At least, three days prior to the test, the patient must consume minimum 150 g of carbohydrates per day. This will help to keep the activity of the enzymes of carbohydrate metabolism at the optimum level so as to elicit a maximum response.
4. Patient should fast for about 8 to 12 hours before the test.

B. Test procedure

1. Individual who is fasting for the past 8 to 12 hours and ready for OGTT is made to sit comfortably in the laboratory.
2. The fasting blood sample is collected.
3. Individual is made to drink **75 g of anhydrous glucose** in about 200 mL of water.

4. Exactly two hours after the glucose is ingested, one more blood sample is taken and it is called 2 hour post glucose load sample.
5. The glucose present in the plasma is estimated by **glucose oxidase peroxidase method**.

C. Interpretation of OGTT Result

	Fasting plasma glucose	*2-hour post glucose load*
1. Normal	<126 mg/dL	<140 mg/dL
2. Impaired tolerance	>110 mg/dL and <126 mg/dL	>140 mg/dL and <200 mg/dL
3. Fasting hyperglycemia	>110 mg/dL and <126 mg/dL	<140 mg/dL
4. Diabetes mellitus	≥126 mg/dL	≥200 mg/dL

Glycated Haemoglobin (HbA1c)

If blood glucose level is increased and remains high for a long time, it will glycate the proteins present in the body. The amino-terminal valine residue of the β chain of haemoglobin is glycated non-enzymatically in diabetes mellitus. It is called **HbA1c**.

Since the lifespan of erythrocytes is on an average 90 days, measurement of glycated haemoglobin gives an idea about the blood glucose level of an individual during the past few months.

QUESTIONS

Long Essay

1. Discuss in detail the regulation of blood glucose level.
2. Discuss in detail glycolysis.
3. What is gluconeogenesis? Mention the substrates for gluconeogenesis, and the key enzymes of gluconeogenesis. Write the reaction sequence of gluconeogenesis from lactate.
4. Describe the reactions of HMP shunt, and write the significance of this pathway.
5. Discuss in detail the citric acid cycle.
6. Discuss in detail the glycogen metabolism. Add a note on glycogen storage disorders.

Short Essay

1. Write short notes on the following:
   (a) Glycogen storage disorders
   (b) Glycogenesis
   (c) Glycogenolysis
   (d) Diabetes mellitus
   (e) Oral glucose tolerance test
2. What you mean by gluconeogenesis? Discuss the key enzymes of gluconeogenesis.

Short Answers

1. Write the defect in:
   (a) Galactosemia.
   (b) Essential fructose intolerance.
2. What is the significance of estimation of glycated haemoglobin?
3. What is the biochemical basis for cataract in galactosemia?
4. Describe von Gierke's disease.
5. Name the regulated enzymes of glycogen metabolism.
6. What is ketoses? Mention any two conditions where ketoses is seen.
7. How will you prepare the patient for oral glucose tolerance test?
8. What is the normal blood glucose level? Mention any two causes for hyperglycemia.
9. How many ATPs are formed when glucose is metabolized by glycolysis under anaerobic condition?

11

Lipid Metabolism

INTRODUCTION

Fats are concentrated form of energy. They are source of energy for many human tissues except RBCs and brain. Fat is chemically triacylglycerol esterified with three fatty acids. When excess amount of calories is supplied to the body, it is converted to fat and stored in the adipose tissue.

METABOLISM OF FATTY ACID

In a well-fed state, when there is a sufficient supply of carbohydrates, proteins, and lipids, the excess carbohydrates and proteins are converted into fatty acids (biosynthesis of fatty acids). These fatty acids are esterified with glycerol to form triacylglycerol (fat) and stored in the adipose tissue. When the body requires energy in the form of fatty acids, they are released from the adipose tissue. Then these fatty acids are carried to various tissues by albumin and are oxidized there.

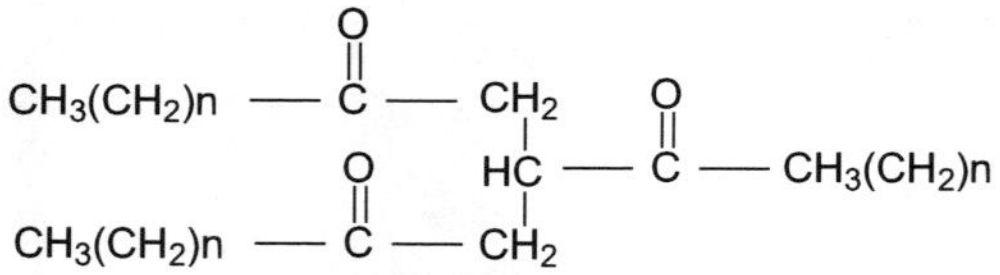

TRIACYLGLYCEROL
(TRIGLYCERIDE)

MOBILIZATION OF FATTY ACIDS

Fat in the adipose tissue is acted upon by an enzyme called **hormone-sensitive lipase (HSL)**. The activity of lipase is regulated by hormones like glucagon, insulin, epinephrine, and cortisol. Hence, it is called hormone-sensitive lipase. HSL acts on fat and hydrolyzes it to release fatty acids and glycerol. Hydrolysis of fat by the HSL is called **mobilization of fatty acids** (Fig. 11.1)

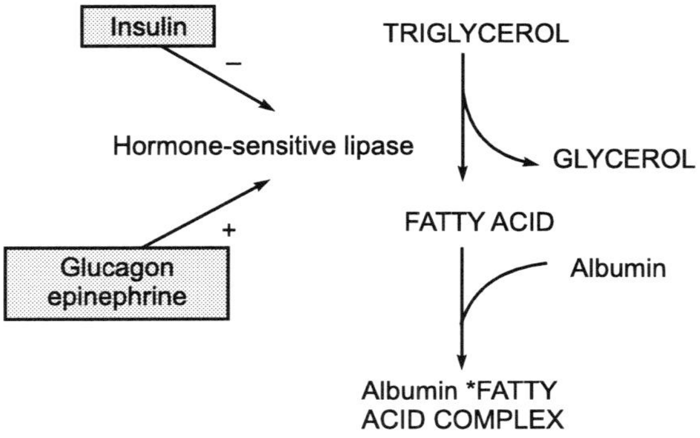

Fig. 11.1 Fatty acid mobilization.

TRANSPORT OF FATTY ACIDS

Fatty acids released from the adipose tissue are called free fatty acids (FFAs). They are hydrophobic in nature. They are insoluble in aqueous phase of plasma. So, the transport of fatty acids requires a carrier molecule. The carrier molecule makes fatty acids relatively soluble in the aqueous phase of plasma by forming a complex with them. Albumin present in the plasma acts as a carrier protein for fatty acids by forming **albumin fatty acid complex**.

INSIDE THE CELL

The albumin fatty acid complex reaches the tissues, which are able to metabolize fatty acids. Fatty acids cross the cell membrane and enter into the cell. Inside the cell, they bind to a carrier protein called fatty acid binding protein (FABP) to reach the subcellular site, where they are metabolized further.

ACTIVATION OF FATTY ACIDS

Fatty acids, which enter into the cells, are activated to form acyl-CoA by the action of the enzyme thiokinase that uses CoA SH and ATP.

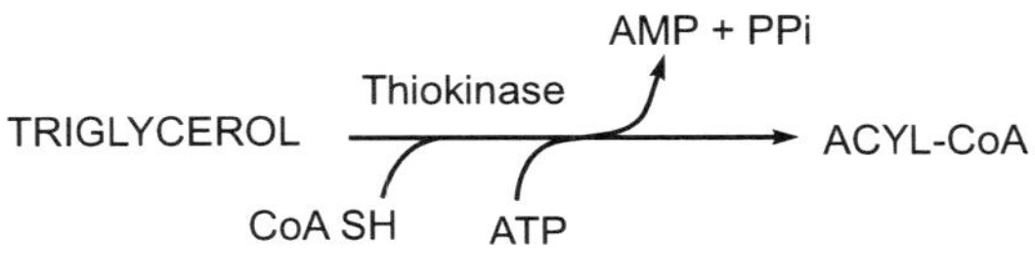

Transport of Long Chain Fatty Acids Across the Mitochondrial Membrane

The oxidation of fatty acids mainly takes place in the liver. Other tissues where fatty acids can be oxidized are the muscles. Short chain fatty acids and medium chain fatty acids can cross the inner mitochondrial membrane freely, and get oxidized within the mitochondria.

Long chain fatty acids cannot cross the inner mitochondrial membrane. For transport across the inner mitochondrial membrane, they require a special biomolecule called **carnitine**. Long chain fatty acids are transported across the inner mitochondrial membrane as carnitine derivatives.

Transport of Long Chain Fatty Acids Across the Inner Mitochondrial Membrane

Acyl-CoA reacts with carnitine in the presence of an enzyme called carnitine acyltransferase I located in the outer mitochondrial membrane. Acylcarnitine is transported across the inner mitochondrial membrane with the help of a translocase protein called carnitine acylcarnitine translocase. This translocase transports acylcarnitine inside the mitochondria and carnitine out from the mitochondria.

Once inside the mitochondria, acyl-CoA is regenerated by the action of carnitine acyltransferase II of inner mitochondrial membrane, releasing free carnitine (Fig. 11.2)

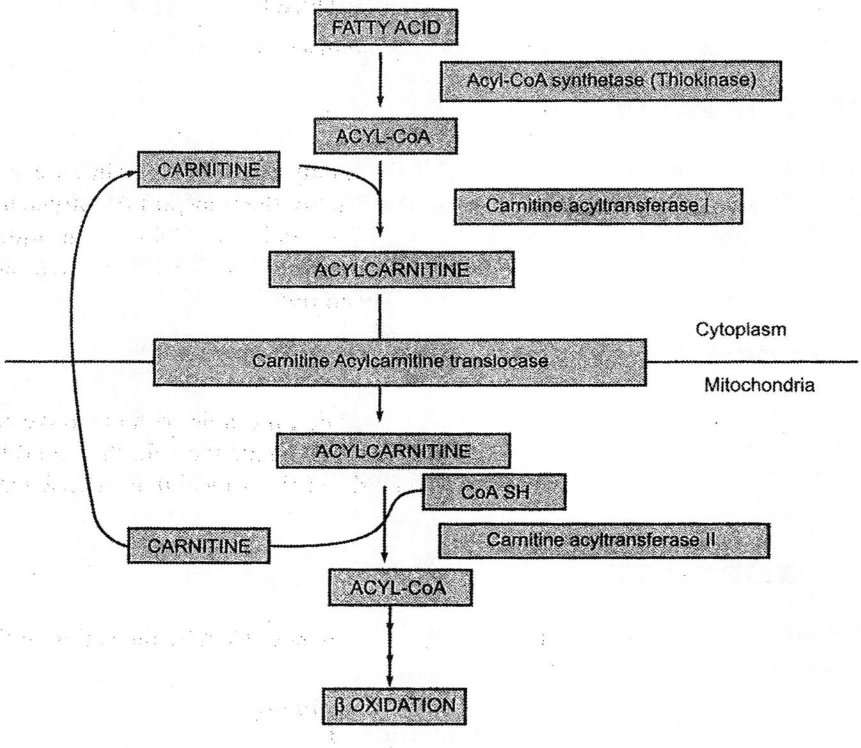

Fig. 11.2 Transport of long chain fatty acids across the inner mitochondrial membrane.

β-OXIDATION OF FATTY ACIDS

Mitochondria contain all the enzymes required to oxidize the fatty acids. These enzymes are present in the mitochondrial matrix. During the oxidation of fatty acids, β-carbon atom of the fatty acid chain is oxidized. Hence, the oxidation is called β-oxidation of fatty acids.

Acyl-CoA is first acted upon by an enzyme called acyl-CoA dehydrogenase. There are different acyl CoA dehydrogenases to act on fatty acids of different chain length. Acyl-CoA dehydrogenase contains FAD as the prosthetic group. This enzyme removes two hydrogen atoms from the 2nd and 3rd carbon atoms of the acyl-CoA to form 2,3 unsaturated acyl-CoA, and simultaneously FAD is reduced to $FADH_2$ (Fig. 11.3).

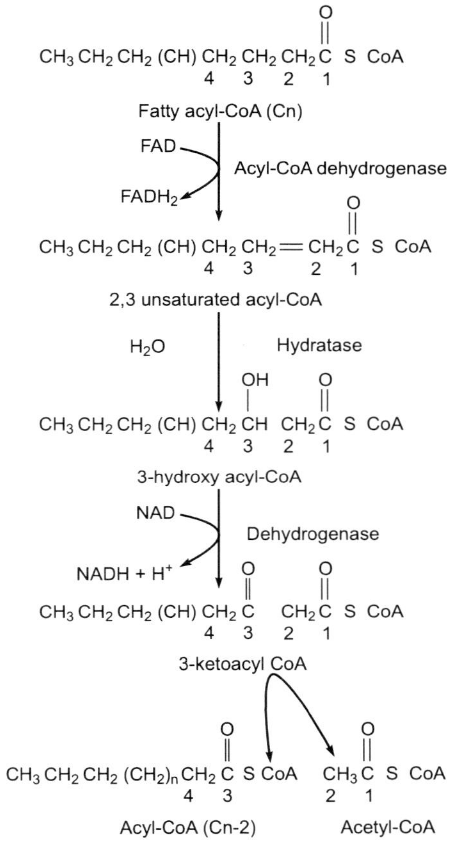

Fig. 11.3 β-oxidation of fatty acids.

In the next reaction, water is added to saturate the double bond present in 2,3 unsaturated acyl-CoA by an enzyme called **hydratase**, leading to the formation of 3-hydroxyacyl-CoA.

Hydroxyacyl-CoA is oxidized by a **dehydrogenase** to form 3-keto acyl-CoA. In this reaction, the hydrogen atoms removed from 3-hydroxyacyl-CoA are accepted by the coenzyme NAD^+. It is reduced to NADH + H^+.

In the last step, the first and second carbon atoms of the fatty acids are removed as acetyl CoA, and the other product is the original fatty acid short of two carbon atoms.

Fatty acids undergo repeated cycles of β-oxidation, and during each cycle, two carbons are removed as acetyl-CoA. Palmitate containing 16 carbon atoms, it is completely oxidized by seven cycles of β-oxidation within the mitochondria to form eight acetyl-CoA. Acetyl-CoA further enters into TCA cycle and is oxidized.

In one β-oxidation cycle, one $FADH_2$ and one NADH + H^+ are formed. $FADH_2$ is equal to 2 ATPs, and NADH + H^+ is equivalent to 3 ATPs. One β-oxidation cycle gives rise to 5

ATP's. Since palmitate undergoes seven β-oxidation cycles, 35 ATP's are formed at the end of β-oxidation.

One acetyl CoA, when oxidized by TCA cycle, gives rise to 12 ATPs. When eight acetyl-CoA produced by the β-oxidation of palmitate are oxidized by TCA cycle, it gives rise to 98 ATPs. So, the total ATPs produced by the oxidation of palmitate are 131.

Two ATPs are utilized for the activation of fatty acids. The net production is 129 ATP's.

KETOGENESIS

When excess of fatty acids are mobilized from the adipose tissue, the blood level of free fatty acids is increased. This increases the uptake of fatty acids by the liver, resulting in the increased oxidation of fatty acids. When the rate of fatty acid oxidation is increased, rate of acetyl-CoA production is also increased. Hence, the concentration of acetyl-CoA within the mitochondria increases. The excess of acetyl-CoA accumulated in the mitochondria forms **ketone bodies.**

The last four carbons of the fatty acid during β-oxidation can enter ketogenic pathway. Alternatively, acetyl-CoA can condense with another acetyl-CoA leading to the formation of acetoacetyl-CoA by reversal of the last step of β-oxidation.

Acetoacetyl-CoA condenses with one more acetyl-CoA, leading to the formation of HMG CoA (3-Hydroxy 4-methylglutaryl CoA) in the presence of an enzyme called HMG CoA synthase (Fig. 11.4).

Another enzyme called **HMG CoA lyase** removes one acetyl CoA from the HMG CoA to form acetoacetate. **Acetoacetate** is the first ketone body to be formed during ketogenesis.

Acetoacetate is later reduced to form 3-hydroxybutyryl CoA, or it is decarboxylated to form actone. Acetone is volatile in nature, so it is expelled through lungs. Acetoacetate and 3-hydroxy butyrate are excreted in urine.

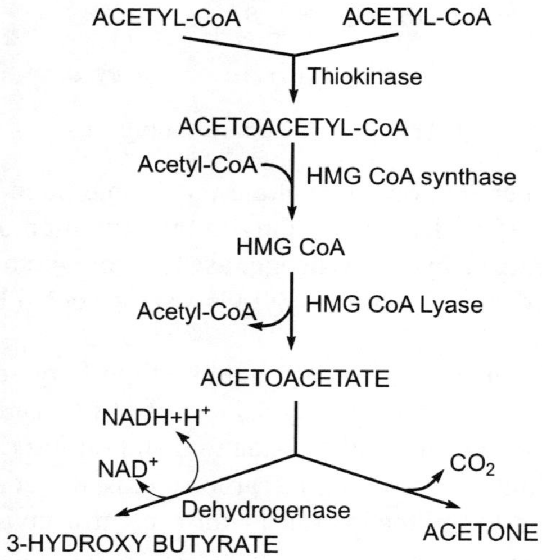

Fig. 11.4 Ketogenesis.

The liver is able to produce ketone bodies, but it cannot utilize it. This is because of the absence of the enzyme required for the catabolism of ketone bodies in the liver. Extra hepatic tissues like the muscle, heart, etc. are able to utilize ketone bodies. So, the ketone bodies formed in the liver are utilized by the extrahepatic tissues for energy.

In the extrahepatic tissues, **3-hydroxybutaryl-CoA** is oxidized back to acetoacetate. Acetoacetate is activated by an enzyme called **thiophorase** to form acetoacetyl-CoA. The CoA donor in this reaction is succinyl-CoA. Acetoacetyl-CoA is converted into two molecules of acetyl-CoA, which enter TCA cycle.

Fatty acids are oxidized in other subcellular sites to a limited extent. They can be oxidized by β oxidation or ω oxidation in the endoplasmic reticulum.

FATTY ACID SYNTHESIS

Fatty acids are synthesized in a well-fed state, that is when surplus amount of calories is supplied in the diet and the blood level of insulin is optimum. Glucose or carbon skeleton of amino acids, which is converted to acetyl-CoA, acts as a source of carbon for synthesizing fatty acid molecules. It is a reductive synthetic pathway, so it requires lot of reducing equivalents in the form of NADPH + H^+.

Fatty acid synthesis takes place in the cytosol with the help of fatty acid synthase complex (Fig. 11.5). Fatty acid synthase complex contains two identical polypeptide chains. It is a dimer with the polypeptide chains having head to tail arrangement. Each polypeptide chain contains seven catalytic activities and one acyl carrier protein (ACP).

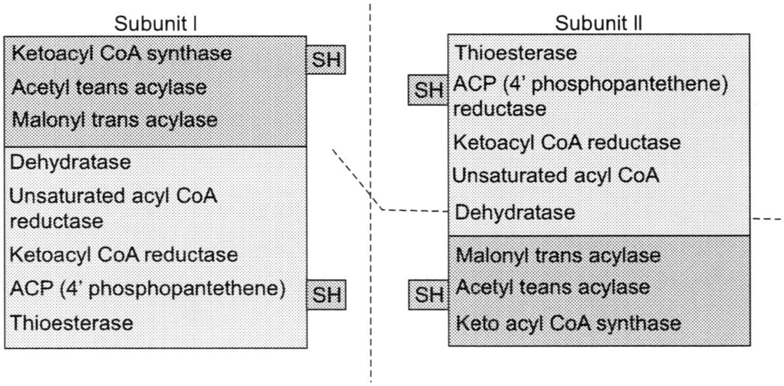

Fig. 11.5 Fatty acid synthase complex.

Acyl carrier protein having the 4′ phosphopantethene has a sulfhydryl group (SH). The newly-synthesized fatty acid is carried by this ACP. The enzyme ketoacyl synthase has cysteine residue. The sulfhydryl group of cysteine of one polypeptide chain is close to the sulfhydryl group of ACP of the other polypeptide chain. Fatty acid synthase is active in dimeric form.

Acetyl-CoA is a carbon donor for the fatty acid biosynthesis. Acetyl-CoA is formed within the mitochondria, but fatty acid synthesis takes place in the cytosol. Acetyl-CoA cannot cross the inner mitochondrial membrane freely. So, it is converted to citrate. The citrate is carried across the mitochondrial membrane with the help of a transporter. In the cytosol, citrate is acted upon by the enzyme ATP citrate lyase to form acetyl-CoA and oxaloacetate.

Reaction 1

The first reaction of fatty acid synthesis is the formation of malonyl CoA from acetyl CoA by **acetyl-CoA carboxylase** (Fig. 11.6).

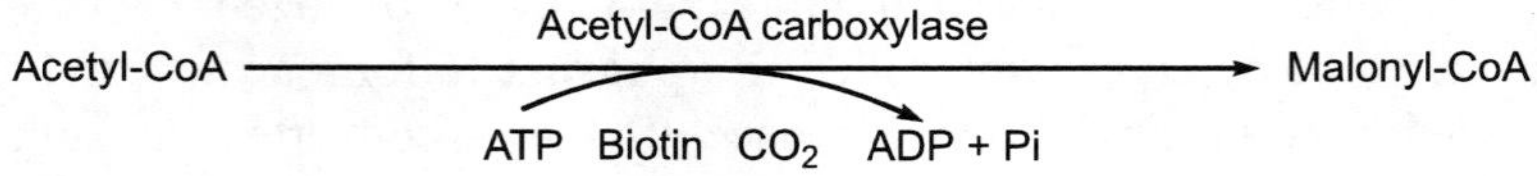

Fig. 11.6

Malonyl-CoA is utilized by the fatty acid synthase complex to construct a new fatty acid molecule.

Reaction 2

Malonyl-CoA and acetyl-CoA react with the sulfhydryl groups of the fatty acid synthase complex to form acetyl malonyl enzyme. Malonyl-CoA reacts with the SH (sulfhydryl) group of ACP, and acetyl-CoA reacts with the SH group of cysteine residue of the ketoacyl-CoA synthase in the presence of an enzyme malonyl transacylase and acetyl transacylase respectively. Since there are two sets of SH groups on either end of the fatty acid synthase complex, so, at a time, two fatty acids can be synthesized by one fatty acid synthase complex.

Reaction 3

The condensing enzyme ketoacyl synthase shifts and condenses acetate present on one polypeptide chain with the malonate present on the other polypeptide chain of the fatty acid synthase complex to form ketoacyl enzyme.

Reaction 4

Ketoacyl enzyme is reduced by using the reducing equivalents from NADPH + H^+ to form hydroxyl acyl enzyme. This reaction is catalyzed by a reductase.

Reaction 5

Hydroxy acyl group present on the enzyme is dehydrated by dehydratase to form unsaturated acyl enzyme.

Reaction 6

Unsaturation is saturated by another reductase to from acyl enzyme by using the reducing equivalents from NADPH + H^+.

Reaction 7

Acyl group is shifted from the acyl carrier protein (ACP) to the sulfhydryl group of the cysteine residue of ketoacyl-CoA synthase. Second malonyl CoA reacts with the acyl enzyme to form acyl malonyl enzyme. One more cycle of reaction by the enzymes of fatty acid synthase complex takes place.

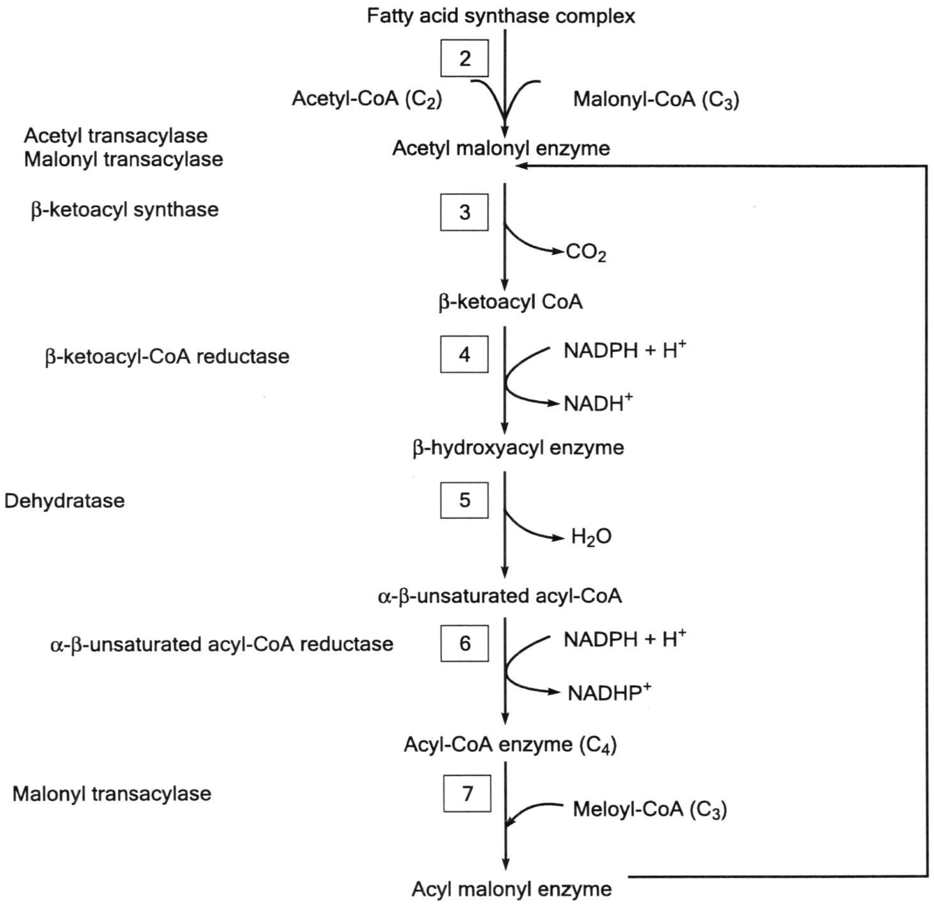

Once the fatty acid of desired chain length is formed, it is acted upon by the thioesterase, and the fatty acid is released in the free form. **Thioesterase** is specific for the chain length of the fatty acid.

The reducing equivalents required for the fatty acid synthesis are mainly derived from the **hexose monophosphate shunt (HMP shunt)**. Some amount of NADPH + H^+ is derived by the activity of the malic enzyme. Oxaloacetate formed by the action of ATP citrate lyase on citrate is converted to malate by the action of **cytoplasmic malate dehydrogenase**. Malate is converted to pyruvate by **malic enzyme** present in the cytosol. During this reaction, $NADP^+$ is converted to NADPH + H^+, which is utilized for fatty acid biosynthesis.

Regulation

The **acetyl-CoA carboxylase activity is regulated** by insulin, glucagon, epinephrine, and citrate. The conversion of acetyl CoA to malonyl CoA is the first step in fatty acid biosynthetic pathway.

The enzyme catalyzing this reaction is acetyl-CoA carboxylase. This enzyme is activated by citrate and inhibited by acyl-CoA.

Insulin favours lipogenesis, and simultaneously inhibits lipolysis. Insulin is the activator of acetyl CoA carboxylase. The hormone glucagon and epinephrine inhibit acetyl CoA carboxylase.

Nutritional status of the organism controls the rate of lipogenesis. If diet contains surplus amount of carbohydrate, it increases the blood glucose level and stimulates the release of insulin. Insulin favours lipogenesis. Glucagon and epinephrine increase the lipogenolysis and blood level of free fatty acids; they inhibit lipogenesis.

CHOLESTEROL

Cholesterol is a lipid that has drawn a lot of attention and concern during the recent past because of its involvement in atherosclerosis and myocardial infarction.

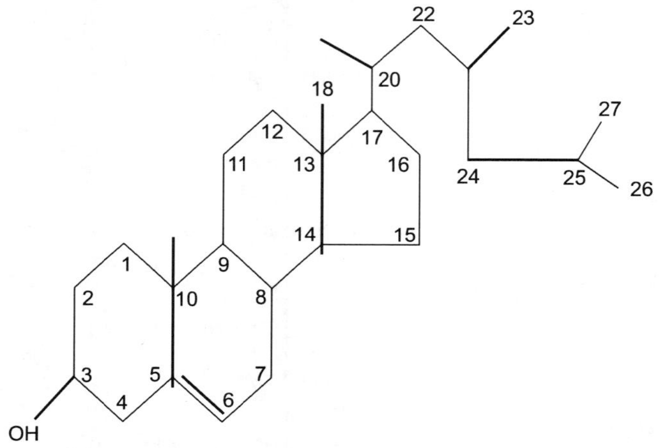

Cholesterol

Cholesterol is a sterol. It contains cyclopentanoperhydrophenanthrene ring, with a hydroxyl group at the third position and a double bond at the fifth position.

The **normal blood level of cholesterol is 150 to 200 mg/dL**. If there is an increase in serum cholesterol level, it is a health hazard as it predisposes for atherosclerosis. Increase in the blood cholesterol is called hypercholesterolemia.

Hypercholesterolemia is seen in conditions like obstructive jaundice, nephrotic syndrome, hypothyroidism, and familial hypercholesterolemia, etc.

Cholesterol can be synthesized in our body. About 50% of the cholesterol present in our body is the result of endogenous synthesis and the remaining is supplied by the diet. The cyclopentanoperhydrophenanthrene ring of the cholesterol cannot be catabolized in our body. It is utilized for the formation of certain physiologically important substances, and the remainder is excreted from the body. Cholesterol is the precursor for the synthesis of steroid hormones, bile acids, and vitamin D.

Metabolism of Cholesterol

Cholesterol Biosynthesis

- Cholesterol can be synthesized by all nucleated cells.
- Subcellular site: Endoplasmic reticulum.
- The carbon atoms required for the synthesis of cholesterol are derived from acetyl-CoA.

The process of cholesterol biosynthesis can be divided into five different stages.

(1) *Formation of mevalonate*

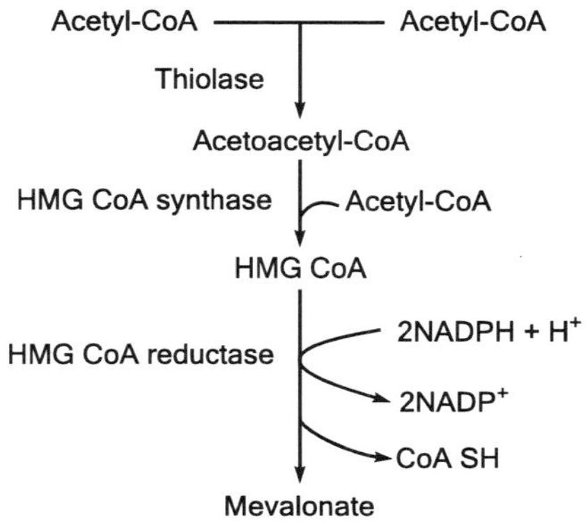

Acetyl-CoA is converted to acetoacetyl-CoA when it condenses with another molecule of acetyl-CoA in the presence of the enzyme thiolase. Acetoacetyl CoA further condenses with acetyl-CoA to form 3-hydroxy-3-methyl glutaryl CoA (HMG CoA). This reaction is catalyzed by HMG CoA synthase. The **rate limiting enzyme HMG CoA reductase**, reduces HMG CoA to form mevalonate.

(2) *Formation of active isoprenoid units*

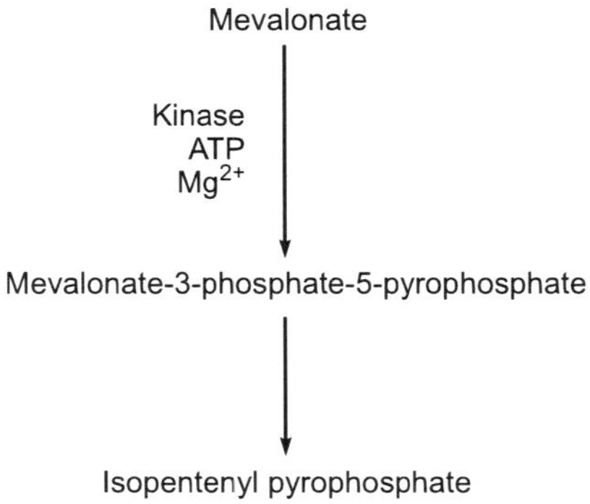

(3) Formation of squalene

Isopentenyl pyrophosphate isomerizes to dimethylallyl pyrophosphate. These two isoprenoid units condense to form geranyl pyrophosphate. One more isopentenyl pyrophosphate condenses with geranyl pyrophosphate to form farnesyl pyrophosphate.

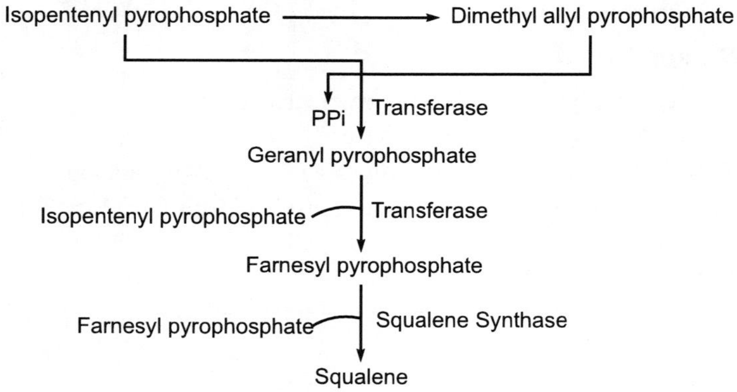

The farnesyl pyrophosphate condenses with another farnesyl pyrophosphate. This reaction takes place in the presence of the enzyme squalene synthase and requires NADPH + H^+. The resulting product is squalene.

(4) *Formation of lanosterol*

The structure of squalene closely resembles the steroid ring structure. Cyclization of squalene by the successive action of endoplasmic enzymes epoxidase and cyclase forms lanosterol. **Lanosterol is the first steroid compound to be formed** during the cholesterol biosynthesis.

(5) *Formation of cholesterol*

Lanosterol is converted to zymosterol, later to desmosterol, and then cholesterol by the removal of few carbon atoms, shifting of the double bonds, and reducing the double bond present in the side chain attached to the steroid ring structure.

Regulation of Cholesterol Biosynthesis

HMG CoA reductase is the regulated enzyme of cholesterol biosynthetic pathway. This enzyme is inhibited by cholesterol. Insulin and thyroid hormones increase the activity of the enzyme, whereas glucagon and glucocorticoids decrease the activity of the enzyme.

Statin group of drugs (mevastatin, lovastatin) inhibit the enzyme HMG CoA reductase, so they are used to decrease the blood cholesterol level.

BILE ACIDS

Bile acids are synthesized from cholesterol. Everyday around 0.5 g of bile acid is synthesized from cholesterol to replace the lost bile acids. Cholic acid and chenodeoxy cholic acid are called primary bile acids. After their metabolism in the intestine, they are converted to secondary bile acids like deoxycholic acid and lithocholic acid respectively.

Synthesis of Bile Acids

- **Organ:** Liver.
- **Subcellular site:** Endoplasmic reticulum.

- **Precursor** for the synthesis of bile acids is cholesterol.
- **Bile acid synthesis** requires molecular oxygen and reducing equivalents in the form of NADPH + H^+ and vitamin C.

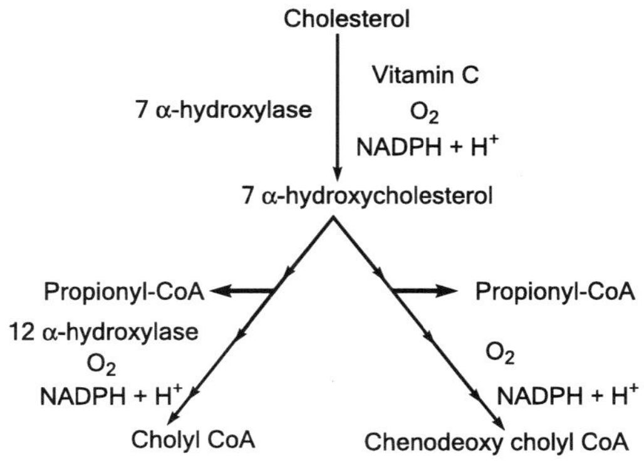

Cholesterol is hydroxylated by **7 α-hydroxylase** to form 7 α-hydroxy cholesterol. This is the rate limiting step as well as committed step in bile acid biosynthesis. The enzyme 7 α-hydroxylase, which **is a microsomal enzyme**, requires vitamin C for its activity.

From this point the pathway bifurcates; one leads to the formation of cholic acid, and the other leads to the formation of chenodeoxy cholic acid. During the formation of cholic acid, 7 α-hydroxycholesterol undergoes an additional hydroxylation at the 12th position by 12 α-hydroxylase.

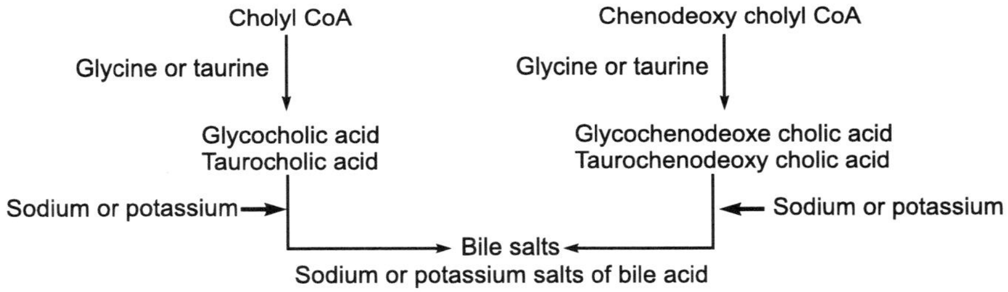

The cholic acid and chenodeoxy cholic acids are called primary bile acids. They combine with glycine or taurine to form glycocholic acid or taurocholic acid and glycochenodeoxy cholic acid or taurochenodeoxy cholic acid. They are the conjugated bile acids. They later combine with sodium or potassium to form their respective salts.

These conjugated bile salts enter the intestine, where they are deconjugated and dehydroxylated to form **secondary bile acids**. Cholic acid is converted to **deoxycholic acid**, and chenodeoxy cholic acid is converted to **lithocholic acid**.

Bile acids are efficiently reabsorbed from the intestine (about 95%). The portal blood brings these bile acids to the liver. Bile acids are taken up by the liver and reconjugated with glycine or taurine to form bile salts, and then secreted into bile. It later reaches the intestine. This is called **enterohepatic circulation of bile acids**.

The **total bile acid pool is about 3 to 5 grams**, but liver secretes about 15 to 30 grams of bile acids into the intestine. This is possible because bile acids undergo enterohepatic circulation for about 8 to 10 times.

Regulation

The bile acid biosynthesis is regulated by regulating the activity of the enzyme, 7 α-hydroxylase. Bile acids act as inhibitor of this enzyme. If the concentration of bile acids increases, the activity of 7 α-hydroxylase is also inhibited. Since 0.5 g of bile acid is excreted, the amount of bile acid produced per day is 0.5 g.

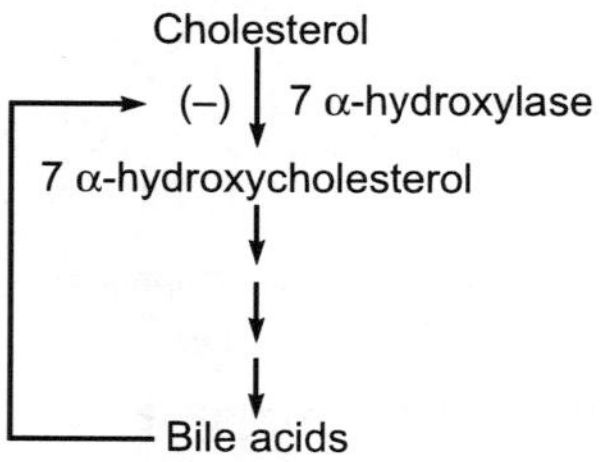

If the excretion of bile acid is enhanced, bile acid pool will decrease. This activates the enzyme 7 α-hydroxylase. It converts more and more cholesterol to bile acid and enhances the excretion of cholesterol from the body. This principle can be used to treat hypercholesterolemia. A resin called cholesteramine can bind to bile acid and decrease its reabsorption. This results in decreased bile acid pool. So, 7 α-hydroxylase becomes active, converts cholesterol to bile acid, and enhances the excretion of cholesterol from the body as bile acid.

Functions

Bile salts are amphipathic in nature, so they help in digestion and absorption of lipids and fat-soluble vitamins. In the bile, bile salts help to keep cholesterol in soluble state. If the bile salt to phospholipid ratio is altered in the bile due to defective bile acid synthesis, cholesterol becomes insoluble in the bile and starts getting precipitated, leading to the formation of cholesterol stones in the biliary tract (cholelithiasis). Cholesterol is excreted from the body in the form of bite acid.

LIPOPROTEINS

Lipoproteins are the complex of lipids and proteins. They are the transport forms of lipids. Lipids are hydrophobic in nature. The exogenous and endogenous lipids can be transported with the help of lipoproteins. Since lipoproteins are made up of proteins and lipids, depending upon the protein and lipid content, they have different densities. Lipoproteins having highest protein content and lowest lipid content have highest density, and lipoproteins with lowest protein content and highest lipid content have lowest density.

Classification

Lipoproteins are classified, based upon their density, into high density lipoproteins (HDL), low density lipoproteins (LDL), very low density lipoproteins (VLDL), and chylomicrons. High density lipoproteins have the highest density, whereas chylomicrons have the lowest density.

Apoproteins (Apolipoproteins)

The protein part of the lipoprotein is called apoprotein (apolipoproteins). There are different types of apoproteins. They are named apoprotein A (apo A), apoprotein B (apo B), apoprotein C (apo C), apoprotein D (apo D), and apoprotein E (apo E). This nomenclature of apoproteins is called ABC nomenclature of apoproteins.

Apo A is further subdivided into apo AI and apo AII. There are two different types of apo B apo B100 and apo B48. Apo C is of three different forms apo CI, apo CII, and apo CIII.

Apoproteins have got variety of functions. They act as cofactors for enzymes involved in the metabolism of lipoproteins, e.g., apo CII acts as cofactor for lipoprotein lipase. Along with phospholipids, in lipoproteins they help to solubilize the lipids. They act as ligands for the receptors during the metabolism of lipoproteins.

Chylomicron transports exogenous lipids, whereas **VLDL and other lipoproteins are involved in endogenous lipid transport**.

Chylomicrons

Chylomicrons are **synthesized in the enterocytes**. The major apoprotein of chylomicron is apo B48. It also contains apo CI, apo CII, apo CIII, and apo E. Lipids present in the diet are digested and absorbed. After absorption inside the enterocytes, the digested products of lipids are converted back to triacylglycerol, cholesterol esters, and phospholipids. Along with the apoproteins synthesized in the intestine, they are packed into a lipid-protein complex called lipoproteins. This lipoprotein is called chylomicron, as it appears in the lymphatic circulation of the intestine at the highest concentration after the ingestion of diet rich in fat, and its diameter is about one micron.

From the lymphatic circulation, chylomicrons enter the systemic circulation through thoracic duct. Different tissues contain an enzyme called **lipoprotein lipase** attached to the wall of the capillary endothelium. These enzymes act on chylomicrons and hydrolyze the ester linkage present in the triacylglycerol, releasing fatty acids and glycerol. Fatty acid is taken up by the tissue, which stores it as triacylglycerol.

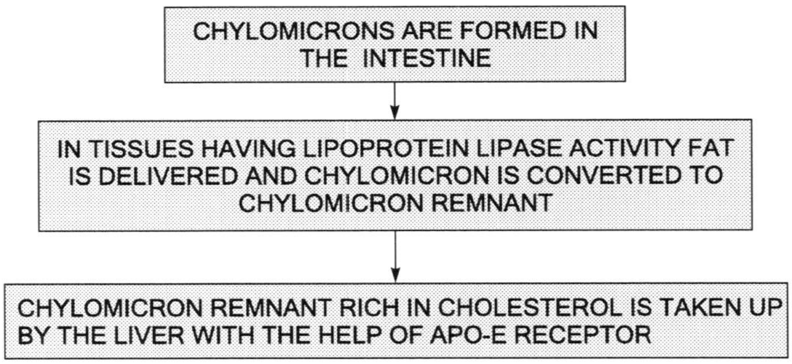

The chylomicron, which is acted upon by the lipoprotein lipase, is called chylomicron remnant. Chylomicron remnant is taken up by the liver and further metabolism takes place in the liver.

Very-Low-Density Lipoprotein (VLDL)

Very-low-density lipoprotein is involved in the transport of endogeneous lipid. It is **synthesized in the liver**. The major apoprotein of VLDL is apo B100. VLDL carries lipids which are synthesized in our body, between different tissues. After it is secreted from the liver, it delivers fat to the tissues having lipoprotein lipase activity. Now, VLDL is converted to LDL.

Liver as well as extrahepatic tissues contain receptors for LDL. These receptors are called LDL receptors. Liver and extrahepatic tissues are capable of taking up LDL due to the presence of LDL receptors.

The LDL is rich in cholesterol. If the level of LDL increases in the blood, more and more cholesterol is deposited in the extrahepatic tissues. This can lead to atherosclerotic changes. Hence, **LDL cholesterol is called bad cholesterol**. The liver takes up LDL and metabolizes it. The cholesterol is excreted as such or in the form of bile acid.

VLDL IS FORMED IN THE LIVER

↓

IN TISSUES HAVING LIPOPROTEIN LIPASE ACTIVITY FAT IS DERIVED AND VLDL IS CONVERTED TO LDL

↓

THE LIVER AND EXTRAHEPATIC TISSUES TAKE UP LDL WHICH IS RICH IN CHOLESTEROL WITH THE HELP OF LDL RECEPTOR

High-Density Lipoprotein (HDL)

High-density lipoprotein is synthesized by the liver. The nascent HDL has disc-like shape. The core of this HDL is empty. The major apoprotein present in HDL is apo-A. It is required for the activity of lecithin–cholesterol acyl transferase (LCAT).

HDL in association with the **enzyme LCAT** and phospholipids, convert cholesterol to cholesterol esters. The cholesterol esters, which are hydrophobic in nature, enter the core of HDL. As HDL receives more and more cholesterol esters, it becomes spherical in shape.

HDL containing cholesterol esters is taken up by the liver and releases cholesterol there. HDL transports cholesterol from extrahepatic tissues to the liver. Hence, HDL prevents the deposition of cholesterol in the extrahepatic tissues. Cholesterol present in HDL is called **HDL cholesterol, which is good cholesterol.**

HDL IS FORMED IN THE LIVER

↓

IN EXTRAHEPATIC TISSUES HDL ALONG WITH THE LCAT ACTIVITY CONVERT CHOLESTEROL PRESENT IN THE EXTRAHEPATIC TISSUE TO CHOLESTEROL ESTER. CHOLESTEROL ESTER WHICH IS HYDROPHOBIC IN NATURE OCCUPIES THE CORE OF HDL

↓

LIVER HDL RELEASE ALL THE CHOLESTEROL SCAVENGED FROM THE EXTRAHEPATIC TISSUE

FATTY LIVER

Liver cell damage and fibrosis of the liver takes place if an abnormal amount of fat gets accumulated in the liver. The change, which takes place in the liver due to the accumulation of abnormal amount of fat, is called fatty change of the liver.

Fatty liver may be due to:

1. Decreased formation of VLDL due to defect in the synthesis of apolipoproteins, and phospholipids;
2. Defect in the synthesis of lipoprotein; and
3. Defect in the secretion of lipoprotein from the liver.

Factors which increase the mobilization of fat from the liver are called lipotropic factors, e.g., methionine, choline, etc.

QUESTIONS

Long Essay

1. Define ketone bodies. Discuss in detail about the metabolism of ketone bodies.
2. Discuss synthesis of fatty acids by fatty acid synthase complex. Add a note on its regulation.
3. Write the reaction sequences of beta-oxidation of fatty acids in the mitochondria. Add a note on the mobilization of fatty acids from the adipose tissue.
4. What are lipoproteins? Classify them. Discuss the metabolism of chylomicron.

Short Essay

Write short notes on the following:

1. Ketogenesis.
2. Fatty acid synthase complex.
3. Reaction sequence of beta-oxidation.
4. Chylomicron metabolism.
5. VLDL metabolism.
6. HDL metabolism.

Short Answers

1. What is fatty liver?
2. What is the normal serum total cholesterol level? Write the special products formed from cholesterol.
3. Mention the normal serum level of total cholesterol and any two causes for hypercholesterolemia.

12

Protein Metabolism

INTRODUCTION

Proteins are utmost important in a cell to keep the life in the dynamic state. The proteins present in a given cell, tissue or organism undergo turnover. Turnover is a process of continuous degradation of the protein followed by its replacement by resynthesis.

There are three major sources of free amino acids found in our body (Fig. 12.1). The degradation of proteins leads to the formation of amino acids. Amino acids are also absorbed from the gastrointestinal tract after the digestion of proteins. Some amino acids (non-essential amino acids) can be synthesized in our body.

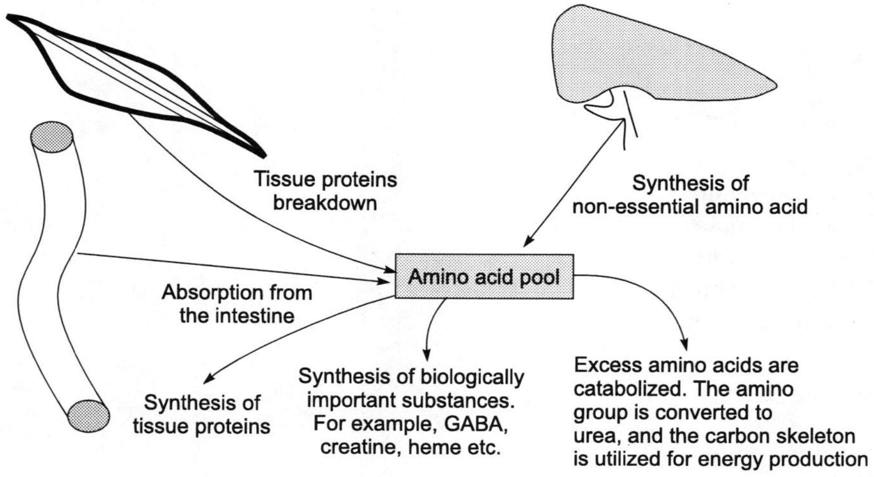

Fig. 12.1 Amino acid pool in the body.

The total amount of amino acids present in our body is called amino acid pool. It is about 100 g at any given time. It is not an anatomical reality. It includes the amino acids present inside and outside the cell.

Amino acids from the amino acid pool can be used for the synthesis of tissue proteins, to form biologically important substances like neurotransmitters catecholamine, etc. The excess amounts of amino acids are not stored in the body; they are degraded.

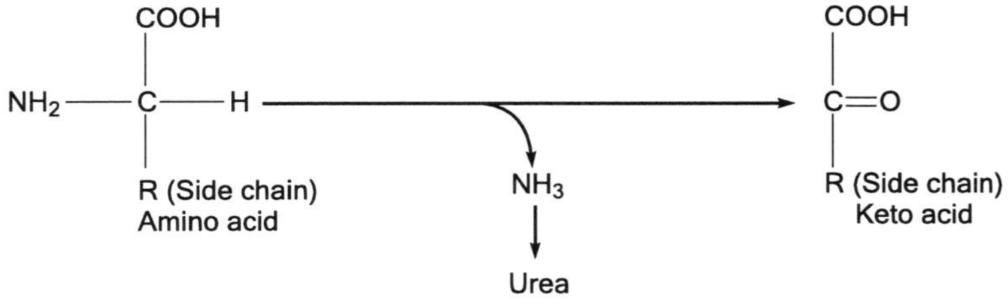

During the degradation of amino acids, the α-amino group is removed as ammonia. It is degraded to form urea, and the carbon skeleton is converted into intermediate of TCA cycle or acetyl CoA. The removal of the amino group of amino acid and converting it into urea constitute the general metabolism of amino acids.

TRANSAMINATION

Transamination reaction is a **reversible reaction** in which the α-amino group of an amino acid is transferred to a α-keto acid, leading to the formation of a new amino acid corresponding to the α-keto acid, and a new keto acid corresponding to the original amino acid. This reaction is catalyzed by **transaminase (aminotransferase)**.

$$\text{AMINO ACID} \xleftrightarrow[\text{PLP}]{\text{Transaminase}} \alpha\text{-KETO ACID}$$

There are two clinically important transaminases:

1. Alanine transaminase (ALT)
2. Aspartate transaminase (AST)

The serum levels of these enzymes are used for the diagnosis of the liver and heart diseases. ALT is specific for liver damage and AST is specific for the heart and liver diseases.

In majority of transamination reactions, α-ketoglutarate acts as a keto acid. During transamination reaction, α-ketoglutarate receives amino group from amino acids and gets converted to glutamate. Because of this, amino group of various amino acids is concentrated as amino group of glutamate.

The transaminated amino acid is converted to a keto acid, and it is further metabolized by TCA cycle to produce energy or enters gluconeogenic pathway to produce glucose.

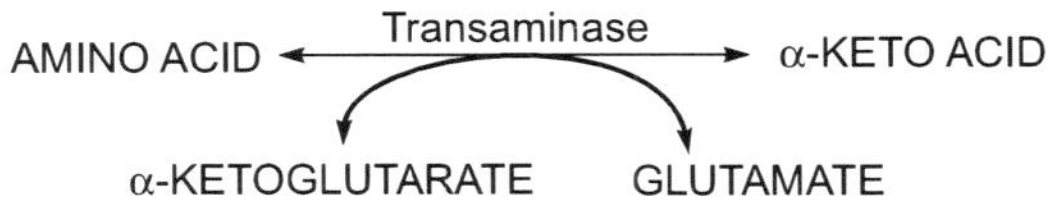

The functions of transamination reaction are as follows:

1. It converts the amino acids into their respective carbon skeleton, which can be further catabolized by amphibolic pathway.

2. It helps in the synthesis of non-essential amino acids.
3. The amino group of amino acids is removed, and it gets concentrated as amino group of glutamate mainly and aspartate partially.

OXIDATIVE DEAMINATION

Oxidative Deamination L-Glutamate Dehydrogenase

The enzyme L-glutamate dehydrogenase acts on glutamate and oxidatively deaminates it to form α-ketoglutarate. In this reaction, amino group is released as ammonia. It is a reversible reaction in which ammonia and α-ketoglutarate can be converted back to glutamate. The enzyme can operate in both directions to degrade glutamate to form α-ketoglutarate or to synthesize glutamate from α-ketoglutarate.

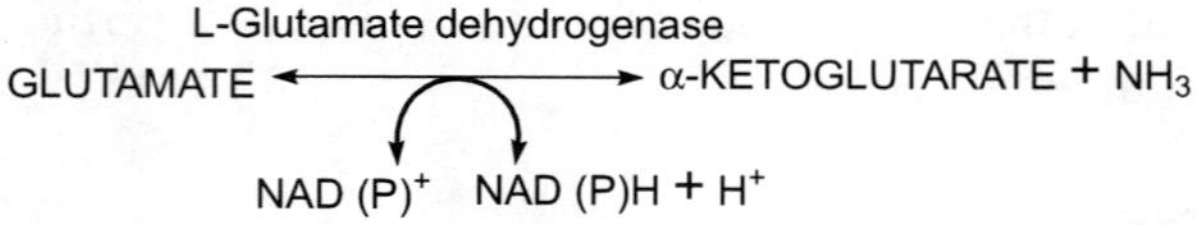

L-glutamate dehydrogenase requires the coenzyme NAD^+ or $NADP^+$. The enzyme requires NAD^+ as a coenzyme to liberate ammonia from glutamate, whereas $NADP^+$ is used as a coenzyme during the synthesis of glutamate. The enzyme is present in the cytosol and mitochondria of the cells. It is allosterically inhibited by ATP and GTP, and activated by ADP and GDP.

OXIDATIVE DEAMINATION BY D- AND L-AMINO ACID OXIDASE

D-Amino Acid Oxidase

D-amino acid oxidase can act on D-amino acid. It oxidatively converts the amino acid into a α-keto acid, and ammonia is released during this process. This enzyme requires FAD as the coenzyme. During the reaction, FAD is converted to $FADH_2$.

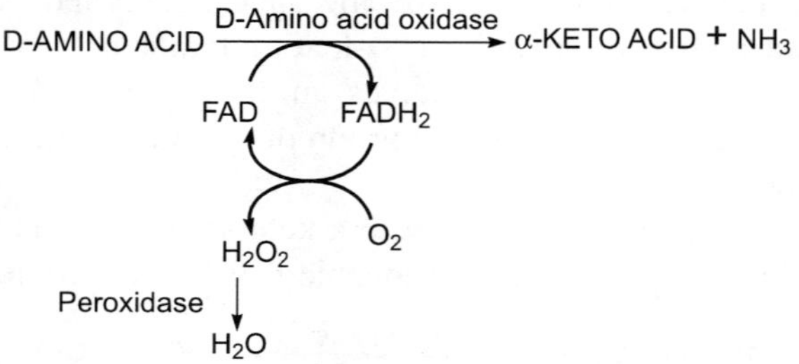

The $FADH_2$ is reoxidized to form FAD by donating its reducing equivalents directly to molecular oxygen, leading to the formation of hydrogen peroxide (H_2O_2). Hydrogen peroxide is later detoxified by the enzyme, peroxidase, to form water.

L-Amino Acid Oxidase

L-amino acid oxidase can act on the L-amino acid. It oxidatively converts the amino acid into α-keto acid, and ammonia is released. This enzyme requires FMN as a coenzyme. During the reaction, FMN is converted to $FMNH_2$.

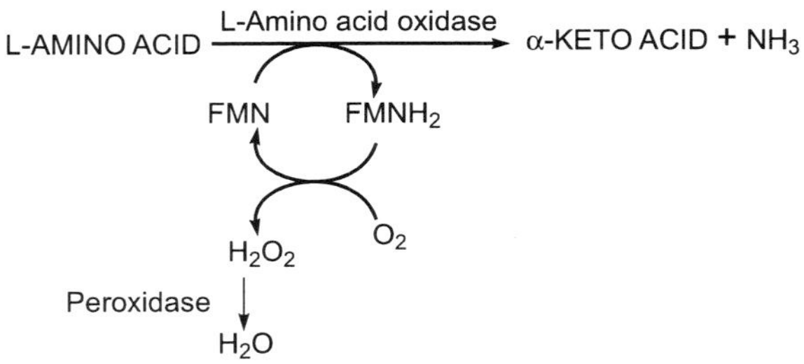

The $FMNH_2$ is reoxidized to form FMN by donating its reducing equivalents directly to molecular oxygen, leading to the formation of hydrogen peroxide (H_2O_2). Hydrogen peroxide is later detoxified by the enzyme, peroxidase to form water.

The sequential action of transaminase and L-glutamate dehydrogenase on the amino acid produces ammonia from the amino acid. This is called **transdeamination**.

In addition to transdeamination reaction and oxidative deamination by D-amino acid oxidase and L-amino acid oxidase, there are other sources of ammonia in our body.

They are:

1. Non-oxidative deamination
2. Ammonia absorbed from the intestine
3. Degradation of nitrogen containing substances

NON-OXIDATIVE DEAMINATION

Many enzymes deaminate amino acids without oxidation. Examples are as follows:

(1) Histidase deaminates histidine to form urocanate

(2) Serine dehydratase converts serine to pyruvate and ammonia

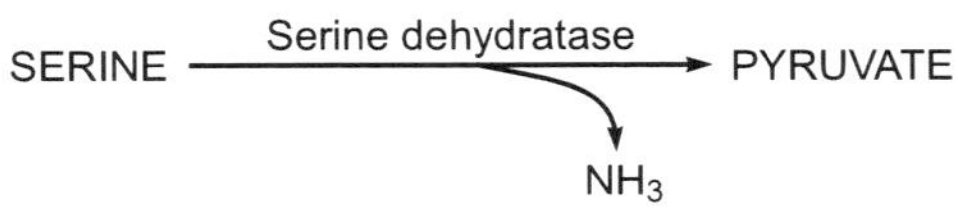

(3) Aspartase converts aspartate to fumarate and ammonia

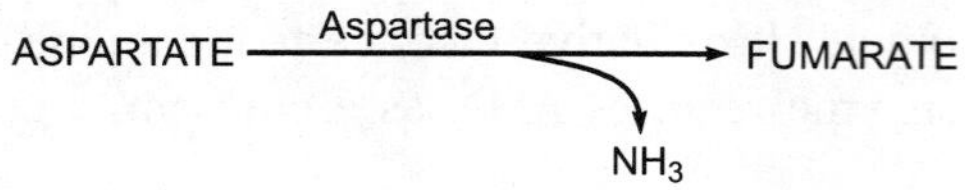

(4) Glutamine is converted to glutamate and ammonia by glutaminase

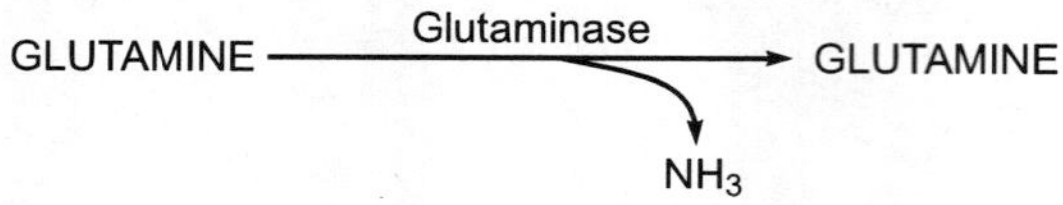

Absorption of Ammonia from the Intestine

Urea is mainly excreted by the kidneys. A small portion of urea is secreted into the intestine. In the intestine, it is acted upon by the enzyme, urease, leading to the formation of ammonia. Ammonia is absorbed by the intestine (one of the major sources of ammonia). From the intestine, it travels along the portal circulation to reach the liver to get detoxified.

Degradation of Nitrogen Containing Substances to form Ammonia

Ammonia is produced during the degradation of various nitrogen-containing substances. For example, catabolism of pyrimidines, catecholamines, etc.

TRANSPORT OF AMMONIA

Ammonia is toxic to the body. It is detoxified in the liver by urea cycle. Therefore, ammonia formed in various parts of the body is transported to the liver.

Glutamine as the Transport Form of Ammonia

The major transport form of ammonia is glutamine. Glutamate reacts with ammonia to form glutamine in the presence of an enzyme called **glutamine synthetase**.

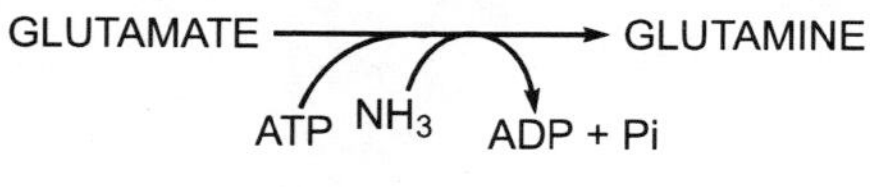

Ammonia is carried to the liver as amide group of glutamine. In the liver, glutamine is acted upon by an enzyme called **glutaminase**. Glutaminase releases ammonia from glutamine, and the other product is glutamate.

Alanine as the Transport Form of Ammonia (Glucose–Alanine Cycle)

Ammonia from the muscle is transported in the form of alanine. In the muscle, pyruvate is formed from glucose by glycolysis. Pyruvate is an α-keto acid, which participates in the transamination reaction and gets converted to alanine. Alanine enters circulation and is transported to the liver. In the liver, it undergoes transamination to form pyruvate which is converted to glucose by gluconeogenic pathway.

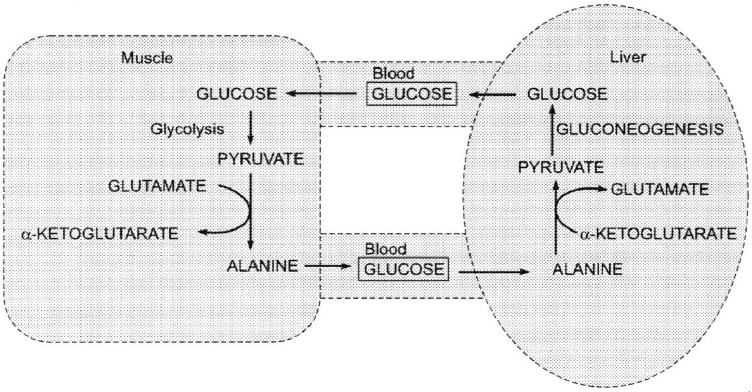

Fig. 12.2 Glucose–Alanine cycle.

Glucose enters circulation and is transported to the muscle. In the muscle, glucose is again converted to pyruvate by glycolytic pathway. Again the same cyclic reaction continues (Fig. 12.2). It is called Glucose – Alanine Cycle.

If ammonia accumulates, it is toxic to the body. Our body has got an efficient mechanism to detoxify ammonia. In the liver, ammonia is converted to urea, which is water-soluble and can be easily excreted from the body through kidney. The biochemical pathway involved in the formation of urea from ammonia is called urea cycle.

Urea Cycle

Urea cycle is the cyclic reaction taking place in the liver. In urea cycle ornithine, which is utilized in the beginning of the pathway, is resynthesized at the end. Urea cycle takes place in the mitochondria and the cytosol of the cell (Fig. 12.3).

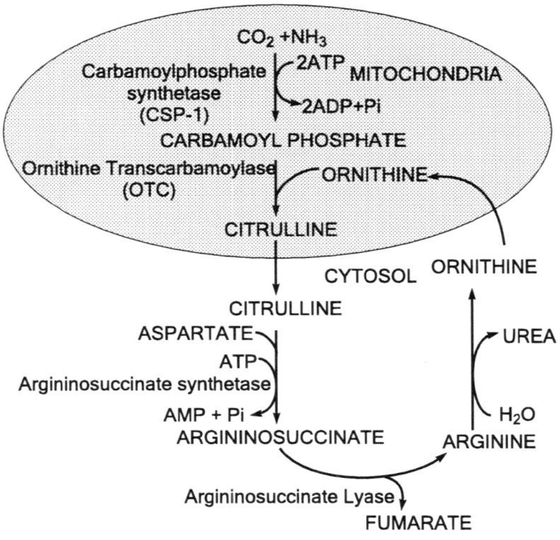

Fig. 12.3 Urea cycle.

Initially, ammonia and carbon dioxide react in the presence of an enzyme called carbamoyl phosphate synthetase to form carbamoyl phosphate. This reaction requires energy in the form of ATP.

Carbamoyl phosphate later combines with ornithine in the presence of an enzyme called ornithine transcarbamoylase to form citrulline. These two reactions take place in the mitochondria of hepatocytes.

Citrulline is transported to the cytosol, and further reactions of urea cycle take place in the cytosol.

Citrulline combines with aspartate to form argininosuccinate in the presence of an enzyme argininosuccinate synthetase. This reaction requires ATP (ATP is converted to AMP and Pi).

Argininosuccinate is acted upon by the enzyme argininosuccinase to release the carbon skeleton of aspartate as fumarate. The amino group of the aspartate is retained on citrulline, which is now called arginine.

Fumarate is converted to oxaloacetate by the reactions of Krebs's cycle. Oxaloacetate is transaminated along with the amino acid glutamate to form aspartate.

Aspartate is again utilized in the reactions of urea cycle. An enzyme called arginase acts on arginine to release urea. The other product formed is called ornithine. Ornithine enters mitochondria and is ready for another cycle of reactions to produce urea.

Regulation

The regulatory enzyme of urea cycle is carbamoyl phosphate synthetase. **N-acetyl glutamate** is the allosteric activator of this enzyme. The level of N-acetyl glutamate increases if the protein intake is high. A regular high protein intake increases the synthesis of the enzymes of urea cycle.

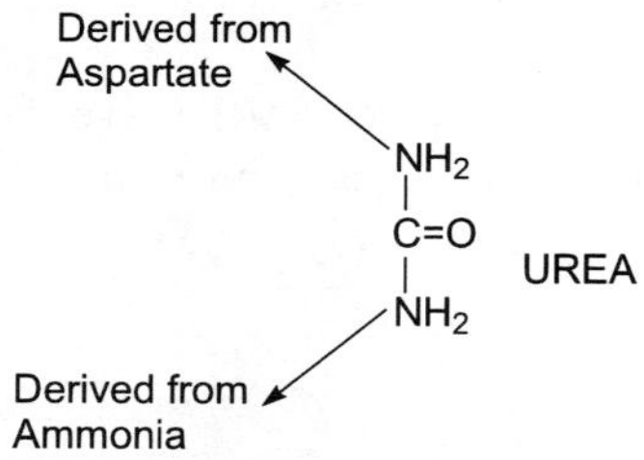

Urea has two nitrogens and one each of carbon and oxygen. One of the nitrogens is derived from ammonia, and the other is derived from the amino group of aspartate. The molecular weight of urea is 60. Urea is water-soluble and freely crosses the cell membrane. It is excreted mainly by the kidney.

The excretion of urea depends upon the glomerular filtration rate. The normal blood urea level is 15 to 45 mg/dL. Blood urea level increases in kidney diseases, for e.g., nephritis. The increase in blood urea level is seen not only in kidney diseases but also in other conditions. Decrease in the blood volume due to any reason (excessive blood loss, severe vomiting and diarrhoea etc.) decrease in renal perfusion that results in decreased glomerular filtration so urea is retained in the blood.

Ammonia Toxicity

Normal blood ammonia level is 10 to 45 μg/L. The accumulation of ammonia in circulation is called hyperammonemia. It gives rise to ammonia toxicity. In ammonia toxicity, the signs and symptoms seen are mostly related to CNS like blurring of vision, slurring of speech, tremors, vomiting, and in severe cases coma and death.

Causes

Normally, ammonia produced in the brain is converted to glutamine by the action of an enzyme glutamine synthetase. When ammonia starts getting accumulated, it reacts with α-ketoglutarate to form glutamate by the action of glutamate dehydrogenase. When the intermediate of TCA cycle — α-ketoglutarate — is diverted for this reaction, the rate of TCA cycle comes down and the energy state of the cell is decreased. Since CNS requires a lot of energy in the form of ATP to maintain the sodium and potassium balance, signs and symptoms related to this tissue are predominantly seen in ammonia toxicity.

Hyperammonemia Due to Defects in Urea Cycle

Defect in any one of the five different enzymes involved in urea cycle can lead to hyperammonemia. It gives rise to mental retardation:

- Defect in the enzyme carbamoyl phosphate synthetase is called **type I typerammonemia**.
- Defect in the enzyme ornithine transcarbamoylase is called **type II typerammonemia**.

Other Inherited Defects in Urea Cycle are

Defect in the enzyme	*Disease*
Argininosuccinic acid synthetase	Citrullinemia
Argininosuccinase	Argininosuccinic aciduria
Arginase	Hyperargininemia

Acquired cause for hyperammonemia is due to liver cell failure as seen in case of cirrhosis of liver due to variety of reasons.

METABOLISM OF INDIVIDUAL AMINO ACIDS

Glycine

Glycine is a non-essential amino acid. It is synthesized in the body from serine, threonine, and glyoxylate.

1. **From Serine**

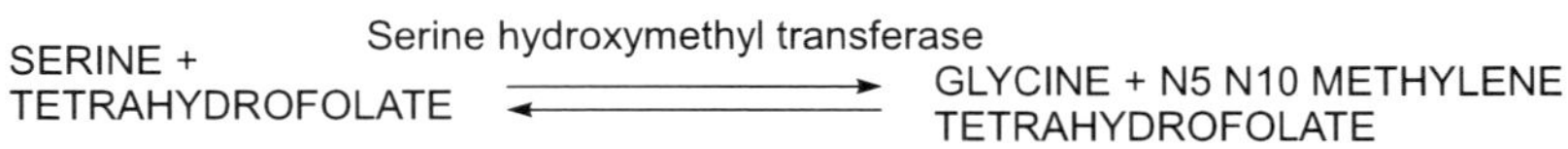

2. **From Threonine**

3. By the Action of Glycine Synthase

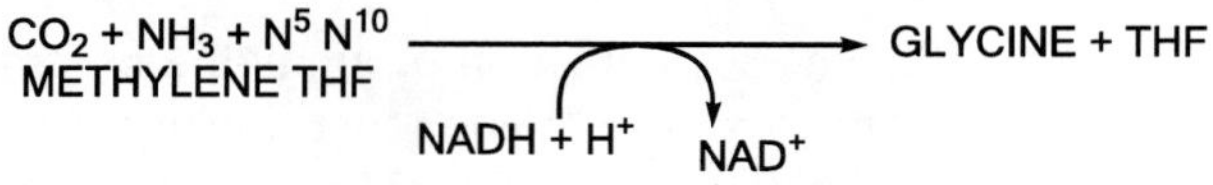

Glycine synthase is involved in the synthesis of glycine from carbon dioxide and ammonia. Glycine cleavage system is involved in catalyzing the reverse reaction.

Metabolic Functions of Glycine

Glutathione synthesis requires glycine. Glutathione is a tripeptide containing glutamate, cysteine, and glycine. It helps to detoxify the free radicals.

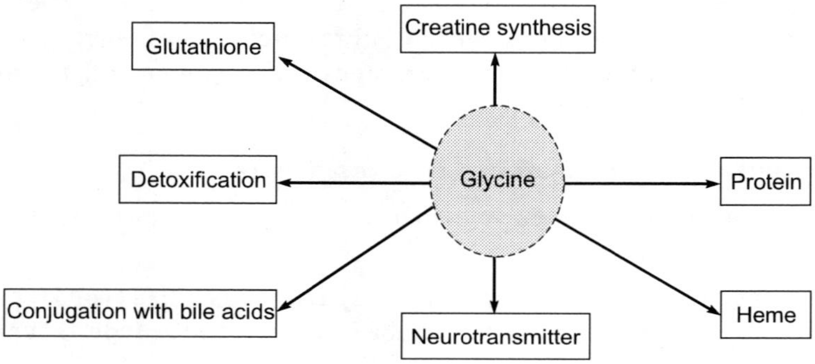

2. Glycine is required for the synthesis of creatine.

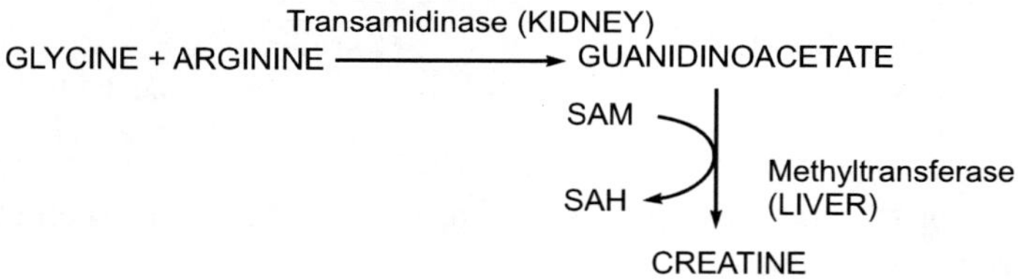

— Creatine formed is transported to skeletal muscle, where it helps to store the energy in the form of creatine phosphate. The reaction involving the conversion of creatine to creatine phosphate is catalyzed by creatine kinase (CK).

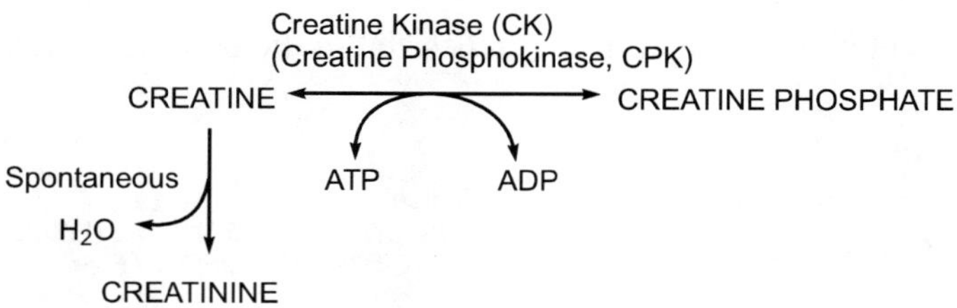

— When energy is required during muscle contraction, creatine phosphate donates its high energy phosphate group and converts ADP to ATP.

— During the metabolism of creatine phosphate, it is spontaneously converted to creatinine. Creatinine is an anhydrous product of creatine. It is a waste product and excreted through kidney.

3. Glycine is required for protein synthesis.
4. Glycine donates C4 and C5 carbon, and N7 nitrogen for the purine ring synthesis.

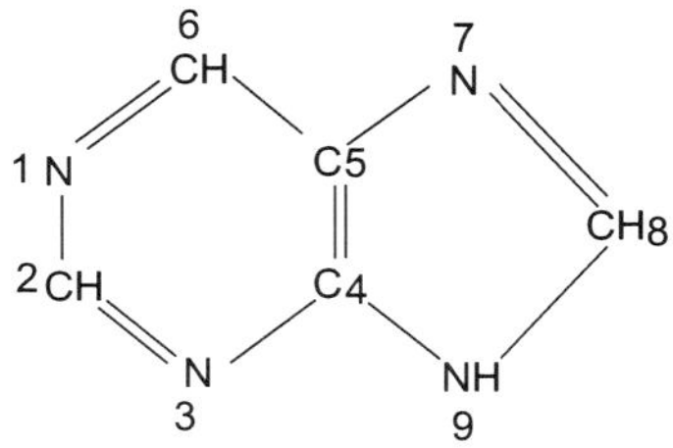

Purine

5. Glycine is required for heme synthesis.

GLYCINE + SUCCINYL CoA ⟶ HEME

6. Glycine participates in detoxification reactions.

BENZOIC ACID + GLYCINE ⟶ HIPPURIC ACID

7. Glycine gets conjugated with bile acids.

CHOLIC ACID + GLYCINE ⟶ GLYCOCHOLIC ACID

8. Glycine acts as a neurotransmitter.

PHENYLALANINE AND TYROSINE

Phenylalanine is an aromatic amino acid. It is a non-essential amino acid. It is both glucogenic and ketogenic amino acid. Phenylalanine can be converted to tyrosine by the action of an enzyme phenylalanine hydroxylase.

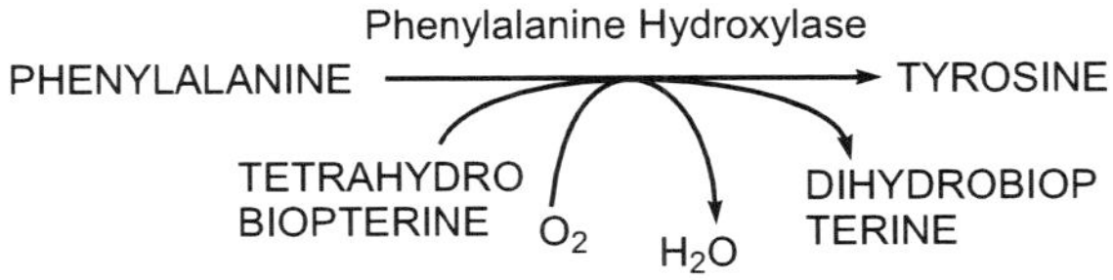

Until phenylalanine supply is adequate in the diet, tyrosine is non-essential.

Phenylketonuria

Defect in the enzyme phenylalanine hydroxylase is responsible for a disease called phenylketonuria. It is one of the commonest genetic diseases, which gives rise to mental retardation in children. Hence, phenylalanine should be restricted in the diet of these children during initial five years of life.

Alkaptonuria

It is a genetic disease related to tyrosine metabolism. There is a defect in the enzyme homogentisate oxidase, which is involved in the oxidation of homogentisic acid— one of the intermediates of tyrosine metabolism.

In this condition, homogentisic acid is accumulated and excreted in the urine. Urine becomes dark-coloured on standing, due to the oxidation of homogentisic acid.

Albinism

It is due to the defect in the enzyme tyrosinase, which converts tyrosine to dopaquinone during melanine synthesis.

Tyrosinemia

Defect in the enzyme transaminase and parahydroxy phenyl pyruvate oxidase causes hereditary tyrosinemia. Tyrosine is converted to various biologically important substances.

TRYPTOPHAN

Tryptophan is an essential amino acid. It can be converted to different biologically important substances. Tryptophan is both glucogenic and ketogenic amino acid. **From 60 mg of tryptophan, 1 mg of niacin can be synthesized**.

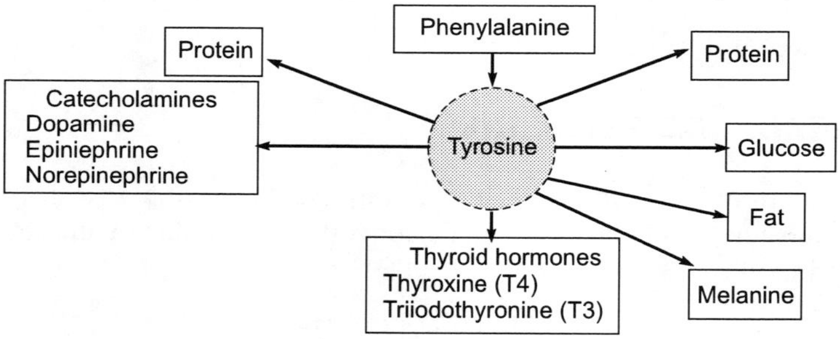

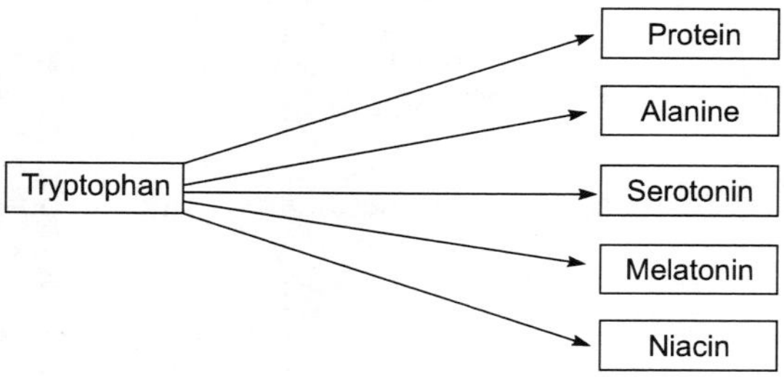

Hartnup's Disease

It is a genetic disease, where the absorption of tryptophan from GIT and renal tubules is defective. In Hartnup's disease, pellagra-like symptoms can be seen.

METHIONINE AND CYSTEINE

Methionine is an essential amino acid. It is both glucogenic and ketogenic amino acid. The active form of methionine is S-Adenosyl methionine. It is an active methyl group donor. The reaction in which a methyl group is added to the substrate is called methylation reaction.

Cysteine is a non-essential amino acid, because it can be synthesized from methionine.

HOMOCYSTEINURIA

During the metabolism of methionine, homocysteine is formed. Homocysteine combines with cysteine to form cystathionine by the action of an enzyme cystathionine synthase, which requires the coenzyme PLP.

If cystathionine synthase is defective or deficiency of vitamins PLP or vitamin B_{12}, homocysteine gets accumulated in blood and is excreted in the urine. An excess of homocysteine in the body interferes with the collagen formation, and is one of the risk factors for atherosclerosis.

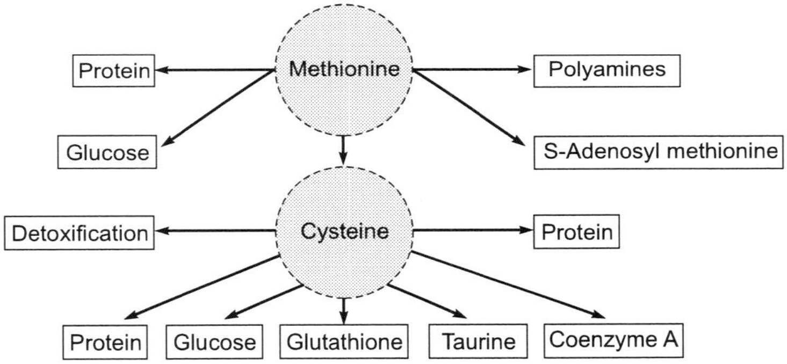

QUESTIONS

Long Essay

1. Write the different sources of ammonia in our body. Discuss in detail the detoxification of ammonia (urea cycle).
2. Write the reactions of urea cycle, and add a note on ammonia toxicity.

Short Essay

1. What is amino acid pool?
2. Describe transamination reaction.
3. Write a note on oxidative deamination reaction.
4. Write a note on non-oxidative deamination reaction.
5. Discuss ammonia toxicity.

Short Answers

1. What is the normal blood urea level? Mention any two conditions where blood urea level is increased.
2. Name two transaminases, which have got clinical importance.
3. Discuss the transport of ammonia from muscle to liver.
4. Describe the transport of ammonia from brain to liver.

13

Heme

INTRODUCTION

Heme is a prosthetic group present in the heme-containing proteins. Haemoglobin is the major heme-containing protein of our body. It is made up of a protein portion called "globin" and a prosthetic group "heme".

Heme has got four pyrrole rings linked to one another (tetrapyrrole), and iron is present at the centre in the ferrous form. Heme is synthesized from glycine and succinyl CoA.

BILIRUBIN METABOLISM

Heme is Catabolized to form Bilirubin

When the heme-containing protein is destroyed, heme and the protein portion are separated. Protein portion is catabolized to form amino acids. Heme is catabolized in hemeoxygenase system of reticuloendothelial system. Heme oxygenase is substrate inducible system (heme concentration will increase the activity of this enzyme system).

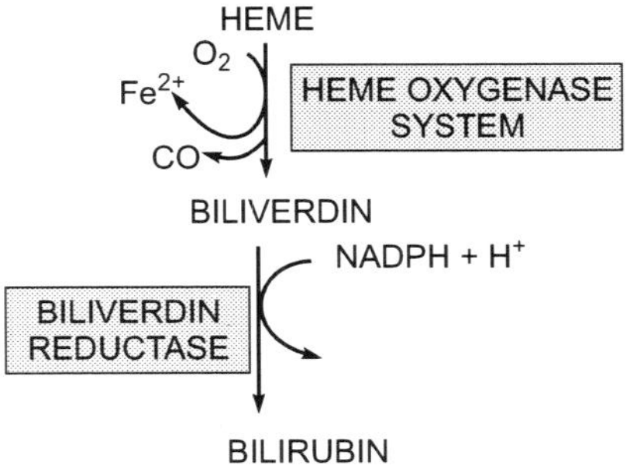

Heme oxygenase system converts heme to biliverdin. In the cytosol, an enzyme called biliverdin reductase converts biliverdin to bilirubin. Bilirubin is a yellow coloured hydrophobic pigment. If it is accumulated in the body, it is toxic. It should be detoxified by making it more hydrophilic.

Bilirubin Metabolism in the Liver

Bilirubin is transported to liver with the help of albumin by forming bilirubin-albumin complex. In the liver, bilirubin is taken up by the hepatocytes. In the cytosol, bilirubin binds to a carrier protein called ligandin and is transported to the site, where it is further metabolized.

In the liver, bilirubin is conjugated with glucuronic acid in the presence of an enzyme called glucuronyl transferase. The conjugated product is bilirubin monoglucuronide and bilirubin diglucuronide.

Bilirubin Metabolism in the Intestine

The conjugated bilirubin enters bile and reaches intestine, where it is deconjugated and converted to form urobilinogen. Urobilinogen is partly reabsorbed and reaches liver. Liver extracts the urobilinogen from portal blood and secrete it into the intestine. A small portion of this urobilinogen reabsorbed from the intestine enters systemic circulation and is excreted through kidney.

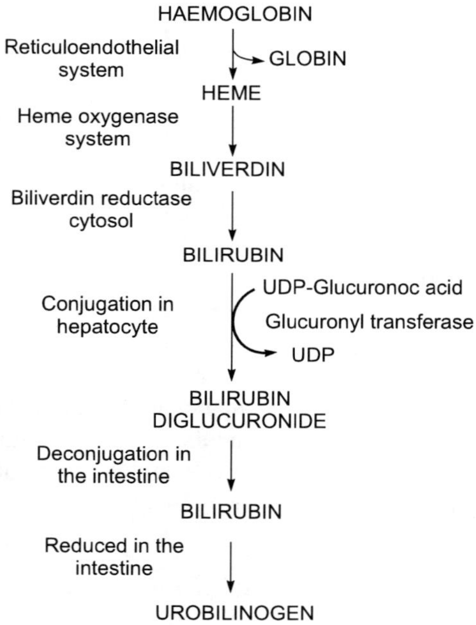

VAN DEN BERGH TEST

When bilirubin is treated with diazotized sulphanilic acid, it is converted to azobilirubin and gives a reddish-purple colour. The same principle is used to estimate the bilirubin in the serum sample. Free bilirubin is insoluble in water, whereas conjugated bilirubin is relatively soluble. When diazotized sulphanilic acid is added directly to the serum, only the conjugated bilirubin, which is soluble in the aqueous phase, will react with the reagent and give colour. This fraction of bilirubin is called direct bilirubin.

The free bilirubin shoud be made soluble by adding methanol before it reacts with diazotized sulphanilic acid. If methanol is added to the serum and treated with diazotized sulphanilic acid, both free as well as the conjugated bilirubin are measured (total bilirubin). If the value for total and conjugated bilirubin is known, unconjugated bilirubin can be calculated by subtracting direct bilirubin from the total bilirubin.

Normal total serum level of bilirubin is less than 1 mg/dL. It ranges from 0.2 to 0.8 mg/dL. Total bilirubin is the sum of direct bilirubin and indirect bilirubin. When bilirubin level increases in the serum, the condition is called hyperbilirubinemia. If bilirubin level goes above 2 or 2.5 mg/dL, it starts diffusing to the tissues and colours them. The sclera of the eye, nail bed, and mucous membrane become yellow in colour. This condition is called jaundice or icterus.

Jaundice is classified into three types:

1. Pre-hepatic jaundice
2. Hepatic jaundice
3. Post-hepatic jaundice

In pre-hepatic jaundice, the liver is normal. The level of unconjugated bilirubin increases due to excess production of bilirubin, as in case of haemolytic jaundice. Since the bilirubin production increases, urobilinogen formation in the intestine also increases. It results in increased excretion of urobilinogen in the urine. But the conjugated bilirubin level and all the enzymes level, which increases due to liver disease, are normal.

In hepatic jaundice there is damages to the hepatocytes. Even though the production of bilirubin is normal, the liver is not able to handle it efficiently due to decreased number of normal hepatocytes. The serum level of conjugated as well as unconjugated bilirubin increases. Since liver is damaged and there is a partial obstruction, serum level of ALT, AST, and alkaline phosphatase increases. Urine level of urobilinogen may be normal or decreased. Bile salt and bile pigments are seen in urine. Since the number of functional hepatocytes decreases, synthesis of proteins required for coagulation also decreases in severe and chronic cases. This results in increased prothrombin time.

Post-hepatic jaundice is due to the obstruction in the flow of bile. Since there is an obstruction, it regurgitates to the blood and blood level of conjugated bilirubin increases. Serum alkaline phosphatase is increased. Bile salts and bile pigments are present in the urine. Urobilinogen excretion in urine decreases or is absent. Since the absorption of fat-soluble vitamin is decreased, prothrombin time is increased.

QUESTIONS

Long Essay

1. Discuss bilirubin metabolism in the intestine.
2. Describe hepatic jaundice.
3. Discuss the catabolism of heme.

Short Essay

1. Define heme.
2. What is van den Bergh test?
3. Discuss the classification of jaundice.
4. What is hyperbilirubinemia?
5. Write a note on pre-hepatic jaundice.
6. What do you mean by direct bilirubin?
7. What is the cause of post-hepatic jaundice?

14

Immunoglobulins

INTRODUCTION

Immunoglobulins are also called antibodies. They occupy the gamma region in the electrophoretogram, so they are also called gamma globulins. They are synthesized by the B-lymphocytes upon stimulation. These are mainly involved in mediating humoral immunity.

STRUCTURE

All immunoglobulins share the same basic structural characteristics. They all differ from one another at the antigen binding region. Immunoglobulins have two identical heavy chains and two identical light chains. The heavy chain and the light chain have amino terminal and carboxy terminal ends. The amino acid sequence of the amino terminal varies in both heavy chain and the light chain. This region is called variable region. Both heavy chain and light chain have variable region at the amino terminal (Fig. 14.1).

Heavy chain and light chain have one variable region. Light chains have one constant region towards the carboxy terminal; heavy chains have three constant regions.

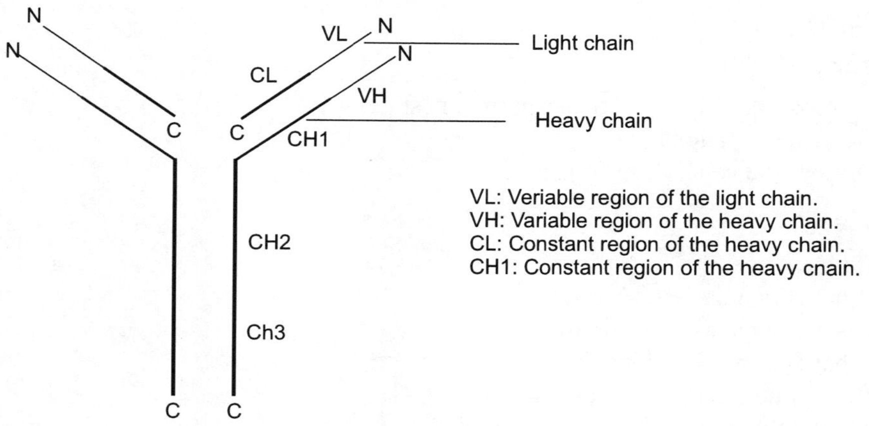

Fig. 14.1 Structure of immunoglobulin.

TYPES OF IMMUNOGLOBULINS

There are different types of heavy chains—α, γ, δ, ε. Based upon the type of heavy chains, there are different immunoglobulins. Immunoglobulin with α heavy chain is called IgA. Similarly, γ heavy chain containing immunoglobulins is called IgG. IgD contains δ heavy chain; IgM contains μ heavy chain; and ε is the heavy chain of IgE.

FUNCTIONS

1. **IgA is secretory immunoglobulin.** It is found in various body secretions. It prevents the invasion of various microorganisms and protects the body.
2. **IgG is the immunoglobulin of secondary immune response.** It is capable of crossing the placental barrier and provides protection to the infant.
3. **IgM is immunoglobulin of primary immune response.** It cannot go out from the vascular compartment. It is involved in protection against the blood-borne microorganisms.
4. **IgE is involved in the hypersensitivity reactions.** The exact function of IgD is not known, but it acts as a receptor on the B-cell membrane.

MAJOR HISTOCOMPATIBILITY COMPLEX (MHC)

Human major histocompatibility complex is also called the human leukocyte antigen (HLA) complex. Gene coding for MHC is present on chromosome 6. HLA class I and HLA class II gene products have got role in immunological reaction of the body and tissue transplantation histocompatibility. HLA plays a major role in susceptibility to a number of autoimmune diseases.

HLA class I and class II molecules have different structures and both are involved in binding antigen and present it to the T lymphocyte (T cell).

QUESTIONS

1. Write the structure of immunoglobulins.
2. Name different types of immunoglobulins and mention their function.
3. MHC.

15

Nucleic Acids

INTRODUCTION

Human body is made up of millions of cells having different structures and functions. What determines the structure and function of these cells? It is determined by nucleic acid. They are macromolecules, made up of building blocks called nucleotide. Nucleotides are linked to one another by phosphodiester linkage to form polynucleotide.

Depending upon the type of sugar present in nucleic acids, there are two different types of nucleic acids:

1. Ribonucleic acids (RNA) : The sugar present is ribose.
2. Deoxyribonucleic acids (DNA) : The sugar present is deoxyribose.

NUCLEOTIDES

Nucleotides are made up of:

1. Heterocyclic nitrogen containing ring (base)
2. Pentose sugar
3. One or more phosphate group (Fig. 15.1).

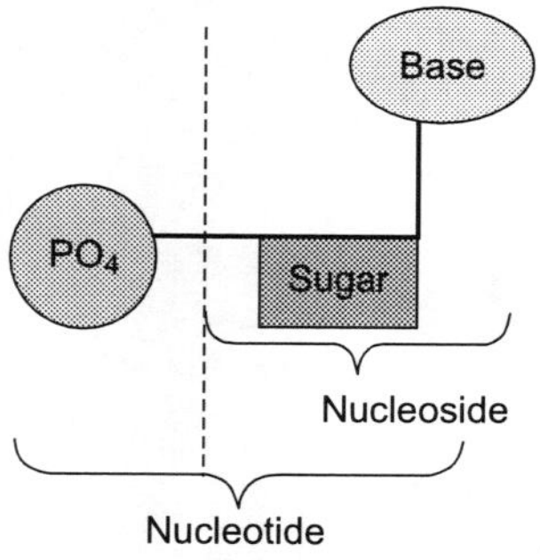

Fig. 15.1 Nucleotide.

NITROGEN CONTAINING BASES

They are of two types—purine and pyrimidine (Fig. 15.2 (a) and (b))

Major purines: Adenine and guanine.

Major pyrimidines: Cytosine, thymine, and uracil.

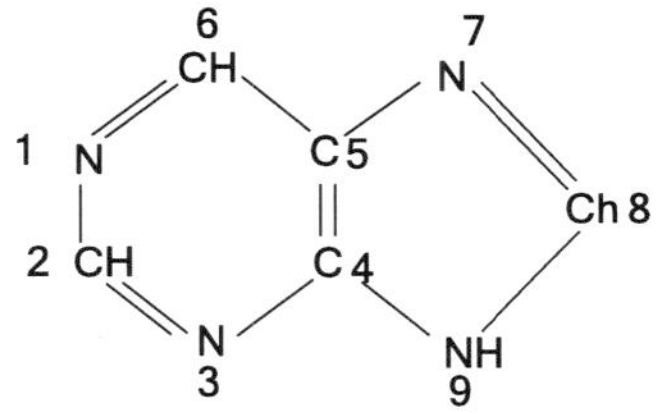

(a) Purine

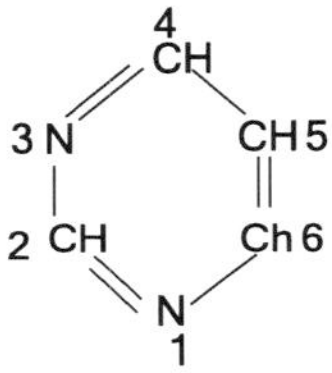

(b) Pyrimidine

Fig. 15.2 (a) Purine (b) Pyrimidine

Purine derived from
(N1) Aspartate
(C2) N10 formyl H4 folate
(C4), C5, N7 Glycine
(C6) Carbon dioxide
(N3) N9 Glutamine

Pyrimidine derived from
(C4, C5, C6, N1) Aspartate
(C2) Carbon dioxide
(N3) Glutamine
(C8) N5 N10 methenyl H4 folate

SUGAR

Nucleic acids contain an aldopentose—ribose. The carbon atoms of ribose are numbered as shown below to differentiate them from the carbon atoms of the nitrogen containing bases.

NUCLEOSIDE

The nitrogen-containing base is connected to ribose by **β-N-glycosidic linkage**. The first carbon atom of ribose is connected to the ninth nitrogen of purine ring, or first nitrogen of the pyrimidine ring. Nitrogen containing base and sugar together constitute the nucleoside.

NUCLEOTIDE

Phosphorylated forms of nucleosides are called nucleotides. Nucleotides may be nucleoside monophosphates or may be nucleoside diphosphates or triphosphates. For example,

1. Adenosine monophosphate (AMP) (adenylate)
2. Adenosine diphosphate (ADP)
3. Adenosine triphosphate (ATP)

Nucleotide form of adenine is adenosine monophosphate (adenylate) and guanine is guanosine monophosphate (guanylate). Similarly, cytidine monophosphate (cytidylate), uridine monophosphate (uridylate), and thymidine monophosphate (thymidylate) are the nucleotide forms of cytosine, uracil and thymine respectively.

FUNCTIONS OF NUCLEOTIDES

1. Nucleotides are the building blocks of nucleic acids—RNA and DNA.
2. They form a portion of several important coenzymes such as NAD^+, $NADP^+$, FAD, and coenzyme A.
3. They serve as mediators of numerous important cellular processes such as second messengers in signal transduction events. The predominant second messenger is cyclic-AMP (cAMP)—a cyclic derivative of AMP formed from ATP.
4. They regulate various enzymatic reactions by their allosteric effects on enzyme.
5. They serve as activated intermediates in biosynthetic reactions. These activated intermediates include S-adenosylmethionine (S-AM) involved in methyl transfer reactions as well as many sugar coupled nucleotides involved in glycogen and glycoprotein synthesis.

BIOLOGICALLY IMPORTANT NUCLEOTIDES

Guanosine Derivatives

Cyclic GMP (cGMP; 3'-5' cyclic guanosine monophosphate) is found in cells. It acts as a second messenger for nitric oxide and atrial natriuretic factor (ANF). The most important cGMP coupled signal transduction cascade is involved in the conversion of light energy into nerve impulse in the retina of the eye.

S-adenosyl Methionine (SAM)

S-adenosyl methionine is an active methionine. It serves as a methyl group donor in methylation reactions, and as a source of propylamine in the synthesis of polyamines.

Cyclic Adenosine Monophosphate (cAMP)

The most common adenosine derivative is cyclic AMP (cAMP; 3′-5′-cyclic adenosine monophosphate). It is a second messenger. Cyclic AMP is also involved in the regulation of ion channels by direct interaction with the channel proteins. Formation of cAMP occurs in response to activation of receptor-coupled adenylate cyclase on the cytosolic surface of the cell.

Phosphoadenosyl Phosphosulphate

Phosphoadenosyl phosphosulphate is abbreviated as PAPS. It is called active sulphate. It acts as a sulphate group donor in the synthesis of sulphated sugar and some detoxification reactions.

Synthetic Nucleotide Analogues

Nucleotide analogues can be synthesized chemically and are used in the treatment of various diseases. The nucleotide analogues can be utilized to inhibit specific enzymatic activities. These analogues are capable of interfering with the synthesis of DNA, and thereby preferentially kill rapidly dividing cells such as tumour cells. So, a number of nucleotide analogues are used in the treatment of cancer. Some of these commonly used in chemotherapy are 5-iodo-2′-deoxyuridine, 6-mercaptopurine, 5-fluorouracil, and 6-thioguanine.

Nucleotide analogues such as AZT (azidothymidine) and ddI (dideoxyinosine) have been used as antiviral agents. Purine analogues are used to treat gout. The most commonly used analogue is allopurinol. Several nucleotide analogues are used after organ transplantation in order to suppress the immune system and reduce the likelihood of transplant rejection by the host.

PHOSPHODIESTER LINKAGE

Polynucleotide is formed when the nucleotides are joined to one another by phosphodiester linkage. The 5′ carbon atom of one nucleotide is connected to the 3′ carbon of the adjacent nucleotide. Hence, the phosphodiester linkage is called 3′5′phosphodiester linkage (Fig. 15.3).

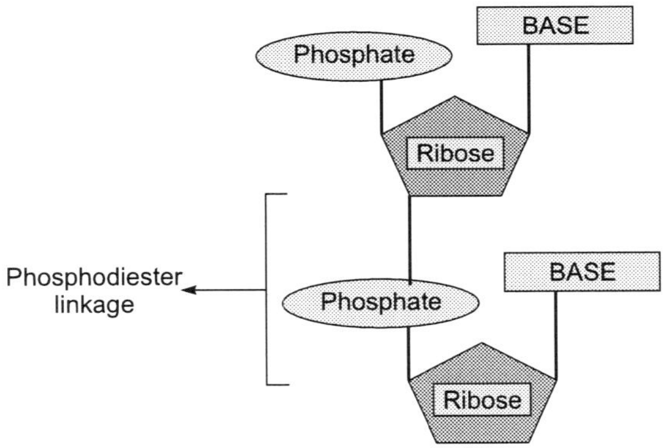

Fig. 15.3 Phosphodiester linkage.

Polynucleotide Chain

The alcohols of 5′-phosphate of one nucleotide and the 3′-hydroxyl of a next nucleotide form a phosphodiester bond. The formation of phosphodiester bonds in DNA and RNA exhibits directionality. One end of the polynucleotide chain, where 3′hydroxyl group is present in free form, is called 3′end, and the 5′end phosphate group, which is in the free form, is called 5′end of the polynucleotide chain. The primary structure of DNA and RNA (the linear arrangement of the nucleotides) proceeds in the 5′ to 3′ direction.

e.g., 5′-pApGpTpC-3′

STRUCTURE OF NUCLEIC ACIDS

Nucleic acids are polymers of nucleotides held together by phosphodiester bonds. Cellular nucleic acids are of two types—deoxyribonucleic acid (DNA) and ribonucleic acid (RNA).

Deoxyribonucleic Acid (DNA)

In 1953, James Watson and Francis Crick deduced the double-helical structure of DNA. This DNA structure is called B-DNA. The salient features of B-DNA can be listed as follows:

1. DNA contains two polynucleotide chains, which wind around each other along a common axis to form a double helix.
2. Polynucleotide chain is made up of four different types of nucleotides, deoxyadenylate, deoxyguanylate, deoxycytidylate, and deoxythymidylate.
3. B-DNA is a right-handed double helix.
4. The two strands are antiparallel to each other, i.e., they run in opposite directions. (3′ end of one strand corresponds to 5′ end of other strand).
5. Both strands are complementary to each other.
6. The strands are stabilized by the hydrogen bonding between purine of one strand and pyrimidine of other strand. Adenine base pairs with thymine, forming two hydrogen bonds, and guanine base pairs with cytosine, forming three hydrogen bonds. This is called Watson and Crick type of base pairing (Fig. 15.4).

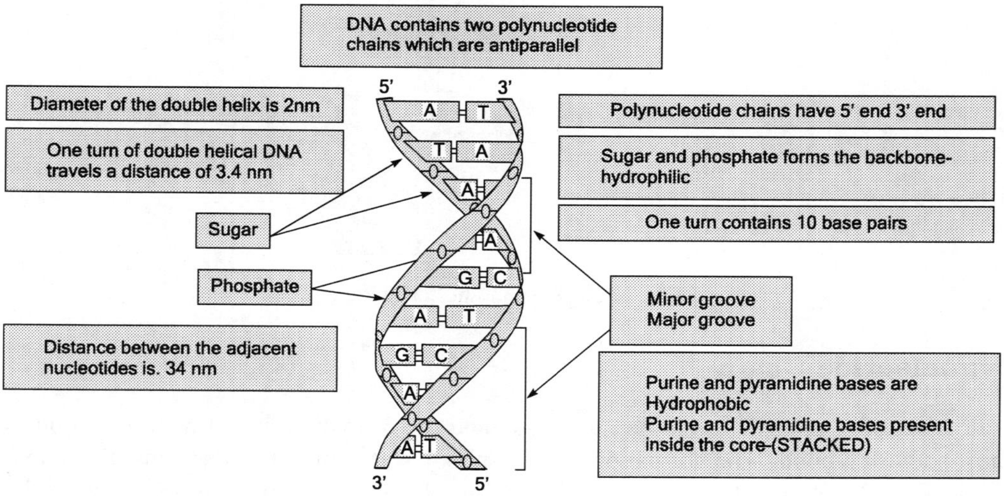

Fig. 15.4 Structure of B-DNA.

7. Erwin Chargaff quantified purines and pyrimidines present in the genome of different species during 1940s, and reported that total number of purines is equal to the total number of pyrimidines in the genome of a given species.
8. The backbone of the double helix is made up of alternate deoxyribose and phosphate.
9. Nitrogen-containing bases are stacked at the core of the double helix. Bases are perpendicular to the long axis of the helix.
10. Diameter of the helix is 2 nm.
11. Each 360° turn of the helix travels a distance of 3.4 nm along the helix axis.
12. Distance between adjacent bases is 0.34 nm; adjacent bases are related to each other by 36°.
13. One full turn contains 10 base pairs.

Ribonucleic Acid (RNA)

There are three types of ribonucleic acids—tRNA, mRNA, and rRNA.

Transfer RNA (tRNA)

Transfer RNA is called tRNA. It acts as an adaptor molecule during translation.

Primary Structure

On an average, tRNA contains about 75 ribonucleotides. Ribonucleotides are linked to each other by phosphodiester linkage. The ribonucleotides present in tRNA are adenosine monophosphate, guanosine monophosphate, cytidine monophosphate and uridine monophosphate. In addition to these, tRNA contains unusual bases like pseudouracil, dihydrouracil and thymine. Polyribonucleotide has a 5′ end and a 3′ end.

Secondary Structure

Certain regions of the tRNA are complementary to each other. These complementary regions base pair to form a clover leaf structure (Fig. 15.5).

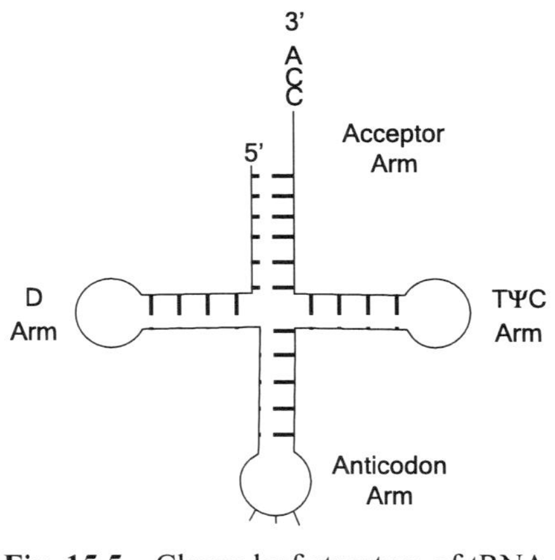

Fig. 15.5 Clover leaf structure of tRNA.

In the secondary structure of tRNA, four different arms can be seen. The arm is made up of a stem and a loop. Stem is the portion, which is formed due to the base pairing of the complementary bases, and the loop is formed by the segment of the nucleic acid where there is no complementary base pairing.

In the secondary structure of tRNA, four different arms can be seen. The arm is made up of a stem and a loop. Stem is the portion, which is formed due to the base pairing of the complementary bases, and the loop is formed by the segment of the nucleic acid where there is no complementary base pairing.

The arm, which contains the **unusual base dihydrouracil** is called D-arm; the arm containing the **unusual base sequence thymine, ψ, C**, is called T ψ C arm. The arm of tRNA containing three nucleotide sequences, which are complementary to the codons present in genetic code, is called anticodon arm. The arm of tRNA which is opposite to the anticodon arm is called acceptor arm. Acceptor arm accepts the amino acid when the tRNA is charged with it.

Tertiary Structure

The **clover leaf structure of tRNA** folds on itself to form the tertiary structure. It is L-shaped (Fig. 15.6). At one end it contains the anticodon, and at the other end it contains the amino acid.

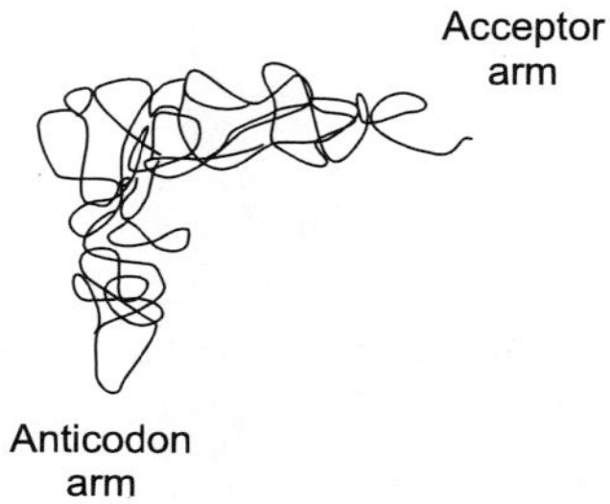

Fig. 15.6 Tertiary structure of tRNA.

Messenger RNA (mRNA)

RNA, which brings message from DNA to the translational machinery for protein synthesis, is called messenger RNA. In case of eukaryotes, RNA produced by transcription is processed by post-transcriptional modification to form messenger RNA.

In prokaryotic cells, the immediate product of translation is messenger RNA. Messenger RNA has a **5′ cap** and a **3′ tail**. Cap is chemically 7-methyl guanosine triphosphate, and the 3′ tail is **'poly A' tail**. The 5′ cap prevents the action of 5′ exonuclease on that end, and helps in the recognition of the mRNA by the ribosome during translation. The 3′ tail helps to prevent the action of 3′ exonuclease on that end.

Ribosomal RNA (rRNA)

It has got extensive stem and loop-like structures due to the base pairing between the complementary bases present within the molecules. Ribosomal RNA is associated with the ribosome. It is required for translation.

Ribosome has smaller and larger subunits. Each of these subunits contains rRNA, which have a role in translation. The exact functions of these RNAs present in ribosome are not yet evident.

Catabolism of Purine

Purines are catabolized to form uric acid (Fig. 15.7). **Normal blood uric acid level is 3 to 7 mg/dL.** Uric acid is capable of donating proton, so it is called an acid. At physiological pH, it gets deprotonated to form urate. The negatively charged urate is associated with positively charged sodium to form sodium urate. In the blood, uric acid is present as sodium urate.

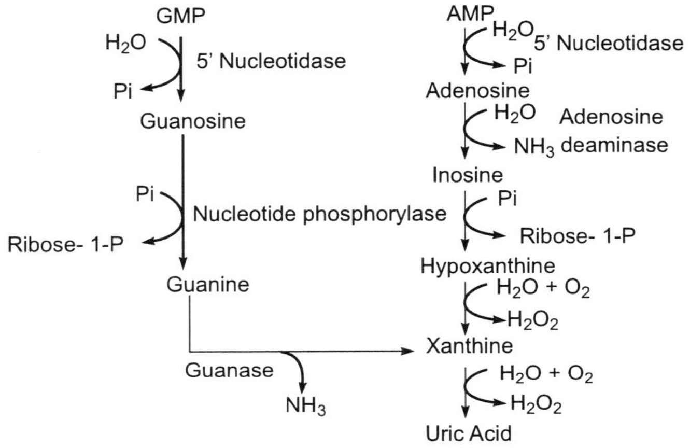

Fig. 15.7 Catabolism of purine.

If the uric acid level increases above 7 mg/dl, it is called hyperuricemia. When the serum uric acid level reaches to 7 mg/dl, blood is said to be saturated with uric acid. If uric acid content of the blood rises above this level, it starts getting precipitated in the soft tissues and joint cavities.

The deposition of uric acid in the soft tissues and joint cavities is called **"tophi"**. In the joint cavity, uric acid crystals evoke inflammatory reaction. Individuals suffer from excruciating pain in the joint (especially small distal joints) due to arthritis (acute or chronic). In gout, excess of uric acid is excreted in the urine. Urinary tract stones can be seen in gout.

Gout is a clinical condition due to hyperuricemia, where clinical manifestations like acute arthritis, chronic arthritis, tophi, or urolithiasis (formation of uric acid stones in the urinary tract) are seen.

Causes of Gout

Primary gout

Defect in the enzyme, PRPP synthetase, HGPRT, glucose-6-phosphatase, genetic defect in the excretion of uric acid.

Secondary gout

Increased rate of turnover of the cells or destruction of large number of tissue cells.

1. Lymphoma
2. Leukemia
3. Psoriasis, etc.

The accumulation of any organic acids like lactic acid, acetoacetic acid that inhibit the excretion of uric acid. Uric acid accumulation may be due to the excess production of purine, which results because of a defect in the biosynthesis of purine nucleotides or increased turnover of cells.

Purines are synthesized starting from phosphoribosyl pyrophosphate (PRPP). If PRPP concentration increases in the cell due to any defect in its metabolic pathway, it will increase the synthesis of purine nucleotide biosynthesis and proportionate increase in its breakdown, leading to an increased formation of uric acid.

PRPP is synthesized from ribose-5-phosphate by the action of an enzyme PRPP synthetase. The activity of PRPP synthetase is increased in the presence of excessive ribose-5-phosphate.

Free bases, adenine and guanine, can be phosphoribosylated to form nucleotides. The phosphoribosyl group donor is PRPP. Enzymes catalyzing these reactions are adenine phosphoribosyl transferase and hypoxanthine guanine phosphoribosyl transferase (HGPRT).

The genetic defect in the enzyme PRPP synthetase may be responsible for the overproduction of PRPP. In case of glucose-6-phosphatase deficiency, glucose-6-phosphate gets accumulated. It enters into HMP shunt pathway and the concentration of ribose-5-phosphate increases. Increase in the concentration of ribose-5-phosphate activates the enzyme PRPP synthetase and concentration of PRPP increases. Defect in the enzyme HGPRT can lead to underutilization of PRPP and increases its concentration.

Accumulation of PRPP leads to increased production of purine nucleotides. The proportionate increase in the breakdown of purines is responsible for hyperuricemia and gout.

Uric acid is excreted by a transporter present in the kidney tubules. The same transporter transports other organic acids like lactic acid, salicylic acid, etc. If the concentration of these organic acids increases in the circulation, they compete with uric acid for the excretion. These transporters prefer other organic acids than uric acid. So, uric acid concentration in the blood will increase.

QUESTIONS

Short Essay

1. Discuss the Watson and Crick model of DNA.
2. Describe the structure of t-RNA.
3. Write a note on gout.
4. Discuss the catabolism of purines.
5. Mention the differences between DNA and RNA.
6. Give some of the functions of nucleotides.
7. Write about some biologically important nucleotides.

Short Answers

1. Mention different types of RNA and their functions.
2. What is the normal serum uric acid level? Mention some of the causes of hyperuricemia.
3. Describe tophi.
4. Indicate the sources of carbon and nitrogen required for the purine ring formation.
5. Mention any four biologically important nucleotides.
6. Write the reaction catalyzed by HGPRT.

16

Protein Biosynthesis

NUCLEOSOME

The DNA present inside the nucleus of a eukaryotic cell when extended is about 1 metre long. Proteins in the nucleus help to pack the long slender DNA inside the nucleus, which is having a volume of few cubic micrometres.

Histones are a family of basic proteins present in the nucleus. They are **rich in basic amino acids arginine and lysine**. Histones have positive charge and bind to the negatively charged backbone of the DNA. This DNA protein interaction helps in the orderly arrangement of the DNA inside the nucleus.

Histones are of five different types–they are H1, H2A, H2B, H3 and H4. Histone H2A, H2B H3 and H4 are core histones. They interact in a specific way as follows:

- 2(H3) and 2(H4) interact to form tetramer
- H2A and H2B interact to form dimer

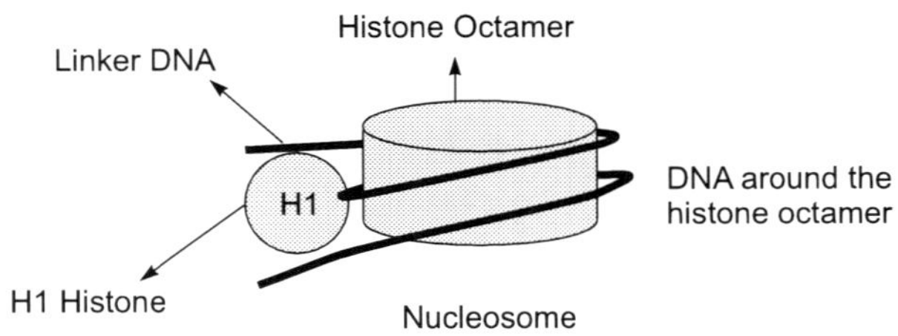

One tetramer and two dimer associate to form histone octamer. Around the histone octamer 1.75 superhelical turn of DNA are wrapped. H1 histone binds to the DNA outside the histone octamer. This structure is called **nucleosome**. Nucleosome is a flat disc-like structure with 10 nm diameter and 5 nm height. Two nucleosomes are linked to each other by a stretch of DNA called linker DNA which has got 30 base pairs (bp) giving rise to 10 nm fibril. It has an appearance of **beads on string**.

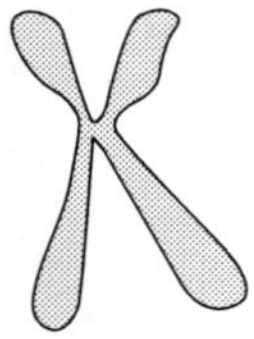

Chromosome

The 10 nm chromatin fibril coils on itself (supercoil) to form 30 nm fibril. It later gets more compactly arranged with the help of other proteins to form chromosome.

REPLICATION

Cells in human body contain 23 pairs of chromosomes (diploid genome). 23 different chromosomes are called haploid genome. The haploid genome contains about 3 × 109 basepairs. Genetic information is present in the nucleotide sequence of the DNA. When cell divides genetic information is transferred from parent cell to the daughter cell. So a copy of DNA should be transferred from parent cell to daughter cell. This requires synthesis of DNA in parent cell which is an exact copy of the DNA present in the parent cell. The process of **synthesis of DNA by using DNA as the template is called replication**.

DNA contains information regarding the structure and function of various proteins. Portion of the DNA specifying the amino acid sequence of the protein is called **gene**. **Gene expression** is the process of converting the nucleotide sequence present in the DNA to amino acid sequence of protein. Gene expression includes transcription and translation. **Synthesis of RNA by using DNA as the template is called transcription**. **Synthesis of protein by using the RNA is called translation**.

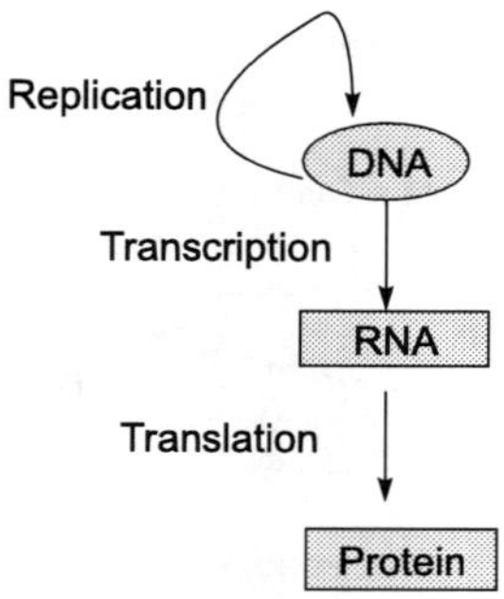

Central dogma of molecular biology

Replication is a process of DNA synthesis using DNA as the template. Replication takes place during the S-phase of cell cycle.

DNA Replication is Semiconservative

Both strands of DNA are used as templates. New strand synthesized is complementary to the existing (parent) strand. The new duplex DNA has one parent strand and one new strand. So replication is semiconservative.

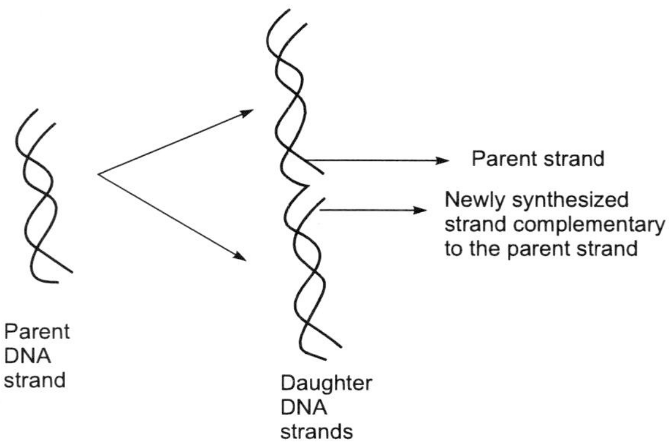

DNA Replication is Bidirectional

DNA replication equires separation of two strands of DNA and polymerization of deoxyribonucleotides to form new strand. Process of replication is well studied in prokaryotes. In *E.coli* the genome is circular. There is a region in this circular DNA which is rich in AT sequence called origin of replication. Since it is rich in AT sequence, initiation of unwinding of DNA is easier in this area.

A protein called dna A-binds to the origin of replication and unwinds a small stretch DNA forming single stranded DNA in that area. The unwinding of a stretch of DNA forms two replication forks.

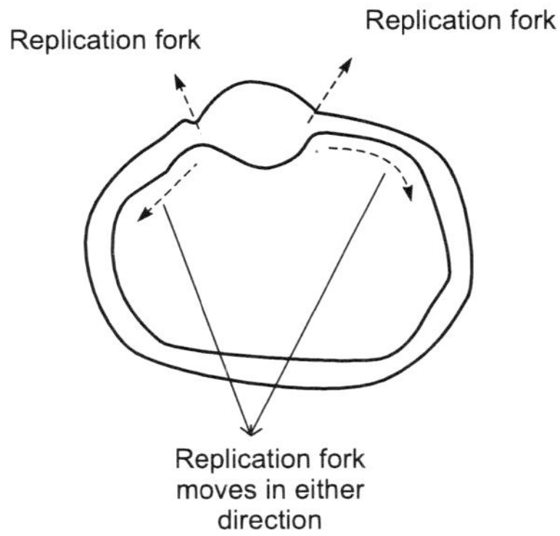

Replication fork is the region where there is active synthesis of DNA occurs. Replication forks move in either direction along the DNA copying the parent DNA strands. So DNA replication is bidirectional.

DNA helicase binds to single stranded DNA and further unwinds the DNA. **Single-strand** binding proteins (SSB) binds to single stranded DNA. These proteins keep two strands separate and prevent the action of nuclease on single-stranded DNA.

DNA Replication is Semidiscontinuous

Polymerization of deoxyribonucleotides complementary to the template strands is catalyzed by the enzyme **DNA polymerase**. DNA polymerase reads the template strand from 3′ to 5′ direction and synthesizes the new strand in the 5′ to 3′ direction. The polymerization of deoxyribonucleotides takes place at the 3′ end of the new strand.

DNA polymerase cannot initiate the process of polymerization of deoxyribonucleotides. It requires a RNA primer with the free 3′ hydroxyl group. The primer is synthesized by an enzyme called **primase.**

In prokaryotes there are three different types of DNA polymerase:

1. DNA polymerase I
2. DNA polymerase II
3. DNA polymerase III

The DNA polymerase moves in one direction along the DNA the strand which is read in the 3′ to 5′ direction is synthesized continuously. It is called **leading strand.** The other strand of DNA is synthesized discontinuously. It is called **lagging strand.** Each fragment of DNA present in the lagging strand contains a small stretch of RNA and a stretch of DNA. These fragments are called **Okazaki fragments.** So DNA replication is semidiscontinuous.

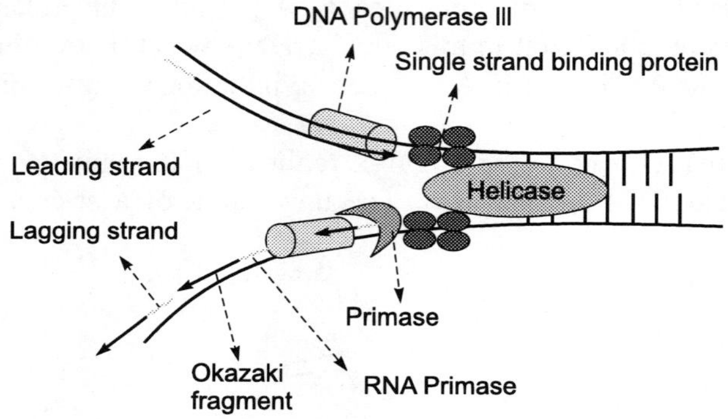

DNA polymerase I has got 5′ to 3′ exonuclease and 3′ to 5′ exonuclease activity. The 5′ to 3′ exonuclease activity removes the RNA primer attached to the Okazaki fragment. The gap is filled up by the polymerase action of DNA polymerase I. The gap between two DNA fragments is sealed by **DNA ligase.**

DNA polymerase III has got 3′ to 5′ exonuclease activity. So it has got **proofreading activity** (mismatched bases are removed during replication).

During replication the unwinding of DNA double strand leads to appearance of **positive supercoils** in the DNA head of replication fork. Accumulation of positive supercoils may stop the process of replication. There are enzymes called **DNA topoisomerase** which helps to release the supercoils. Topoisomerase are of two types.

1. **Topoisomerase I**
   It cuts one strand of DNA and releases the supercoil. It does not require ATP.
2. **Topoisomerase II**
   It cuts both strands of DNA and releases the supercoil. It requires ATP. Type II topoisomerase found in the bacteria is called DNA gyrase. DNA gyrase is **inhibited by antimicrobial agent, quinolones** (e.g., Ciprofloxacin).

TRANSCRIPTION

Transcription is the process of RNA synthesis by using DNA as the template. Transcription requires an enzyme **RNA polymerase**. This enzyme polymerizes ribonucleotides which are complementary to a stretch of DNA called gene. Based on the sensitivity to a mushroom poison **α-amanitin**. There are three different types of RNA polymerase in mammals.

1. **RNA polymerase I**
   It is insensitive to α-amanitin. It is required for synthesis of 28s, 18s and 5.8s ribosomal RNA (rRNA).
2. **RNA polymerase II**
   It is highly sensitive to α-amanitin. It is required for the synthesis of messenger RNA (mRNA).
3. **RNA polymerase III**
   It is partially sensitive to α-amanitin. It synthesizes tRNA and 5srRNA.

RNA polymerase requires DNA as the template. It reads the template strand from 3′ to 5′ direction and synthesizes the RNA in 5′ to 3′ direction. RNA polymerase lacks exonuclease activity. So it does not have proofreading activity. RNA polymerase does not require primer to begin polymerization of ribonucleotides.

The process of transcription includes three different steps:

1. Initiation
2. Elongation
3. Termination

Initiation

A segment of the DNA strand contains information regarding the amino acid sequence of the polypeptide chain. This portion of the DNA is transcribed during transcription. There are certain nucleotide sequences on the DNA adjacent to the transcribed region of DNA called promoter region.

RNA polymerase binds to the promoter region along with a protein called **sigma (σ) factor**. RNA polymerase has got two α subunits and one each β and β' subunit. It is called core polymerase. Sigma factor helps RNA polymerase to identify and bind to the promoter. Core polymerase with the sigma factor is called holopolymerase.

In prokaryotes on the upstream side of the transcription start site, there are two consensus regions

1. Tata box (Pribnow box)
   10 basepairs upstream sequence 5′-TATAAT-3′
2. –35 sequence
   35 basepairs upstream of transcription start site sequence 5′-TTGACA-3′

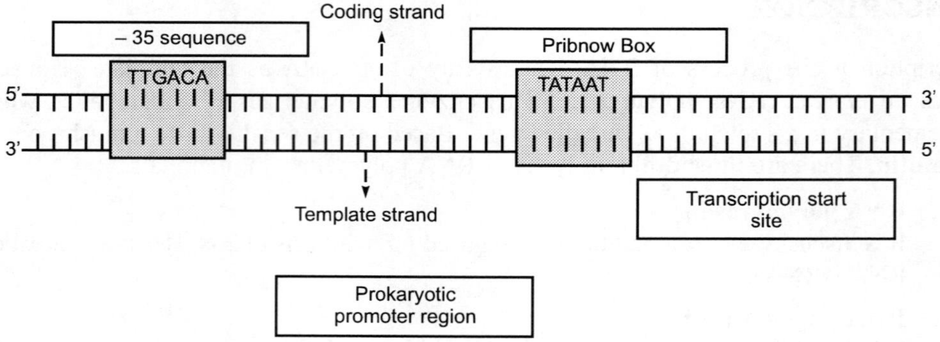

In eukaryotes the corresponding consensus sequences are called

1. Goldberg-Hogness box
2. CAAT box

In addition to the promoter region, there are different nucleotide sequences which increase or decrease the rate of transcription called enhancers and silencers.

Elongation

Once holopolymerase binds to the promoter region and unwinds a small stretch of DNA, RNA polymerase starts transcribing the DNA leading to formation of several short stretches of RNA having less than 11 nucleotides. These stretches of RNA are discarded.

The elongation of transcription begins when the transcript length exceeds 10 nucleotide. Now the sigma factor is released and the RNA polymerase leaves the promoter region and moves along the DNA template and polymerizes ribonucleotides complementary to it.

The site where transcription begins is called transcription initiation site. The first nucleotide present in the transcription start site is purine.

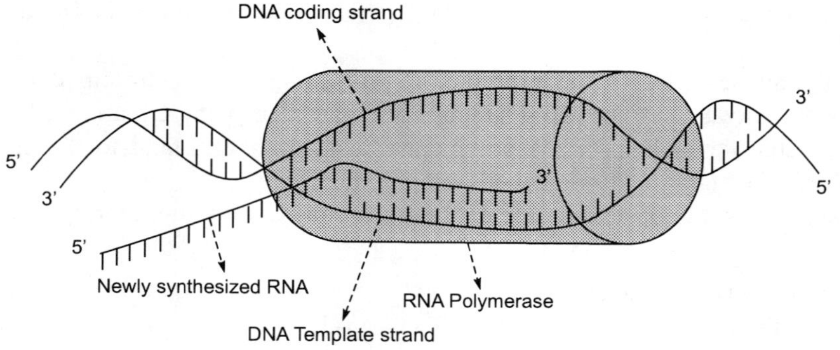

Transcription

Termination

Transcription of the DNA template continues until the termination signal is reached. Termination may be by two different methods.

1. **Rho (ρ) dependent**
   Protein having ATPase and helicase activity called rho factor is required. It binds to the C rich site near the 3′ end of the newly synthesized RNA and moves along the RNA and reaches RNA polymerase. With the help of its helicase activity, separates RNA and DNA hybrid leading to release RNA.
2. **Rho independent**
   The 3′ end of the newly synthesized RNA has self complementary region which is rich in GC bases. The self-complementary region forms hairpin. Beyond the hairpin there is a uracil rich region in the 3′ end of the RNA. The weak base pairing between the uracil and adenine facilitates the separation of the RNA from DNA.

Inhibitors of Transcription

Actinomycin D binds to DNA template and prevents the transcription by interfering the movement of the RNA polymerase. Rifampin binds to the β-subunit of RNA polymerase and prevents transcription.

Post-transcriptional Modifications

Messenger RNA is formed by extensive processing of the immediate product of transcription called **primary transcript**. The primary transcript has got noncoding regions (nucleotide sequence does not code for the protein) called **introns**. The coding regions are called **exons**.

Post-transcriptional modification involves

1. Removal of extrons and splicing of introns
2. Addition of 5′ cap (7 methyl gaunosine)
3. Addition of 3′ tail (polyadenylate tail)

Removal of extrons and splicing of introns is carried out by **spliceosomes**. Spliceosomes contain small nuclear RNA (SnRNA) and protein. The SnRNAs have enzymatic activity.

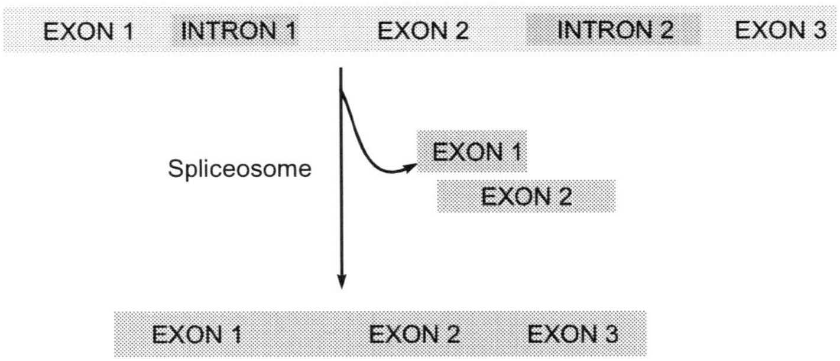

7-Methyl gaunosine is added to the 5′ end of the RNA. It protects the RNA from 5′ exonuclease and is required for translation. About 40 to 200 adenylate (AMP) is added to the 3′ end of the RNA. It is called poly A tail. Poly A tail protects RNA from the action of 3′ exonuclease.

Ribosomal RNA (rRNA) and transfer RNA (tRNA) are modified by different mechanisms.

TRANSLATION

Information regarding the amino acid sequence of protein is present in the nucleotide sequence of a stretch of DNA. This nucleotide sequence is used to synthesize messenger RNA by a process called translation. Messenger RNA is sent to the cytosol. The nucleotide sequence of messenger RNA expressed to produce specific amino acid sequence of the protein and the process is called translation.

GENETIC CODE

Three nucleotides sequence present in mRNA constitute **codon**. Each codon specifies an amino acid. There are 4 different nucleotides and each codon is made up of 3 nucleotides. The total number of specific codons is 64 (43).

5'	U		C		A		G		3'
	UUU	Phe	UCU		UAU	Tyr	UGU	Cys	U
U	UUC		UCC	Ser	UAC		UGC		C
	UUA	Leu	UCA		UAA	Ter	UGA	Ter	A
	UUG		UCG		UAG		UGG	Trp	G
	CUU		CCU		CAU	His	CGU		U
C	CUC	Leu	CCC	Pro	CAC		CGC	Arg	C
	CUA		CCA		CAA	Gln	CGA		A
	CUG		CCG		CAG		CGG		G
	AUU	Ileu	ACU		AAU	Asn	AGU	Ser	U
A	AUC		ACC	Thr	AAC		AGC		C
	AUA		ACA		AAA	Lys	AGA	Arg	A
	AUG	Met	ACG		AAG		AGG		G
	GUU		GCU		GAU	Asp	GGU		U
G	GUC	Val	GCC	Ale	GAC		GGC	Gly	C
	GUA		GCA		GAA	Glu	GGA		A
	GUG		GCG		GAG		GGG		G

Out of 64 codons, 61 codes for specific amino acids and 3 codons are called **stop codons or nonsense codons**. They are UAG, UGA, and UAA. Initiation of translation starts from the codon **AUG (initiation codon)**.

The **collection of codons is called genetic code**. The knowledge of genetic code is responsible to understand the relationship between the nucleotide sequence of nucleic acids and the amino acid sequence of the proteins.

Characteristics of Genetic Code

1. **Specificity:** A given codon always codes for specific amino acid.
2. **Universality:** It is same for all species of plant and animals (exceptions are seen in mitochondria).

3. **Non-overlapping and commaless:** Each codon is a three-nucleotide sequence read from a fixed point continuously—there is no punctuation.
4. **Degeneracy:** A given amino acid may have more than one codon but a given codon specifies a single amino acid.

REQUIREMENTS FOR PROTEIN SYNTHESIS

1. mRNA
2. tRNA containing amino acids
3. Ribosome
4. Protein factors

mRNA

The mRNA is produced by the process of translation. In prokaryotes the primary transcript itself is mRNA. Prokaryotic messenger RNA contains information from more than one gene. So it is called **polycistronic mRNA**. In eukaryotes mRNA contains information from one gene. It is called **monocystronic mRNA**.

tRNA

Each tRNA has got three nucleotide sequences in the anticodon arm complementary to a given codon called **anticodon**. Each tRNA molecule carries specific amino acid on the acceptor arm. tRNA carrying amino acid is said to be charged and the tRNA without amino acid is said to be uncharged. The tRNA is capable of reading the nucleoside sequence of mRNA and brings the specific amino acid to ribosome during protein synthesis. It is known as adaptor molecule.

Amino acid is added to tRNA by an enzyme called **aminoacyl-tRNA synthetase**. There are specific aminoacyl-tRNA synthetase enzymes for charging specific tRNA.

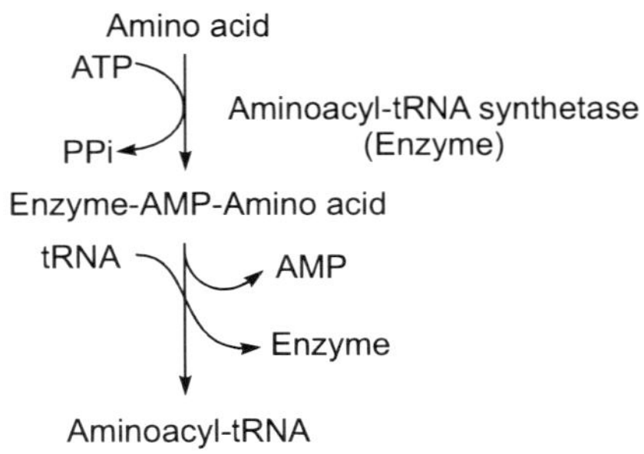

Ribosome

Ribosome is required for protein synthesis. **Cytoplasmic proteins are synthesized by free ribosome. The secretory proteins and membrane proteins are synthesized by ribosomes present on the endoplasmic reticulum.**

Ribosome

In **prokaryotes ribosome is 70S ribosome**. It contains ribosomal RNA and protein. Ribosome has two subunits, larger subunit and smaller subunit. Larger subunit is 50S and the smaller subunit is 30S. Larger subunit has 5S, and 23S rRNA and the smaller subunit has 16S rRNA.

In **eukaryotes ribosome is 80S ribosome**. The larger subunit is 60S. It contains 28S, 5S and 5.S rRNA. The smaller subunit is 40S and it contains 18S rRNA.

Protein Factors

Different protein factors are required for initiation, elongation and termination translation. The process of translation takes place in cytoplasm. The mRNA is translated from 5′ to 3′ end. The protein chain is synthesized from amino terminal and it proceeds towards carboxyl terminal. The process of translation is divided into three different phases.

1. Initiation
2. Elongation
3. Termination

Initiation

Assembly of the tRNA, initiation factors, GTP, mRNA and subunits of ribosome, etc. before the formation of peptide bond is called initiation of translation. There are three initiation factors in prokaryotes and more than ten in eukaryotes

During the process of initiation smaller subunit of ribosome tRNA containing N-formyl methionine (methionine in eukaryotes) and mRNA combine to form 30S preinitiation complex.

Messenger RNA contains consensus sequence near the 5′ end which helps to locate the exact site where the process of translation begins. In prokaryotic messenger RNA consensus sequence is called **Shine-Dalgarno sequence (Kozak sequence in eukaryotes)** is situated near the codon AUG (imitator codon).

50S subunit of ribosome combines with 30S preinitiation complex of ribosome to form 70S initiation complex. Initiation factors and GTP help in the assembly of ribosomal subunits and mRNA. The P site of ribosome is now occupied by methionine containing tRNA (Met-tRNA).

Elongation

New amino acyl tRNA containing anticodon complementary to the codon aligned with the A site of ribosome binds to ribosome. Entry of new aminoacyl-tRNA requires elongation factors EF-Tu and EF-Ts (in eukaryotes EF-1α and EF-1$\alpha\gamma$) and requires GTP hydrolysis.

The **larger subunit of ribosome has peptidyltransferase activity**. It forms peptide bond between the amino acids present in the A and P site. The tRNA present in the P site is devoid of amino acid and it is ejected from the ribosome.

The ribosome moves towards the 3′ end by three nucleotides. The tRNA containing the peptide occupies the P site of ribosome and the A site of the ribosome becomes vacant. The process is called translocation. Translocation requires EF-G (in eukaryotes EF-2) and hydrolysis of GTP. The vacant A site is occupied by the another tRNA containing anticodon complementary to the codon aligned with A site, peptide bond formation and then translocation. These sequences of reaction repeats and the peptide chain grows in length.

Termination

When the termination codon enters the A site it is recognized by the releasing factors. The releasing factors enter the A site and bind to termination codon, induces the peptidyltransferase to hydrolyze the bond connecting the peptide to the tRNA. Release of the peptide chain requires hydrolysis of GTP.

Inhibitors of Translation

1. **Streptomycin** binds to 30S subunit of ribosome and inhibits initiation of translation
2. **Tetracycline** interacts with 30S subunit of ribosome and blocks the entry of aminoacyl-tRNA into the mRNA – ribosome complex.
3. **Chloramphenicol** inhibits the peptidyltransferase activity of the ribosome.
4. **Puromycine** inhibits the elongation of translation. It can inhibit translation both in prokaryotes and eukaryotes.
5. **Erythromycin** inhibits the process of translocation.
6. **Diphtheria toxin** binds to eukaryotic elongation factor 2 and inhibits translocation.

Post-translational Modification of Protein

Polypeptide chains are modified after the synthesis by:

1. Proteolytic cleavage
2. Covalent modification of the peptide chain

Proteolytic Cleavage

Most of the proteins are synthesized in the precursor form which is inactive. Proteolytic cleavage releases the functionally active form of the protein.

Example, secretory proteins are synthesized as a precursor molecules, which contain extra amino acid sequences. These extra amino acid sequences are removed and the functional protein is released.

Insulin is synthesized as a precursor molecule called proinsulin which is converted to insulin by proteolytic cleavage. Zymogen activation is also an example for posttranslational modification. Collagen is released from the cell as procollagen. In the extracellular matrix the N and C terminals of the procollagen molecules are cleaved and tropocollagen is formed.

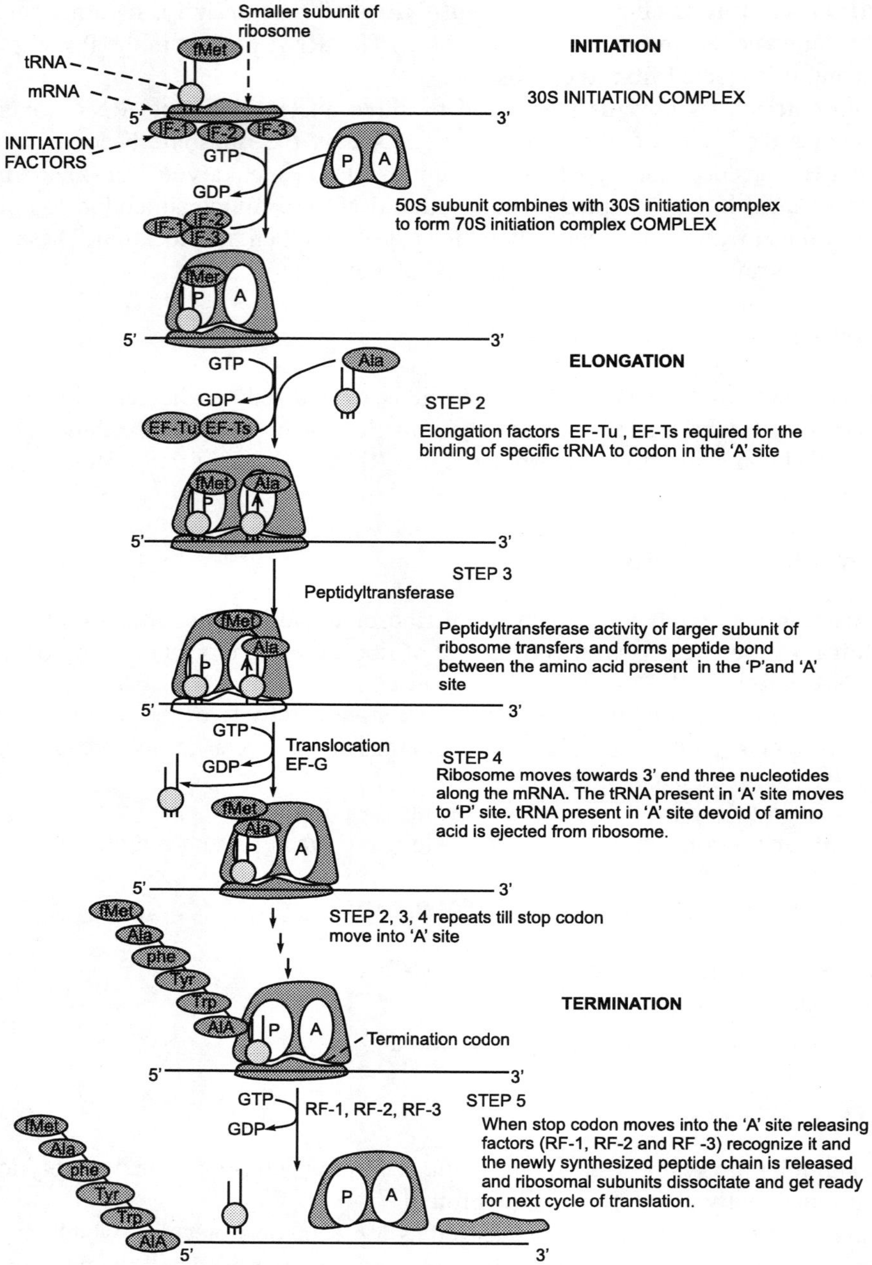

Covalent Modification of the Peptide Chain

Addition of different chemical groups to the protein leads to activation or inactivation of the proteins.

Phosphorylation is addition of phosphate group to the protein. Phosphorylation of the protein is catalyzed by a group of enzymes called protein kinase. Many enzymes activity is regulated by phosphorylation. Phosphorylation may increase the activity of certain enzymes and decreases the activity of some other.

Glycosylation is addition of carbohydrate to the protein. Glycosylation of proteins takes place in endoplasmic reticulum and Golgi apparatus. Many proteins present in human body are glycoproteins.

Hydroxylation is addition of hydroxyl group to the proteins. The proline and lysine residues of collagen are hydroxylated in the Golgi apparatus.

γ-Carboxylation (additional carboxyl group is added) of glutamate residue of various calcium binding proteins takes place after the protein synthesis.

Protein Targetting

Protein synthesis takes place in the cytosol. Many of the proteins are transported to different subcellular organelles where they perform specific function. Peptides destined to different subcellular organelles have got certain signals residing in the polypeptide.

Example, the amino terminal amino acid sequence of the secretory and membrane proteins direct the newly synthesized protein precursor to the endoplasmic reticulum. The proteins destined to the nucleus and mitochondria are targeted by 'nuclear localization' and 'mitochondrial entry sequence' respectively. The enzymes of the lysosome contain mannose-6-phosphate which acts as a signal for targeting them to the lysosome.

QUESTIONS

Long Essay

1. What do you mean by replication? Explain the process of replication.
2. Explain the process of translation and add a note on inhibitors of translation.

Short Essay

1. Genetic code
2. Transcription
3. Replication is semidiscontinuous
4. Post-translational modification
5. Protein targetting
6. Post-transcriptional modification reactions.

17

Acid-Base and Water Electrolyte Balance

INTRODUCTION

The pH of the arterial blood ranges from 7.35 to 7.45 (average 7.4). The pH of blood should be maintained within this narrow range to facilitate all biomolecules, especially proteins, to work at their optimum capacity.

Various biochemical processes produce large amounts of volatile as well as non-volatile acids. All these acids should be efficiently handled by the body to maintain the pH of the body fluids within the normal range. For this purpose, various defence mechanisms are present to defend the body against the deviation of pH from normal.

DEFENCE MECHANISMS

1. Buffer systems
   (a) Bicarbonate buffer system (HCO_3^-/H_2CO_3)
   (b) Protein buffer system (plasma protein in the vascular compartment and haemoglobin buffer in the RBC are the important ones)
   (c) Phosphate buffer system
   (d) Ammonia buffer present in the kidney
2. Respiratory mechanism
3. Renal mechanism

BUFFER SYSTEM

Bicarbonate Buffer

Bicarbonate buffer is the major buffer of the extracellular fluid. The components of bicarbonate buffer are carbonic acid (H_2CO_3), which is acid component, and bicarbonate (HCO_3^-), which is the base. When Henderson Hasselbalch equation is applied to the bicarbonate buffer system, we get the following expression:

$$pH = pKa + \log^{10} \frac{[HCO_3^-]}{[H_2CO_3]}$$

$[HCO_3^-]$ and $[H_2CO_3]$ are the concentrations of bicarbonate and carbonic acid respectively. Carbonic acid is the dissolved form of carbon dioxide produced by the action of the enzyme **carbonic anhydrase** on carbon dioxide and water.

$$CO_2 + HO_2 \xrightarrow[Zn^{2+}]{\text{Carbonic anhydrase}} H_2CO_3$$

The concentration of the carbonic acid can be obtained as follows:

$$pCO_2 \times \text{Solubility coefficient of carbon dioxide} = \text{Carbonic acid}$$

pKa value of bicarbonate buffer is 6.1. If the pH and pCO_2 level of the blood is known, then the bicarbonate level can be calculated by using Henderson Hasselbalch equation.

In arterial blood gas analyzers (ABG), pH and pCO_2 are measured by using electrodes. By using these values, bicarbonate concentration is calculated. Hence, it is a calculated parameter.

In a normal individual, the ratio between the bicarbonate to carbonic acid is 20:1. Any change in the concentration of bicarbonate or carbonic acid will change the value of the ratio between the acid and the base component of the buffer system. If the value of the ratio increases, pH will increase; if the value of the ratio decreases, pH will decrease.

Protein Buffer

The histidine residue of the plasma proteins or haemoglobin has a pKa value of 6.8. This amino acid residue is mainly involved in buffering action. Since the pKa value is 6.8, it acts as an efficient buffer at a physiological range.

Phosphate Buffer

Phosphate buffer is the major buffer inside the cell. The components of phosphate buffer are sodium hydrogen phosphate and sodium dihydrogen phosphate (NaH_2PO_4/Na_2HPO_4). The pKa value of phosphate buffer is 6.8, so it is an efficient buffer at physiological pH.

Ammonia Buffer

The components of ammonia buffer are NH_3/NH_4^+. It is present in the kidney. Ammonia can freely cross the cell membrane, whereas NH_4^+ cannot. Ammonia is produced in the kidney by the action of the enzyme glutaminase on glutamine.

$$\text{Glutamine} \xrightarrow[\searrow NH_3]{\text{Glutaminase}} \text{Glutamate}$$

Respiratory Mechanism

Tissues utilize oxygen and produce carbon dioxide (CO_2). Haemoglobin transport oxygen from lungs to tissues, and carbon dioxide from tissues to lung. At normal state, there is no change in the pH when the pCO_2 is normal.

The alteration in the elimination of carbon dioxide affects the pH of blood. The rate of respiration is regulated by the respiratory centre present in the medulla. It is sensitive to the change in pCO_2 and pH. If pH decreases, the respiratory centre is stimulated. Rate of respiration increases and there is an increased expulsion of CO_2. If pCO_2 is increased immediately, the rate of respiration is increased to increase the expulsion of CO_2.

At the tissue level, carbon dioxide is hydrated to form carbonic acid in the RBC. It dissociates to form hydrogen ion and bicarbonate. Hydrogen ion is buffered by the histidine residue of haemoglobin. Bicarbonate moves out of the RBC, and to maintain the electrical neutrality, chloride enters the RBC. This is called **"chloride shift"** (Fig. 17.1). In the lungs, opposite reactions takes place, this is known as **"Bohr's effect"**.

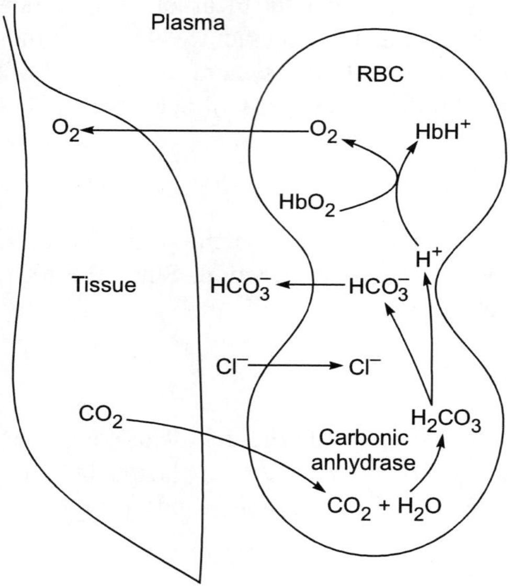

Fig. 17.1 Chloride shift.

Renal Regulation of Acid-Base Balance

Kidney plays a major role in acid-base regulation. When there is an excess of acid production in the body, H^+ are excreted in urine as **titrable acid** and **ammonia**. Kidneys help to reclaim the bicarbonate ions present in the ultrafiltrate. When the acid-base balance is disturbed, kidneys come into action late, but they provide a long-term compensatory mechanism to maintain the pH of the blood.

1. Reclamation of bicarbonate.

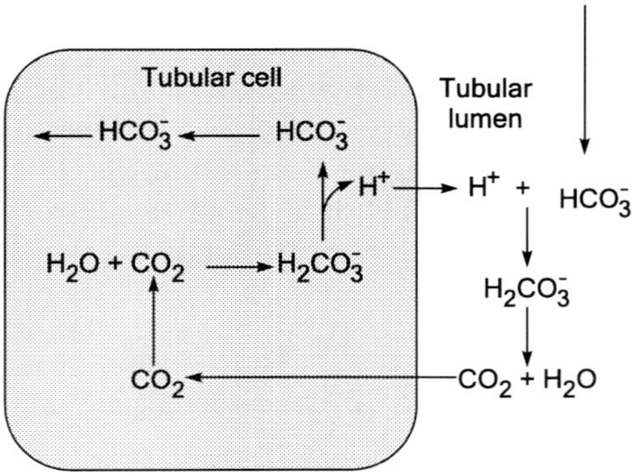

2. Secretion of acids in proximal tubules can be excreted as titrable acid or ammonia in the urine

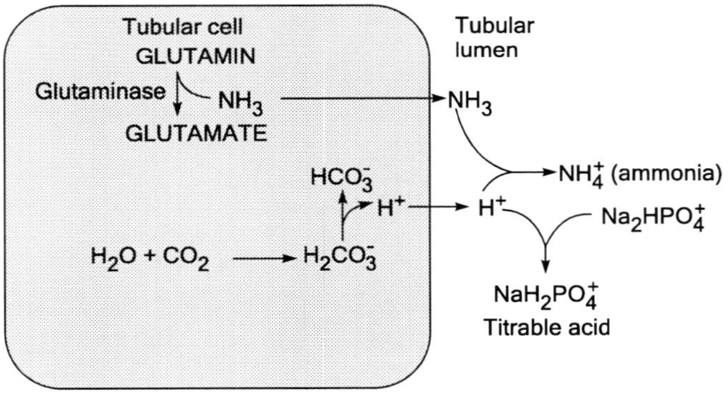

DISTURBANCES IN ACID-BASE BALANCE

Acids, produced in our body during various metabolic activities, are efficiently buffered by various mechanisms described above. Bicarbonate buffer is the major buffer of the extracellular fluid.

Components of bicarbonate buffer system are:

- Bicarbonate (HCO_3^-)/Carbonic acid (H_2CO_3)

Bicarbonate is a metabolic component, and its concentration is mainly regulated by the kidney. Carbonic acid is a respiratory component, and its concentration is regulated by the lungs. Since, the metabolic component (HCO_3^-) and the respiratory component (H_2CO_3) are in communication with the kidney and lungs respectively, it is an open buffer system.

Decrease in the pH of blood is called acidosis, and increase in the pH of blood is called alkalosis. Decrease in the concentration of bicarbonate (metabolic component) or increase in the concentration of carbonic acid (respiratory component) will decrease the value of the ratio between acid and base component of the buffer, and hence the pH.

If acidosis is due to **decrease in the concentration of bicarbonate** (metabolic component), it is called **metabolic acidosis**. If acidosis is due to **increase in carbonic acid** (respiratory component), it is called **respiratory acidosis**.

Increase in the concentration of bicarbonate (metabolic component) or **decrease in the concentration of carbonic acid** (respiratory component) will increase pH blood.

Alkalosis due to **excessive accumulation of bicarbonate is called metabolic alkalosis**, and if it is due to **decrease in the carbonic acid, it is called respiratory alkalosis**.

The parameters to assess the acid-base status are pO_2, pCO_2, pH, and bicarbonate.

Acidosis (decrease in blood pH)

1. Metabolic acidosis (primary decrease in bicarbonate).
2. Respiratory acidosis (primary increase in carbonic acid due to retention of CO_2).

Alkalosis (increase in blood pH)

1. Metabolic alkalosis (primary increase in bicarbonate).
2. Respiratory alkalosis (primary decrease in carbonic acid).

Metabolic Acidosis

Primary decrease in the bicarbonate concentration is called metabolic acidosis.

Causes

1. Diabetic ketoacidosis
2. Starvation ketoacidosis
3. Lactic acidosis
4. Salicylate poisoning
5. Ethylene glycol poisoning
6. Renal failure
7. Excess loss of bicarbonate, as seen in severe diarrhoea
8. Renal tubular acidosis

Compensatory Mechanisms

Buffers and respiratory mechanism

This is the first line of defence against the changing pH in case of metabolic acidosis. Acids, which are produced in excess during metabolism, are immediately buffered by the bicarbonate buffer system present in the extracellular fluids.

During this process, the concentration of bicarbonate is decreased, while that of carbonic acid is increased.

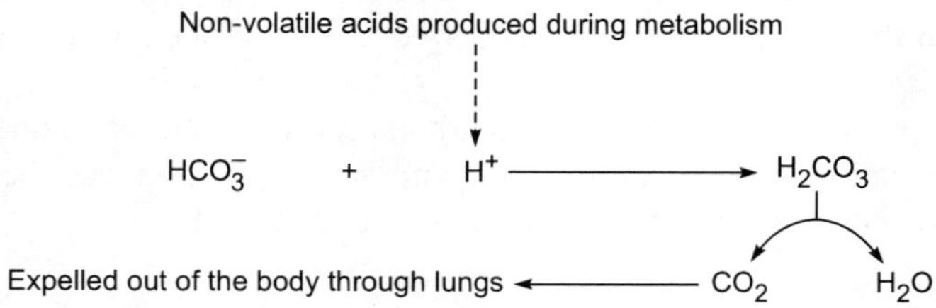

In the lungs, carbonic acid is converted to carbon dioxide and water. Carbon dioxide is expelled out of the body during expiration. As the accumulation of H^+ increases in the body, the bicarbonate level goes on decreasing, and the ratio between the bicarbonate to carbonic acid will also decrease. pH of the blood decreases. The decrease in the pH stimulates the respiratory centre, which leads to increase in the rate and depth of respiration.

Renal mechanism

Renal mechanism comes into action late, but it is a long-term compensatory mechanism. Kidney helps to excrete excess of acid and reabsorb maximum amount of bicarbonate present in the ultrafiltrate.

Laboratory findings

When the arterial blood gas analysis is done in case of metabolic acidosis, pH decreased, pCO_2 is normal, and bicarbonate level in the blood is decreased. When the compensatory mechanism comes into action to bring back the pH to normal, blood pH is still low; pCO_2 is decreased because of the rapid breathing due to stimulation of the respiratory centre by the hydrogen ions; bicarbonate level is still low.

When the renal mechanism also comes into action and the condition is fully compensated, blood pH becomes normal, pCO_2 decreased, and blood level of bicarbonate is also decreased.

Table 17.1 Metabolic acidosis

Parameters	*Uncompensated*	*Partially fully compensated*	*Compensated*
pH	Decreased	Decreased	Normal
pCO_2	Normal	Decreased	Decreased
Bicarbonate	Decreased	Decreased	Decreased

Respiratory Acidosis

The primary increase in the carbonic acid level in blood is called respiratory acidosis.

Causes

1. Acute severe asthma
2. Severe pneumonia
3. Depression of respiratory centre due to various causes (e.g., head injury, drugs, etc.)
4. Respiratory distress syndrome
5. Injury to the thoracic cage or paralysis of respiratory muscles due to neurotoxins
6. Tetanus

In these conditions, due to retention of CO_2, pCO_2 increases.

Compensatory mechanisms

The kidneys come into action. They reabsorb more bicarbonate and excrete more H^+ ions.

Laboratory findings

In uncompensated respiratory acidosis, blood level of bicarbonate is normal, but there is an increase in pCO_2 and pH. When the compensatory mechanism starts (partially compensated) blood level of bicarbonate starts increasing, and pCO_2 level and pH is also high (Table 7.2). In fully compensated respiratory acidosis, bicarbonate level is increased, pCO_2 level is high, and pH is normal.

Table 17.2 Respiratory acidosis

Parameters	*Uncompensated*	*Partially fully compensated*	*Compensated*
pH	Increased	Increased	Normal
pCO_2	Normal	Increased	Increased
Bicarbonate	Increased	Increased	Increased

Metabolic Alkalosis

The primary increase in the bicarbonate level in blood is called metabolic alkalosis.

Causes

1. Severe vomiting
2. Prolonged nasogastric suction
3. Administration of alkali etc.

Compensatory mechanisms

Initially, respiratory mechanism comes into action. The rate of respiration decreases, so that carbon dioxide is retained in the circulation (carbonic acid concentration increases). Later, the kidneys come into action. They retain H^+ ions and excrete bicarbonate.

Laboratory findings

In uncompensated metabolic alkalosis, blood level of bicarbonate increases, pCO_2 level is normal, and pH also increases. When the compensatory mechanism starts (partially compensated) blood level of bicarbonate increases, pCO_2 level increases, and pH is also high (Table 17.3).

In fully compensated metabolic alkalosis, bicarbonate level increases, pCO_2 level is high, and pH is normal.

Table 17.3 Metabolic alkalosis

Parameters	*Uncompensated*	*Partially compensated*	*Fully compensated*
pH	Increased	Increased	Normal
pCO_2	Normal	Increased	Increased
Bicarbonate	Increased	Increased	Increased

Respiratory Alkalosis

The primary decrease in the carbonic acid level in blood is called respiratory alkalosis.

Causes

1. Commonly seen in hysterical overbreathing
2. Diseases of the nervous system like encephalitis
3. At high altitude
4. Sepsis
5. In case of assisted breathing etc.

Compensatory mechanisms

The kidneys retain H^+ ions and excrete bicarbonate.

Laboratory findings

In uncompensated respiratory alkalosis, blood level of bicarbonate is normal, pCO_2 level decreased, and pH increased. When the compensatory mechanism starts (partially compensated) the blood level of bicarbonate starts decreasing, pCO_2 level becomes low, and pH is high (Table 17.4). In fully compensated respiratory acidosis, bicarbonate level decreases, pCO_2 level is decreased, and pH is normal.

Table 17.4 Respiratory alkalosis

Parameters	*Uncompensated*	*Partially compensated*	*Fully compensated*
pH	Increased	Increased	Normal
pCO_2	Decreased	Decreased	Decreased
Bicarbonate	Normal	Decreased	Decreased

Water and Electrolyte Balance

Water is the universal solvent. It is the solvent present in the biological fluids in which the solutes are dissolved. It makes up about 60% of the body mass. Water is distributed between the intracellular and extracellular compartments. Water present in the intracellular compartment makes up about 40% of the total body mass, and that present in the extracellular compartment makes up about 20% of the total body weight.

We drink water, and some amount of water is produced in the body by the oxidation of food stuffs. It has been calculated that proteins, lipids, and carbohydrates, sufficient of generating about 2000 kcal of energy, will generate about 300 ml of water. Water is lost from the body through the skin, lungs, faeces and kidneys.

Normally, about 1500 ml of water is lost from the body. This is called **daily obligatory loss**. In this, 600 ml of water is lost through the skin, 400 ml through the expired air, and about 500 ml through the urine. If the water intake is above this obligatory volume, it will appear in the urine. If the daily water intake is less than the obligatory volume, the total body water level will decrease.

Regulation of Water Content of the Body

Regulatory centre (water intake area and water output area) is mainly located in the hypothalamus. The stimulation of water intake area of the hypothalamus increases thirst, so the free water intake is increased. The stimulation of water output area leads to the secretion of antidiuretic hormone (ADH). This will decrease the water excretion by the renal tubules. The regulatory centres are stimulated by:

(a) The osmolarity (the number of moles of solute per litre);
(b) Blood volume; and
(c) Angiotensin II.

If the osmolality of the extracellular fluid is increased, it will stimulate the regulatory centres of hypothalamus. If the osmolality of the extracellular fluid is decreased, thirst and ADH secretions are inhibited. Hence, loss of free water brings back the osmolality of the extracellular fluid.

Decrease in the Blood Volume Will

(a) Decrease the activity of the distension receptors located in the atria, inferior vena cava etc;
(b) Reduce the activity of the blood pressure receptor in the aorta and the carotid arteries. This signal will stimulate water intake area and suppress the water output area of the hypothalamus; and
(c) Decrease the renal blood flow (renal perfusion). This will lead to the release of **renin** by the juxtaglomerular cells of the kidneys.

Renin is a Proteolytic Enzyme. It is Released in Response to:

1. Decreased renal perfusion;
2. Stimulation of sympathetic nerve supply to kidney; and
3. Decreased sodium concentration in the fluids of the distal tubule.

Renin acts on angiotensinogen (a polypeptide synthesized and secreted by the liver), which is present in the blood and convert it to angiotensin I. Angiotensin I is converted to angiotensin II in the vasculature of lungs and kidneys by the action of **angiotensin converting enzyme (ACE)**. This mechanism helps to increase the water content of the body as shown below.

Aldosterone increases the reabsorption of sodium in the distal tubule. Along with the sodium, water is also reabsorbed and retained in the body. Aldosterone increases the excretion of potassium from the body.

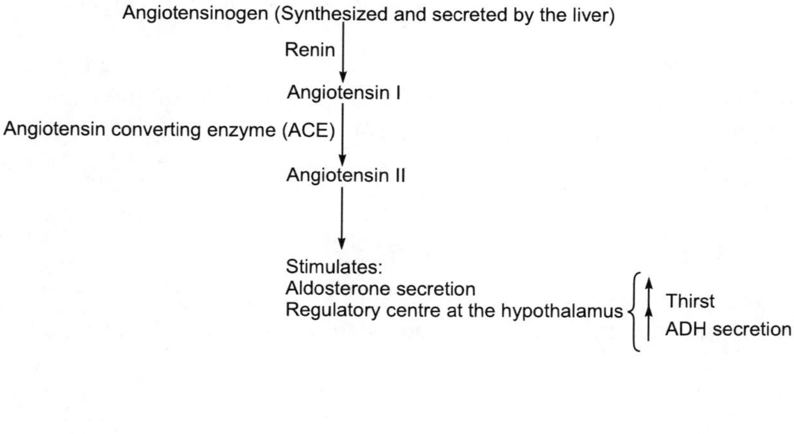

When the blood volume is increased, the distension of the atria will release a peptide called **atrial natriuretic factor** (ANF). Atrial natriuretic factor decreases the reabsorption of sodium. This leads to increased water and salt excretion by the kidneys.

Imbalances in the Water Metabolism

Dehydration

Water content of the body may be decreased with the sodium content of the body normal or both water as well as sodium content of the body may be decreased.

Causes

1. Vomitting and diarrhoea
2. Excessive sweating
3. Osmotic diuresis
4. Adrenocortical insufficiency.

Symptoms

Thirst, dry mucous membranes, decreased skin turgor, decreased urine output, and in severe cases, weakness, hypotension, and shock.

Water intoxication

Water excess may be caused due to:

1. Consumption of excess of water, as in case of certain psychiatric disorder;
2. Excessive ADH secretion, as in the case of ectopic secretion of ADH by bronchogenic carcinoma, adenosarcoma of pancreas, or sometimes in pulmonary tuberculosis; and
3. Secondary to decreased venous return to the heart with no decrease in the total blood volume as in the case of asthma, right sided heart failure, positive pressure ventilation, etc.

ELECTROLYTES

Substances that are capable of dissociating in the water are called electrolytes. Electrolytes are charged species present in the extracellular and intracellular fluid. The composition of electrolytes in the intracellular and extracellular fluid is different.

Electrolytes may be anions or cations (Table 17.5). At any given compartment, the total number of anions should be equal to total number of cations to maintain the electrical neutrality.

Table 17.5 Composition of extracellular fluid and intracellular fluid

Extracellular fluid	*Intracellular fluid*
Cations	**Cations**
Sodium	Sodium
Potassium	Potassium
Calcium	Calcium
Magnesium	Magnesium

Contd.

Anions	Anions
Chloride	Chloride
Bicarbonate	Bicarbonate
Phosphate	Phosphate
Sulphate	Sulphate

The major cation of extracellular fluid is sodium. The major cation of intracellular fluid is potassium. Chloride is the major anion of extracellular fluid, and organic anions are the major anions of the intracellular fluid.

In the hospital practice, the term "electrolytes" includes only sodium, potassium chloride and bicarbonate. In most of the biochemistry laboratories attached to the hospital, sodium, potassium, chloride, and bicarbonate are estimated under the heading "electrolytes".

Importance of maintaining the electrolyte balance

The concentration of electrolytes should be maintained within the acceptable range.

They have a role in:

(a) Transmission of nerve impulse
(b) Neuromuscular excitability
(c) Transport of gases
(d) Blood clotting
(e) Enzyme activity
(f) Bone and teeth formation
(g) Muscle contraction

Sodium

The concentration of sodium is highest in the extracellular fluid. The plasma concentration is 136 to 145 mmol/L. Bone has got a large reserve of sodium.

Functions

(a) Helps to maintain the osmotic pressure. The water content of the body is mainly dependent upon the sodium content.
(b) Required for nerve impulse transmission.
(c) Has role in neuromuscular excitability.
(d) Required for the absorption of glucose and amino acids.
(e) Required for the formation of bile salt.

RDA

- 1 to 3 gm/day.

Regulation

1. Angiotensin II favours the retention of sodium by secretion of aldosterone.
2. Aldosterone increases the reabsorption of sodium in the renal tubules.
3. Atrial natriuretic factor (ANF) and prostaglandins increase the excretion of sodium.

4. Sodium potassium ATPase helps to keep the concentration of sodium low inside the cell.

Disorders related to sodium balance

1. **Hypernatremia**
   The increase in the sodium content of the body is called hypernatremia.

Causes

(a) Excess secretion of aldosterone, as in the case of primary hyperaldosteronism.
(b) Excess of glucocorticoids, as in Cushing's syndrome.
(c) In cases of dehydration where water loss is much more than sodium, as in case of diabetes insipidus and excessive sweating (hypernatremia is due to hemoconcentration).

2. **Hyponatremia**
   Decrease in the serum sodium concentration.

Causes

(a) Increased loss of sodium, as in Addison's disease.
(b) Severe vomiting and diarrhoea, as seen in gastroenteritis.
(c) Renal failure.
(d) When dehydrated patients are rehydrated with salt free water.

Potassium

Potassium is the major cation of the intracellular space. The concentration of potassium in serum is 3.5 to 5 mmol/L.

Functions

(a) Required for the conduction of nerve impulse.
(b) Helps to maintain the osmotic pressure within the cell.
(c) Required for the bile salt formation.
(d) Required as coenzymes for enzymes.

RDA

- 2 to 3 gm/day.

Disorders related to potassium balance

1. **Hyperkalemia**
   Hyperkalemia is increased in the serum potassium concentration. Care must be taken to avoid haemolysis when the blood is collected for the estimation of potassium. Potassium is an intracellular cation present in a highest concentration within the cell. If there is haemolysis, intracellular potassium is poured to the extracellular fluid of blood leading to false high values, which may not be clinically acceptable. In such situations, second blood sample should be sent to the laboratory for estimation of potassium.

Causes

(a) Addison's disease.
(b) Dehydration due to vomiting and diarrhoea.
(c) Renal failure.

2. **Hypokalemia**
   Decreased level of serum potassium is called hypokalemia.

Causes

(a) Hyperaldosteronism.
(b) Prolonged use of diuretics.
(c) During the treatment of diabetic ketosis with insulin and glucose infusion without supplementing with potassium.

Chloride

Chloride is the major anion of the extracellular fluid. Normal serum level is 100 to 108 mmol/L.

Functions

(a) Required for the formation of hydrochloric acid in the stomach.
(b) Helps to maintain the colloidal osmotic pressure of the extracellular fluid.
(c) Has a role in the maintenance of pH of the blood by a mechanism called chloride shift.

Disorders related to the chloride balance

1. **Hyperchloremia**
   Increase in the serum chloride level.
   When bicarbonate is lost from the body due to any cause, chloride is retained in the body to maintain the electrical neutrality. It is responsible for hyperchloremia. Hyperchloremia is seen in metabolic acidosis.
2. **Hypochloremia**
   Decrease in the serum chloride level.
   Chloride is lost from the body, as in the case of excessive vomiting and diarrhoea, alkalosis, prolonged gastric aspiration, etc.

QUESTIONS

Long Essay

1. Describe in detail the various defence mechanisms involved in maintaining the pH of the body.
2. What is metabolic acidosis? Describe its causes and laboratory findings.
3. Discuss respiratory acidosis. Write about its causes and laboratory findings.
4. Describe atrial natriuretic factor.

Short Essay

1. Define electrolytes. Write a note on various electrolytes.
2. Write a note on disorders related to sodium balance.
3. Discuss acidosis and alkalosis.
4. Describe metabolic alkalosis.
5. What is respiratory alkalosis? Write a note on its laboratory findings.
6. Write a note on disorders related to chloride balance.
7. Describe hyperkalemia and hypokalemia.

18

Renal Function Tests

INTRODUCTION

Kidney is mainly involved in filtering the blood and removal and excretion of the toxic and waste products. Kidneys synthesize **erythropoietin** (a hormone required for the formation of erythrocytes). In kidney, 25-hydroxy cholecalciferol is hydroxylated to form **1, 25 dihydroxy cholecalciferol**, which is the active form of vitamin D. Kidney has got major role in **acid base and water electrolyte balance**.

Nephrons are the structural and functional units of the kidney. Each kidney contains about one million nephrons. The functions of these nephrons are filtration at the glomerulous, reabsorption, and secretion of substances at the tubules.

Renal function tests consist of groups of tests, which help to assess the functions of the glomerulous and the tubules. In renal function tests, **urine volume** is determined, presence of various **abnormal constituents in the urine** is analyzed, and **various substances are estimated in the blood and urine**.

Renal function tests can be classified into three major groups:

1. **Tests to Assess the Glomerular Function**
    - (a) Inulin clearance test
    - (b) Creatinine clearance test
    - (c) Urea clearance test
    - (d) 51 Cr-EDTA clearance test
    - (e) 99m Tc- DTPA clearance test
2. **Tests to Assess the Renal Plasma Flow**
    Para amino hippuric acid (PAH) clearance
3. **Tests to Assess the Tubular Function**
    - (a) The concentration and dilution tests
    - (b) Test to assess the ability of the kidney to excrete acidic urine
    - (c) Test to assess the secretory capacity of the tubules (Pheolsulphonephthalein Test)

Tests to Assess the Glomerular Function

Blood is filtered when it is passed through the glomerulous. The filtrate formed at the glomerulous has the same composition as that of protein-free plasma. To assess the glomerular function, estimation of various substances present in the plasma and urine are done. The kidney function can be assessed depending upon the concentration of these substances. For example, serum urea and creatinine.

Urea and creatinine are excreted by the kidney. If the renal function is compromised, excretion of these substances is affected and they get accumulated in the blood. Their concentration in the blood also increases. Serum urea level can sometimes increase in conditions other than the kidney diseases. But serum creatinine is mainly elevated in the kidney diseases. The concentration of these substances starts increasing in blood only in the later stage of the renal disease when more than 50% of the nephrons have lost their function.

Renal Clearance Tests

To detect the renal disease in its early stage, another group of tests called clearance tests can be used. **Clearance of a substance is expressed as mL/minute**. It is the **amount of plasma, which contains the substance that is excreted in the urine per minute**. Clearance value of a substance, which is completely filtered at the glomerulous, not reabsorbed or secreted at the tubules and not metabolized in the body, gives the **glomerular filtration rate**.

Clearance value of a substance, which is totally filtered at the glomerulous and partly reabsorbed by the tubules, is less than the actual glomerular filtration rate. If a substance is filtered and not reabsorbed, but secreted by the renal tubules, it has a clearance value more than the glomerular filtration rate.

Clearance can be calculated by the formula—

$$\text{Clearance (C)} = \frac{\text{Urine concentration of the substance (U)} \times \text{Urine volume (V)}}{\text{Plasma concentration of the substance (P)}}$$

Clearance value varies according to the body surface area, sex, and age.

Inulin clearance

Inulin (polymer of fructose) is a substance, which is completely filtered at the glomerulous, and is neither reabsorbed nor secreted by the renal tubule. It is metabolically inert. Clearance value of inulin gives the glomerular filtration rate. Normal inulin clearance is 110 – 125 mL/min.

Creatinine clearance

It is convenient to measure the creatinine clearance in a clinical set-up. Creatinine is a normal metabolite of the body. Estimation of the serum and urine creatinine is a simple procedure. Creatinine clearance may be endogenous creatinine clearance or exogenous creatinine clearance.

Calculation of endogenous creatinine clearance doesn't require the administration of any exogenous substances orally or through the intravenous route. The clearance value of the creatinine produced in our body can be measured. It requires a 24-hour urine collection, and estimation of creatinine in the urine and serum. By using these values, creatinine clearance can be calculated by the formula

$$C = \frac{UV}{P}$$

Normal endogenous creatinine clearance is about **85 to 120 mL/min**. It is closer to the glomerular filtration rate. Any decrease in the creatinine clearance indicates decrease in the renal function. It is most commonly used in the clinical practice.

Urea clearance

Excretion of urea depends upon the glomerular filtration rate. Urea clearance depends upon the rate glomerular filtration. If the rate of excretion of urine is 2 mL/minute or more, then the normal urea clearance value is about 75 mL/minute in normal individuals. It is called maximum urea clearance. If the rate of urine excretion decreases below 2 mL/minute, urea clearance value will also decrease. It becomes 54 mL/minute when the rate of urine excretion is 1 mL/minute. This is called **standard clearance of urea**.

51Cr-EDTA clearance

51Cr labelled Ethylenediaminetetraacetate (EDTA) is excreted by glomerular filtration. The plasma clearance of 51 Cr-EDTA gives the glomerular filtration rate. This test is suitable for the measurement of GFR in children because urine collection is not required.

99m Tc- DTPA clearance

Clearance of 99m Tc-DTPA can be used to measure the GFR of right and left kidney separately.

Tests for Renal Blood Flow

Para-aminohopuric acid (PAH) is filtered and secreted by the nephrons. When the blood level of PAH is low, the concentration of PAH in the urine becomes a measure of the plasma cleared of PAH per minute if the urine flow rate is known. The PAH clearance, when the blood level of PAH is low, gives the renal plasma flow.

PAH clearance doesn't give the actual renal plasma flow because kidneys are able to clear only about 91% of PAH in a single circulation through the kidneys. Diodrast clearance is another test used to measure the renal plasma flow.

Tests to Assess the Renal Tubular Function

Renal tubules are involved in reabsortion and secretion of certain substances. As the glomerular filtrate passes through the tubules, its composition is altered. Glomerular filtration rate is about 125 mL per minute; urine flow rate is only about 1 mL per minute. This shows the tremendous capacity of the tubules to reabsorb water and other constituents of ultrafiltrate entering the tubule.

Various tests to assess the tubular function can be listed as follows:

1. The concentration and dilution tests
2. Test to assess the ability of the kidney to excrete acidic urine
3. Test to assess the secretory capacity of the tubules (Pheolsulphonephthalein Test).

Concentration and dilution test

Impaired tubular function leads to decrease in the ability of the kidneys to concentrate the urine. Ability to concentrate the urine is more important than the secretion of dilute urine. As the impairment of the tubular function progresses, the diluting ability of the kidney is also affected.

Test to check the concentrating capacity (fishberg concentration test)

After the overnight fasting and fluid restriction, urine sample is collected the following morning. Its specific gravity is measured. If the specific gravity is below 1.022, it indicates decrease in the concentrating ability of the kidney.

Test to check the diluting capacity

A person has to fast overnight with water restriction after midnight. Next morning, about 1200 mL of water is given to him. Urine sample is collected every hour for four hours. Any one of the specimen of urine should show the specific gravity less than 1.003. If there is any impairment of the renal function, specific gravity may not decrease; it might get fixed at 1.010.

Test to assess the ability of the kidney to excrete acidic urine

Kidney plays a role in maintaining the acid base balance. It is able to excrete acid. Acids are excreted as titrable acid and ammonia. To check the ability of the kidney to excrete acid, 0.1 g/kg of gelatin coated capsules of ammonium chloride are given to the individual and the bladder is emptied after an hour and the urine specimen is discarded. After 6 hours, all the urine voided is collected carefully and its pH and ammonia content is measured.

If the pH of any one specimen of urine is 5.3 or less and ammonia excretion is 30 to 90 μ mol/min, the individual is normal. In case of chronic renal failure, ability of the kidney to excrete acid is not impaired, but ammonia excretion is diminished.

Test to assess the secretory capacity of the tubules (Pheolsulphonephthalein [PSP] Test)

Pheolsulphonephthalein is secreted by the kidney. After overnight fast and empting the bladder, I.V injection of Pheolsulphonephthalein is given to the individual. Urine sample is collected every 15 minutes for about 2 hours. About 35% of the PSP is excreted in the first 15 minutes, and about 70% is excreted at the end of 2 hours. Any alteration in the excretion of PSP indicates the impairment of tubular secretory function.

Tests for Abnormal Constituents of Urine

In diseases of the kidney or other organs, urine contains certain substances in excess. These substances may be present in very small amount in the urine of normal individuals, but the sensitivity of the qualitative tests may not be sufficient to detect them. When they are present in excess, they are called abnormal constituents and can be demonstrated by qualitative tests. (Tests for normal constituents of urine are given at the end of the chapter.)

Normal urine contains organic and inorganic constituents (Table 18.1).

Table 18.1 Normal constituents of urine

Inorganic constituents	*Organic constituents*
(a) Calcium	Urea
(b) Sulphate	Creatinine
(c) Ammonia	Uric acid
(d) Phosphate	Urobilinogen
(e) Sodium	Ethereal sulphate
(f) Potassium	

The abnormal constituents seen in urine are as follows:
(a) Glucose
(b) Ketone bodies
(c) Protein
(d) Blood
(e) Bile pigments
(f) Bile salts
(g) Excess of urobilinogen

The following tests are done to detect the abnormal constituents of urine.

SUGAR

Sugars, which can appear in different disease conditions, are glucose mainly in diabetes mellitus; fructose in fructose intolerance and essential fructosuria; and galactose in galactosemia.

Benedict's test (test for reducing substance)

Benedict's reagent consists of copper sulphate, sodium citrate, and sodium carbonate. Copper sulphate provides cupric ions; sodium citrate helps to keep cupric ions in solution; and sodium carbonate acts as mild alkali.

Procedure

To perform Benedict's test, take 5 mL of Benedict's reagent to which add 8 drops of sugar solution. Mix well and boil for 2 minutes. If the blue colour of the Benedict's reagent changes to green, yellow, orange, or brick-red colour at the end of 2 minutes of boiling, it indicates that the given sugar is a reducing sugar. It is a test for reducing sugars.

Principle

When sugars containing free aldehyde or keto groups are treated with Benedict's reagent, sugars are converted to enediols. Enediols are powerful reducing agents. They reduce cupric ions to cuprous ions, which is responsible for the change in the colour of the Benedict's reagent.

Interpretation

This test is used in clinical practice to detect the reducing substance present in the urine. It is a semi-quantitative test. When sugars are present at different concentrations, different colours are developed, which is interpreted and reported as follows:

Table 18.2 Interpretation of Benedict's test

Colour	*Approximate amount of sugar*	*Reported as*
Green colour	0.5 %	+
Yellow	1.0 %	++
Orange	1.5 %	+++
Red	2.0 %	++++

Benedict's test is not a specific test for glucose. The presence of reducing sugar (glucose, galactose, fructose, etc.) or any reducing substance (vitamin C, excess amount of uric acid) in urine can give positive Benedict's test.

Glucose oxidase (Glucose strips)

Glucose oxidase containing strips are now available to detect glucose present in the urine. Glucose oxidase acts specifically on glucose and produces hydrogen peroxide. Hydrogen peroxide is acted upon by the enzyme known as peroxidase to liberate oxygen. This oxygen oxidizes a colourless chemical substance to coloured product, indicating the presence of glucose. It is a specific test for glucose.

Ketone Bodies

Rothera's test

Take about 5 mL of urine. Saturate it with sodium sulphate. Add a drop of sodium niroprusside solution and mix well. Now add liquor ammonia slowly along the side of the test tube. If a purple ring appears at the junction of two solutions, it indicates the presence of ketone bodies.

Acetone, acetoacetate, and 3-hydroxy butyrate are called ketone bodies. Their concentration in blood is increased in uncontrolled diabetes mellitus and starvation ketosis, where the rate of fatty acid oxidation is increased.

PROTEIN

Heat coagulation test

Take about 10 mL of urine in a clean and dry test tube. Heat the upper portion of the test tube. Lower portion is used as a control. Add 2 to 3 drops of 1% acetic acid. If a coagulum is formed, it indicates the presence of proteins in the urine.

Protein is present in the urine in kidney diseases where there is damage to the filtering membrane of the kidney as in case of Nephrotic syndrome. In this condition, more than 3 gm of proteins per day are lost in the urine.

BLOOD

Benzidine test

Take a clean and dry test tube. Mix two drops each of benzidine and hydrogen peroxide. Add a drop of this mixture to the urine. If blue-green colour is formed (which is stable for only a few minutes), it indicates the presence of blood in the urine sample. This test can be used to detect the presence of blood in the faeces (occult blood).

Heme of haemoglobin decomposes hydrogen peroxide and liberates oxygen. Oxygen combines with the benzidine to form a coloured complex. Presence of blood in the urine is called hematuria. The commonest cause of hematuria is injury caused to urethra during the introduction of catheter. Other causes are urinary tract infection, stones in the urinary tract, glomerular nephritis, tuberculosis of the urinary tract, etc.

Bile Salt (Hay's Test) and Bile Pigment (Fouchet's Test)

Hay's test

Take 3 mL of urine and sprinkle a little of sulphur powder. If sulphur sinks to the bottom of the test tube, it indicates the presence of bile salt in the given urine sample. Bile salts reduce the surface tension, so sulphur sinks to the bottom of the tube.

Fouchet's test

To 5 mL of urine, add 2 mL of barium chloride and a little of magnesium sulphate and mix well. Barium sulphate is precipitated. The bile pigments get adsorbed to the barium sulphate and are present in the precipitate. Filter the precipitate by using a filter paper. Add a few drops of Fouchet's reagent to the precipitate.

Green or blue colour formation indicates the presence of bile pigment. Fouchet's reagent oxidizes the bile pigment to form blue-coloured bilicyanine or green-coloured biliverdin.

Bile salts and bile pigments are present in the urine in the case of obstructive jaundice and hepatic jaundice.

QUESTIONS

Long Essay

1. Discuss the various tests used to assess the glomerular function.
2. Describe in detail about the tests used to assess the renal tublar function.
3. Discuss the tests used to assess the abnormal constituents of urine.

Short Essay

1. Write a note on Benzidine test.
2. Write briefly about tests for renal blood flow.
3. Write about the various normal constituents of urine.
4. Write a note on Benedict's test.
5. Mention the procedure of specific test for glucose.
6. What is Rothera's test?
7. Discuss briefly about Fouchet's test.

19

Specimen Collection

INTRODUCTION

The biochemistry laboratory attached to a hospital has got a major role in the diagnosis of disease, treatment, and follow-up of the patients. This laboratory is called clinical biochemistry laboratory. It performs analysis and/or estimation of various constituents of the body fluids. The estimation of body fluids, tissues, faeces, or any other material requires a small volume/quantity of these materials to be collected under specified conditions and sent to the clinical biochemistry laboratory. These are called samples or specimens. The specimen collected from the patient is precious and hazardous, so it should be collected and handled carefully.

Sample collection
↓
Storage and transport
↓
Processing
↓
Analysis

To get a clinically useful, reliable report, it is important to have a good skill in collecting the specimen and knowledge about the interfering factors. It is mandatory to follow the well-accepted procedures to collect, transport and store the specimens to ensure the quality and integrity of the specimen.

The person, who is involved in the specimen collection, always comes in contact with the patient. If the collection of specimen involves any minor procedure, it should be explained to the patient in a non-threatening language. The person involved in the collection of the specimen should dress neatly and should have good communication skills.

The room, where the specimen is collected, should be neat and well organized. Before the collection of any specimen, the patient's identity should be verified and it should be checked whether the right patient has been selected for specimen collection.

SPECIMENS

A variety of specimens are received in the clinical biochemistry laboratory. Blood is the commonest specimen received in addition to urine, plural fluid, ascitic fluid and C.S.F.

Specimens Received in the Clinical Biochemistry Laboratory

Blood	Urine
Plural fluid	Ascitic fluid
Synovial fluid	C.S.F
Pericardial fluid	Faeces
Saliva	Gastric juice

Blood Collection

Blood is present in the vascular compartment. Blood can be collected from the vein, artery, or capillaries. Based upon the vessel from which it is collected, there are three types of blood specimens:

1. Venous blood
2. Arterial blood
3. Capillary blood

Venous Blood

Venipuncture is a procedure for the collection of venous blood, and the trained individual, who performs the venipuncture, is called phlebotomist.

Procedures

1. **Identify the correct patient.**
   If fasting specimen is required, verify whether the patient is fasting or not.
2. **Calculate the approximate amount of blood required** based upon the tests requested. Select the container and the syringe. Keep all the supplies required for the venipuncture and blood collection ready. Label the containers.
3. **Select the vein**
   The vein commonly used to draw the blood specimen is medial cubital vein preset in the antecubital fossa. If this vein is not visualized properly, veins on the dorsum of the hand or veins in the ankle can be used. But it should be avoided as far as possible in diabetic individuals.

   Never select a vein from the arm containing cannula. If an intravenous infusion is being given, it should be shut for about 3 minutes before the blood specimen is collected from the opposite arm. If blood has to be taken from the same arm, the site selected should be below the level of the infusion.

If a woman has undergone unilateral mastectomy, arm on that side should be avoided.

4. **Find vein by touch, not by sight.**
   Vein is dense and rubbery to feel. It springs back when pressed.
5. **Area around the intended puncture site should be cleaned** by using 70% isopropanol by circular motion from the site outward. The skin should be allowed to dry before the venipuncture (a trace of alcohol can give rise to haemolysis).
6. When the blood specimen is collected for the estimation of ethyl alcohol, isopropanol or benzalkonium chloride should be used to avoid the false high value.
7. Cleaned area should not be touched until the specimen collection is over.
8. **Select syringe of appropriate volume and needle of appropriate bore size**
   Gauge of the needle indicates its bore size. As the gauge number increases, bore size deceases. For example, 20 gauge needles have a larger bore size than 23 gauge needles. Normally, 20 gauge needle is used. In case of collapsible veins, 21 gauge needle is used. If large volume of blood is required, 18 gauge needle can be used.
9. **Put the patient and his arm in position.**

   If possible, patient should be in the sitting position and his arm should be extended and well-supported. Patient should be in that position for quiet some time before the venipunture is done.
10. **Tourniquet is applied about 3 to 4 cm above the decided puncture site.**
11. **Tourniquet should not be kept in place for more than one minute.**

    If it is kept for more than one minute, it leads to stasis of venous blood and the concentration of the analyte is altered.
12. When the vein is punctured, the bevel of the needle should face upwards making an angle of 15° with the surface of the arm.
13. Once the tip of the needle is inside the vein, a drop of blood will appear at the tip of the syringe where the needle is fixed to it.
14. Draw the blood slowly by gently pulling the plunger.
15. Once the desired amount of blood is drawn, keep a dry gauze pad at the site of puncture. Then withdraw the needle and put pressure on the puncture site. Hold the gauze pad for few minutes till the bleeding stops. If the bleeding does not stop after few minutes, inform the physician.
16. Remove the needle from the syringe and slowly displace the blood from the syringe to the containers, which are already labelled. Record the time and date of specimen collection.
17. Send the specimen promptly to the lab. If it is not possible, store it as per the specifications. Blood can be collected in the vacutted tubes. Blood collection tubes that are vacuumed inside are called vacutted tubes. They can aspirate specific volume of blood. They are commercially available. Tubes with different additives are also available. They can be identified by the colour of the rubber stopper.

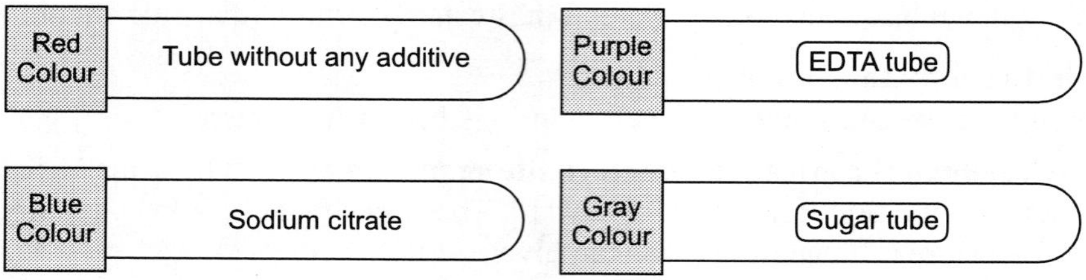

Plasma and Serum

If blood is not allowed to coagulate after it is collected by using an anticoagulant, the supernatant obtained by centrifuging it is called plasma. If blood is collected on a plain tube and allowed to coagulate, the supernatant obtained after centrifuging it is called serum. Serum does not contain fibrinogen and other coagulation factors as compared to plasma.

Anticoagulants

Anticoagulants are used to collect specimen for those tests, which require whole blood or plasma. There are different types of anticoagulants:

1. **Heparin**
   It is a heteropolysaccharide. It is commercially available as sodium, potassium, lithium, or ammonium salt. Approximately, about 2 mg of heparin is used per 10 mL of blood.
2. **Oxalates and citrates**
   They precipitate calcium and act as anticoagulants.
3. **EDTA**
   It chelates calcium and acts as an anticoagulant. EDTA blood is required for hematological studies.
4. **Sodium citrate**
   Citrate does not precipitate calcium, but converts it to a nonionized form. Citrated blood is mainly used for the coagulation studies.
5. **Sodium fluoride**
   At higher concentration it acts as an anticoagulant, but inhibits the activity of several enzymes. So, it is not suitable as an anticoagulant.

Collection of blood for plasma glucose estimation: Blood is collected in a tube or bottle containing potassium oxalate as anticoagulant and sodium fluoride as the antiglycolytic agent. These tubes are known as sugar tubes.

Most of the routine investigations in biochemistry laboratory are done on serum. Whole blood is required for the arterial blood gas analysis (heparin is the anticoagulant) and for the estimation of glycated Hb (EDTA is used as an anticoagulant).

Arterial Blood

It is the blood collected from the artery. Radial artery is the preferred artery. Other arteries used are brachial or femoral arteries. Arterial blood is used for the analysis of pH and arterial blood gas (pCO_2, pO_2). For the arterial blood collection, glass syringes are preferred compared to the plastic syringes. Gases from the blood can be exchanged with the external environment in case of plastic syringes.

The anticoagulant used is heparin. The syringe is flushed with the heparin, so that the air, which is collected in the dead space of the syringe, is replaced by the heparin solution. The amount of heparin collected in the dead space of the syringe is sufficient to act as an anticoagulant. It also prevents the contamination of the blood with the atmospheric air.

Capillary Blood

It is collected by skin puncture. In adults, pulp of the finger is pricked; in older children, ear lobe can be pricked; and in younger children, lateral side of heel is the preferred site for skin puncture (Fig. 19.1).

Before puncturing the skin, the intended area of puncture should not be rubbed or squeezed to increase the flow of blood. This action will increase the efflux of tissue fluids and has effect on the composition of the blood. To enhance the blood flow, warm cotton gauze can be placed on the intended site of puncture. Before the skin puncture, the area should be properly cleaned.

After the puncture of the skin, the first drop of blood should be discarded and the subsequent drops should be collected. Gravity should facilitate the flow of blood.

Urine Collection

Urine examination in the clinical laboratory would help in diagnosis or confirmation of the disease. Urine examination may be done for the detection of:

1. Abnormal constituents (chemical examination);
2. Bacteria or microorganisms (bacteriological examination) and
3. Epithelial cells, pus cells, R.B.Cs, crystals of different salts, or cholesterol, etc. (microscopic examination).

Urine collection is done in different ways:

1. A random urine sample can be collected for qualitative or quantitative tests. A clean, early morning fasting specimen is the preferred specimen. It is the concentrated specimen and the suitable specimen for microscopic examination and also for detection of abnormal constituents. The specimen should be collected in a clean and dry container, and before sending it to the lab, it must be well sealed and labelled.
2. Timed urine sample is the sample that is collected over a specific period of time. It may be 2 hours urine collection, 4 hours collection, 8 hours collection, 12 hours collection, or 24 hours collection. In timed urine collection, clear instructions should be given to the patients (better in writing). If there is any dietary restriction, appropriate written instructions should be given to the patients. For example, urine collection for the estimation of vanillyl mandelic acid, 5 hydroxytryptomine, and 5 hydroxy indole acetic acid.

Patient should strictly adhere to the instructions.

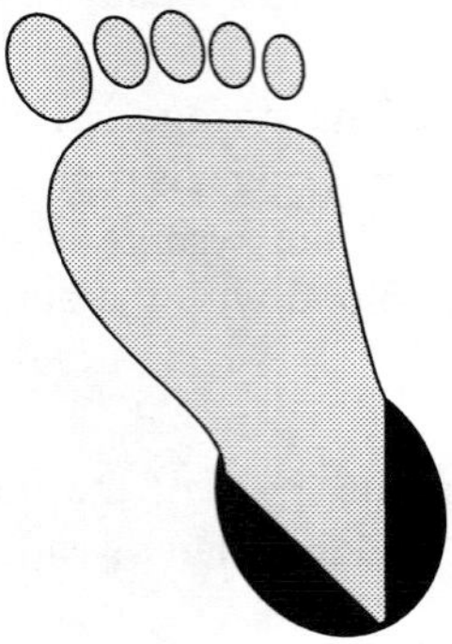

Fig. 19.1 In young children, the lateral side of the heel is the preferred site for the skin puncture.

Before the collection of the urine, the bladder is emptied and the urine is discarded. Thereafter, all urine should be collected until the end of the scheduled time. If collection has to be made over several hours, the container should be refrigerated at 4°C. The voided urine should be collected in a separate container and then poured into the large container.

If the patient has to pass the stool during the collection period, proper precaution should be taken to prevent the contamination of the urine with the faeces. The bottle of specific capacity should be given to the patients. For example, for 2 to 4-hour collection, 1 L bottle; for 12 hours collection, about 2 L bottle; and for 24 hours urine collection, about 3 to 4 L bottle would be sufficient.

Urine can be collected by using catheter in critically ill patients. In children, after cleaning the genitalia and the perineal regions, a plastic bag is placed around the genitalia and urine is collected. If required, urine can be collected from the bladder by saprapubic puncture.

PRESERVATION

Proper preservation is required to reduce the decomposition of the chemicals, the action of the bacteria, and to solubilize the constituents that might otherwise precipitate. Urine specimen can be refrigerated immediately following collection.

Chemicals can be used to preserve the urine specimen. If the specimen is acidified to pH below 3, it will prevent the bacterial growth and the analysts are well preserved. This urine is most suitable for the analysis. About 10 mL of HCl 6 mol/L per 24 hours urine collection is appropriate.

It is advisable to collect all the urine in a single container. Once the collection is over, urine present in the container should be sent to the laboratory immediately. The urine present in the container is thoroughly mixed and then the volume is measured. A small portion of it is kept separately for analysis.

For chemical and microscopic urine analysis, freshly voided urine is the specimen of choice.

Cerebrospinal Fluid (CSF)

Spinal fluid is obtained from the lumbar area by lumbar puncture. CSF examination is done in patients with cerebrovascular accidents, meningeal involvement in the malignant disease, demyelinating disease, meningitis, etc.

The lumbar area, where the lumbar puncture has to be done, is cleaned thoroughly. If there is any sepsis in this region, the lumbar puncture is not done. These measures help to prevent the introduction of infection in the CNS.

If the CSF pressure is more than normal (90 to 180 mm of Hg), the procedure should be done only if the analysis of CSF will definitely help to improve the treatment. In adults, about 20 ml of CSF can be withdrawn, but minimum amount should be withdrawn in case of children.

During collection, three aliquots are collected in separate containers. The first aliquot, which may be contaminated with the tissue debris and the bacteria, is used for the chemical analysis. Second aliquot is sent for microbiological examination, and the third aliquot is used for microscopy and cytological examination. All the containers should be labelled and sealed properly and sent to the laboratory immediately.

Plural Fluid, Pericardial Fluid, Ascitic Fluid

Fluids found in the plural space and pericardial spaces are called plural fluid and pericardial fluid respectively. The excess fluid accumulated in the peritoneum is called ascitic fluid. In normal individuals, a small amount of serous fluid is present in these potential spaces that help to reduce the friction between the parietal and visceral membrane surfaces.

The technique of collecting these fluids is called "Paracentesis". These procedures should be done by the physicians.

Thoracentesis—collection of plural fluid
Pericardiocentesis—collection of pericardial fluid
Peritoneocentesis—collection of ascitic fluid

In some disease conditions, excess amount of these fluids can get accumulated. These fluids are aspirated and analyzed, which would help to determine whether the accumulated fluid is transudate (low protein content of the fluid) or exudates (inflammatory origin; rich in protein) and helps in the diagnosis of disease.

Collection of faeces

Faeces are tested in the laboratory to detect the presence of blood or to estimate the fat content (in steatorrhoea). Blood present in the faeces is called occult blood. Faecal specimen is collected with the help of a spatula, from the faeces excreted into the toilet bowl or bed pan. Faeces present on the gloves of the physician, who has done the rectal examination, should not be used; since this may be contaminated with the blood oozing out from the injuries caused during the procedure and can give false positive results.

Faecal fat is estimated as a part of gastrointestinal function test to assess the malabsorption. It requires the collection of faeces for about 72 hours without adding preservatives. Proper precautions should be taken to avoid contamination with urine. Before and after collection, the weight of the container should be measured to determine weight of the faeces collected. Throughout the period of collection, the container with the faeces should be refrigerated.

QUESTIONS

Long Essay

1. Describe the procedure for the collection of venous blood from the patient.
2. Describe the procedure for the collection of urine from the patient.
3. Discuss the procedure for the collection of faeces from the patient.

Short Essay

1. How is urine sample preserved in the laboratory?
2. Mention the different tests done for the examination of urine.
3. How is cerebrospinal fluid collected from the patient?
4. Write a short note on various anticoagulants used to collect the specimens.

Short Answer

1. Define EDTA.
2. What is the function of sodium citrate as an anticoagulant?
3. Mention various samples which can be taken from the patients in order to diagnose the disease.

20

Vitamins

INTRODUCTION

Vitamins are accessory food factors that are organic in nature and are required in small quantities for normal health, growth, and reproduction. Earlier separated factors were amines in nature, but as they were essential for health, they were called vitamins (vital amines). Later, it was found that all these factors are not amines, so the last letter 'e' of vitamine was omitted, and hence the present name "vitamin" (last letter 'e' is eliminated). Deficiency of vitamins causes various diseases.

CLASSIFICATION

Based upon their solubility, vitamins are classified as (Table 20.1):

1. Fat-soluble vitamins
2. Water-soluble vitamins

Table 20.1 Classification of vitamins

Fat Soluble vitamin	Water soluble vitamin
Vitamin A Vitamin D Vitamin E Vitamin K	Thiamine Riboflavin Niacin Biotin Pantothenic acid Folic Acid Vitamin B_{12}

Fat-soluble vitamins are soluble in fat solvents (Table 20.2). As they are hydrophobic in nature, they require a special carrier protein for transport. Fat-soluble vitamins are stored in the body, so it takes longer time for the deficiency manifestations to appear. They tend to accumulate in the body if ingested in excess, and give rise to toxicity. Most of the fat-soluble vitamins have isoprenoid units.

Table 20.2 Differences between fat-soluble and water-soluble vitamins

	Fat Soluble Vitamins	Water soluble vitamins
1. Solubility	Soluble in fat solvents	Soluble in water
2. Structure	Made up of isoprenoid units	Made up of diverse functional groups
3. Absorption	Require bile salts absorbed along with lipids	Do not require bile salts
4. Transport	Require carrier proteins	Do not require carrier proteins
5. Storage	Stored in liver and adipose tissue	No significant storage
6. Excretion	Not excreted in urine	Excrete d in urine since it is water soluble
7. Frequency	Single dose sufficient for weeks or months	Frequent supply is required
8. Deficiency	Deficiency symptoms appear weeks or months after	Deficiency symptoms appear earlier
9. Hypervitaminosis	Excess intake leads to toxicity Since they are insoluble in aqueous phase, excess tends to accumulate	Usually may not be seen. Since it is water soluble, excess is excreted in urine

FAT-SOLUBLE VITAMINS

VITAMIN A

Vitamin A is a fat-soluble vitamin. It is present in the animal tissues. Plants contain the provitamin form (precursor) of vitamin A called carotene. Compounds with vitamin A activity are called retinoids. The term "retinoids" includes both naturally occurring and synthetic substances with vitamin A activity.

Chemistry

Vitamin A contains polyisoprenoid units joined to β-ionone ring. Vitamin A may be present in three forms depending upon the functional groups present on the polyisoprenoid side chain.

1. Retinol—Vitamin A alcohol
2. Retinoic acid—Vitamin A acid
3. Retinal—Vitamin A aldehyde

Retinol and its two derivatives—retinal and retinoic acid—are responsible for the functions of vitamin A. The polyisoprenoid side chains of vitamin A contain many alternate double bonds. At these double bonds, vitamin A exhibits geometric isomerism. Retinal can exist in all 'trans' form where it shows trans configuration. All trans form of retinal is also called vitamin A1. It is the most common form. Retinal can exhibit 'cis' configuration at 11th position of the isoprenoid side chain. It is called 11-cis retinal.

β-carotene found in plants and plant products is a provitamin form of vitamin A. Carotene does not possess vitamin A activity. Upon hydrolysis of β-carotene in the intestinal epithelial cells, two molecules of vitamin A are released. In humans, this conversion is inefficient, so β-carotene is only about one-sixth as effective as source of vitamin A (as retinol) (6 microgram of beta-carotene amounts to 1 microgram of retinol).

Vitamin A requirement is expressed as retinol equivalents (RE) in order to take into account the variability of the efficiency of conversion of carotenoids into retinol.

1 RE = 1 microgram retinol
= 6 microgram of β-carotene
= 3.3 IU of retinol

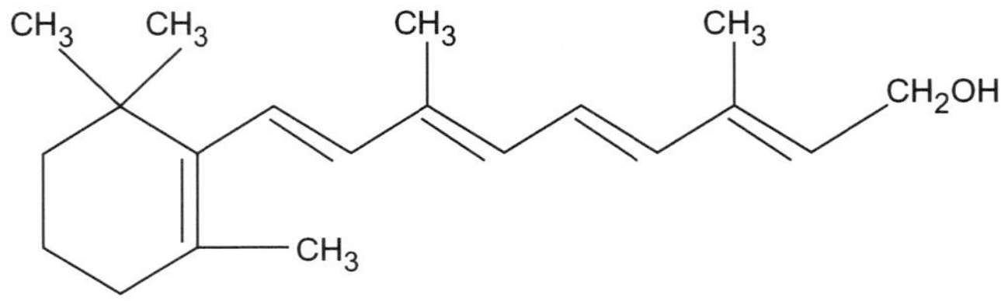

Vitamin A

Retinol can be reduced to retinal by retinal reductase. This reaction is readily reversible. Retinal is oxidized to retinoic acid, which cannot be converted back to retinal.

$$\text{RETINOL} \xrightarrow{\text{Retinal reductase}} \text{RETINAL} \longrightarrow \text{RETINOIC ACID}$$

Sources

Vitamin A is present in its active form only in the animal tissues, but a provitamin form, β-carotene, is present in plant tissues.

Sources of vitamin A are liver oil, butter, milk, cheese, and egg yolk. Richest dietary sources of preformed vitamin A (retinol) are fish liver oils such as those derived from cod and halibut. Animal livers are also good sources. Richest dietary source of carotene is palm oil. Good sources are yellow and red fruits and vegetables such as carrots, tomatoes, and peaches.

Vitamin A is heat stable, and the losses occurring in normal cooking are not serious. It is, however, sensitive to UV light, and for this reason, vitamin A preparation should be kept in dark or opaque containers.

RDA

Men and women need about 750 μg (2500 I.U.) of vitamin A.

Absorption

Retinol is present in animal tissues as retinyl ester with long chain fatty acids. Dietary retinyl esters derived from foods of animal origin are digested by the hydrolases of pancreatic or intestinal origin. They convert retinyl esters to retinol and free fatty acids.

RETINYL ESTER → RETINOL + FREE FATTY ACID

Carotene present in the foods of plant origin is acted upon by the enzyme carotene 15-15'-dioxygenase present in the epithelial cells lining the intestine, and which releases two molecules of retinal. Some amount of carotene is absorbed as such.

Transport and Storage

After the absorption, retinol is reesterified with long chain fatty acids. Retinol esters are incorporated into chylomicron and transported to the liver. They are stored in the liver as retinyl ester with fatty acid. When required, they are released from the storage site.

From the liver, vitamin A is transported to extrahepatic tissues with the help of retinol binding proteins (RBP). The retinal binding protein–vitamin A complex binds to prealbumin. This prevents the RBP–vitamin A complex getting filtered through the filtering membrane of nephron.

Functions

- **Required for normal vision:**
  Retina of the eye contains two types of cells—rod cells and cone cells. The cell membrane of these cells contains photopigments, rhodopsin and photopsin respectively. Cone cells are required for bright light vision and colour vision. Rod cells are involved in dim light vision.

The photopigments contain 11 cis retinal as the prosthetic group. Rhodopsin, the pigment of rod cell, is made up of protein portion opsin and the prosthetic group 11 cis retinal. The 11 cis retinal plays a major role in vision. When light falls on the rhodopsin, 11 cis retinal is converted to all trans retinal and opsin though the series of intermediates (Fig. 20.1).

All trans retinal can be converted to all cis retinal by the action of an enzyme called retinal isomerase. 11 cis retinal combines with opsin to form rhodopsin. Alternatively, some amount of all trans retinal is transported to liver, where it is converted to all trans retinol by alcohol dehydrogenase. Later, all trans retinol is isomerized to 11 cis retinol. Oxidation of 11 cis retinol leads to the formation of 11 cis retinal. It is transported to retina. This combines with opsin to form rhodopsin. These sequences of cyclic reactions were proposed by George Wald, hence it is called Wald's visual cycle.

One of the intermediates formed during the photolysis of rhodopsin activates a G protein called transducin. Activated form of transducin in turn activates an enzyme called phosphodiesterase. Active phosphodiesterase decreases the cGMP level and closes the Na^+ channels. This causes hyperpolarization of the plasma membrane and generation of nerve impulse (Fig. 20.1).

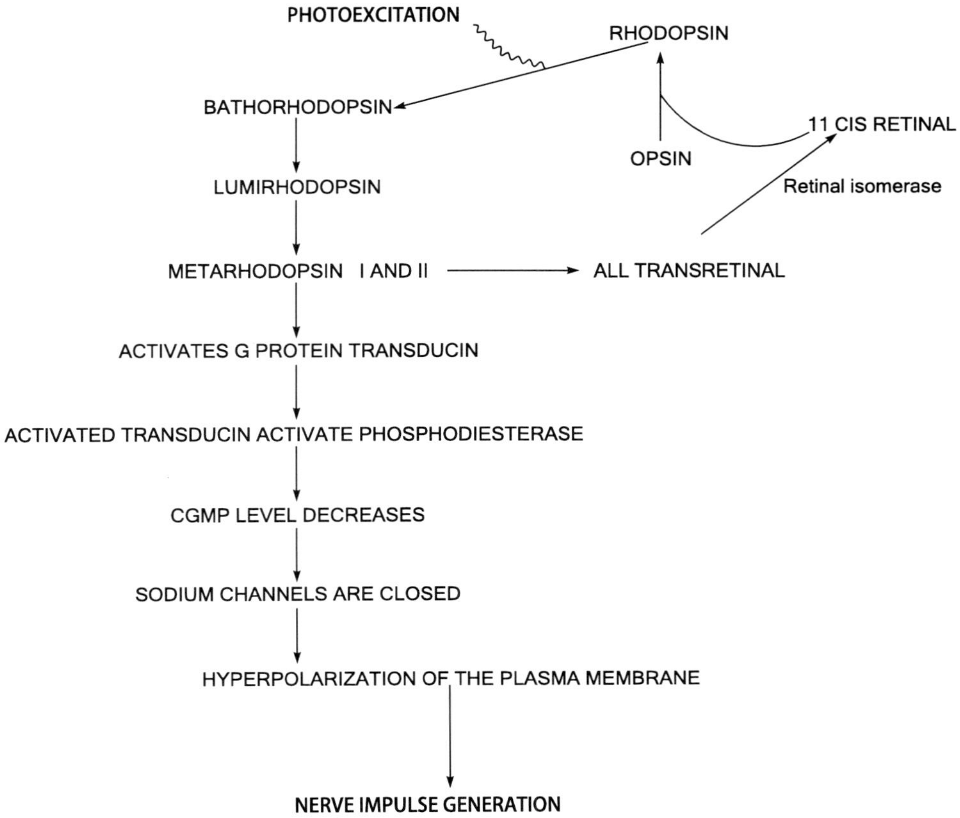

Fig. 20.1 Photoactivation of photopigment and generation of nerve impulse.

Mechanism of action of retinoic acid is similar to steroid hormone. It supports growth and differentiation, and is required to maintain the epithelia. It is also required for the synthesis of glycoproteins and mucopolysaccharides. Retinoic acid is not useful in supporting the normal vision.

— **Anticancer action**
  Mainly provitamin form of vitamin A (β-carotene), acts as an antioxidant and protects the cells against the deleterious effects of powerful oxidants.

— **It is required for the normal growth**
  e.g., bone growth, normal growth of the central nervous system, etc.

— **It is used in the treatment of various skin diseases like acne and psoriasis.**

Deficiency

1. Dark blindness (night blindness) or nyctolopia.
   Due to the defective resynthesis of rhodopsin during visual cycle, a person, who has vitamin A deficiency, may not be able to see in the dim light or darkness. This is known as night blindness.

If the condition is not treated by the supplementation of vitamin A, it leads to dryness and keratinization of the conjunctiva and cornea. When keratinization takes place, the conjunctiva appears dry, wrinkled, and pigmented (xeropthalmia).

Later bitot spots appear, which are triangular-shaped white plaques on the conjunctiva. Xeropthalmia, if not treated, leads to keratomalasia, (degeneration of corneal epithelium), corneal ulcer and permanent blindness.

About 500,000 million children become blind worldwide each year due to the deficiency of vitamin A. It is called preventable blindness, since vitamin supplementation to the children prevents blindness.

2. Epithelium of the respiratory tract, reproductive gland, lachrymal gland, and salivary gland undergo hyperkeratosis, predisposing them for infection. It gives rise to phrynoderma.
3. Skeletal and nerve growth is affected.

Hypervitaminosis A

Intake of vitamin A more than 7.5 mg/day should be avoided. Initial signs of hypervitaminosis are reflected in the skin; skin becomes dry. Liver is enlarged and cirrhotic changes can take place. There may be increase in the intracranial pressure.

During pregnancy, excess intake of vitamin A should be avoided because of its potential for causing congenital malformation in the developing foetus.

VITAMIN D

Antirichatic Factor

Vitamin D is a fat-soluble vitamin. Since the mechanism of action of the active form of this vitamin is similar to that of steroid hormone, the active form of vitamin D (calcitriol) is considered a hormone.

Cholecalciferol, which can be synthesized in the human skin, is also present in the fish liver, is designated as vitamin D_2. Ergocalciferol which is formed by the UV irradiation of ergosterol present in the yeast is called vitamin D_3. Vitamin D_2 and vitamin D_3 are inactive forms. Active form of vitamin D is called calcitriol (1,25-dihydroxy cholecalciferol).

Sources

The richest natural source of vitamin D is fish liver oil, butter, eggs, etc. Since milk is a source of calcium, it was chosen for fortification with vitamin D in the daily diet. The addition of vitamin D to the milk helps to meet the need of the body. This prevents rickets in the growing children.

Synthesis

1. Vitamin D can be synthesized from 7-dehydrocholesterol by the UV irradiation. When we are exposed to sunlight, 7-dehydrocholesterol is converted to cholecalciferol (vitamin D_2).

(7-dehydrocholesterol and ergosterol are the precursors for the synthesis of vitamin D. Hence, they are called provitamin forms of vitamin D.)

RDA

- One international unit of vitamin D = 0.025 μg of cholecalciferol.
- About 200 to 400 IU (2.5 to 5 μg) of vitamin D is required daily. Requirement of vitamin D depends upon the exposure to sunlight.

Activation

Cholecalciferol is present in the inactive form. It is activated by the serial addition of hydroxyl group in liver and kidney (Fig. 20.2). Cholecalciferol is converted to 25-hydroxycholecalciferol in the liver by the action of microsomal enzyme 25-hydroxylase (calcidiol).

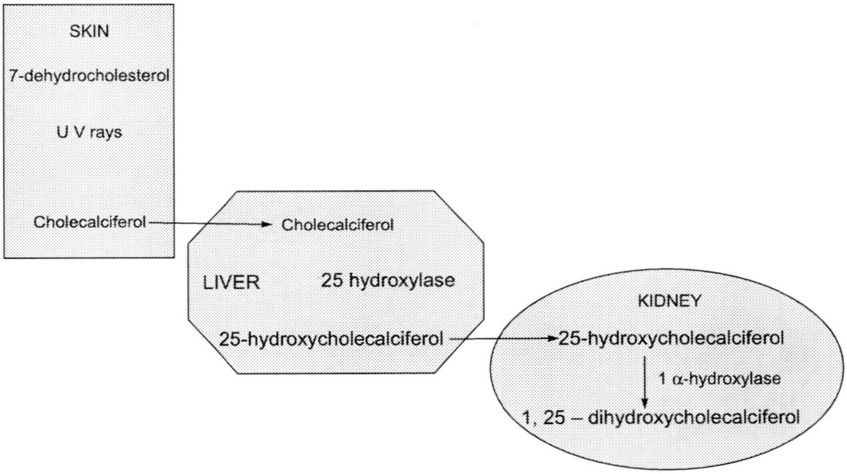

Fig. 20.2 Activation of vitamin D in liver and kidney.

Calcidiol is transported from the liver to kidney bound to the plasma protein called vitamin D binding protein. In the kidney, 25-hydroxycholecalciferol is converted to 1,25 dihydroxycholecalciferol (calcitriol) by mitochondrial 1 α-hydroxylase. Calcitriol is the active form of vitamin D. Since calcitriol is considered a hormone, cholecalciferol and ergocalciferol can be called prohormone forms.

Calcitriol formation is regulated by the blood level of calcium and phosphate. The activity of the enzyme 1 α-hydroxylase present in the kidney is increased directly by plasma phosphate and indirectly by the blood calcium level. Excess of calcitriol inhibits the activity of the enzyme 1 α-hydroxylase. When the activity of 1 α-hydroxylase is suppressed, another enzyme called 24-hydroxylase becomes active. It converts 25-hydroxycholecalciferol to 24, 25-dihydroxy cholecalciferol.

In the kidney, conversion of 25-hydroxycholecalciferol to 24, 25-dihydroxycholecalciferol can happens when there is a decrease in parathyroid hormone level or increase in the calcium and calcitonin level in the blood.

Transport and Storage

Vitamin D present in the diet is absorbed in the small intestine with the help of bile salt. In the enterocytes, vitamin D is incorporated into the chylomicrons and enters the circulation. In the circulation, it binds to vitamin D binding protein (DBP).

The vitamin D binding protein is a globulin present in plasma that can bind to all different forms of vitamin D. Liver, kidney, adrenal gland, and adipose tissues store vitamin D to a significant extent.

Mechanism of Action

Calcitriol binds to its receptor present in the cytosol of the cell. The calcitriol receptor complex then moves to the nucleus. In the nucleus it interacts with the DNA and increases the expression of the gene coding for the calcium binding proteins.

The mechanism of action of calcitriol is similar to that of steroid hormone. So, calcitriol is considered a hormone (Fig. 20.3).

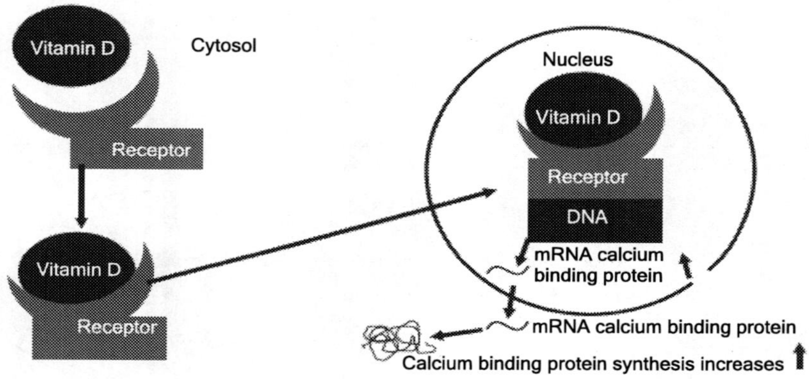

Fig. 20.3 Mechanism of action of calcitriol.

Functions

Active form of vitamin D is mainly involved in blood calcium and phosphate homeostasis.

The major **three target tissues** for calcitriol are:

1. Intestine—increases the absorption of calcium.
2. Kidney—decreases the loss of calcium by the kidney.
3. Bone—increases the bone resorption when necessary.

Action on Intestine

Calcitriol increases the absorption of calcium and phosphate from the intestine.

It increases the concentration of calcium binding protein in the epithelial cells lining the intestine and increases the absorption of calcium from the intestine. This helps to increase the blood level of calcium.

Action on the Kidney

Calcitriol increases the reabsorption of calcium and phosphate by the kidney.

Action on the Bone

Calcitriol stimulates the mobilization of calcium and phosphate from the bone and increases the blood level of calcium and phosphate. When adequate amount of calcium and vitamin D are available, it promotes the mineralization of bone.

Deficiency

The deficiency of vitamin D gives rise to rickets in children and osteomalacia in adults. Vitamin D is required for the absorption of calcium and phosphate. In response to decreased blood level of calcium, parathyroid hormone is secreted. This hormone increases the mobilization of calcium from the bone. This helps to maintain the blood level of calcium.

In children, when there is a deficiency of vitamin D, blood level of calcium decreases. The mineralization of bone is incomplete, but there is continued synthesis of bone matrix. This results in the soft and pliable bones which are deformed.

In adults, osteomalacia occurs due to the deficiency of vitamin D, which leads to decreased plasma level of calcium. In response to this, parathyroid hormone is released. It mobilizes the calcium from the bone. Due to this, bones become weak and susceptible to fracture.

In kidney disease, the conversion of 25-hydroxycholecalciferol to 1, 25-dihydroxycholecalciferol is defective. Defective bone formation is seen. The condition is called renal rickets (renal osteodystrophy). Condition can be resolved promptly by replacing the calcitriol.

Hypervitaminosis D

Vitamin D is a fat-soluble vitamin. It can be stored in the body. Excess of vitamin D gives rise to toxic signs and symptoms. In hypervitaminosis D, person experiences loss of appetite, thirst, neuromuscular depression, irritability, and excessive mobilization of calcium from bones, this gives rise to pathological fracture.

VITAMIN E

Antisterility Factor

It is a fat-soluble vitamin. It contains 6-hydroxy chromane ring with a isoprenoid side chain. Depending upon the number of methyl groups and their position, there are different types of tocopherols—a, b, g, d, and z.

Sources

Good sources of vitamin E are cottonseed oil, safflower oil, wheat germ oil, corngerm oil, mustard oil, soya bean oil, etc.

RDA

Adults need 8 to 10 mg/day. During pregnancy and lactation, the requirement goes up to 12 mg/day.

Absorption, Transport and Storage

It is absorbed from the small intestine with the help of bile salts. Once inside the intestinal lining epithelial cells it is incorporated into the chylomicrons and transported to the liver. From the liver it is distributed to muscle and adipose tissue. They are the sites of storage.

Functions

Vitamin E is one of the natural antioxidants. It acts as chain breaking antioxidant. During metabolism, various free radicals are generated in our body. These free radicals are compounds that contain one or more unpaired electrons in their outermost orbit. Oxygen-based free radicals are the major free radicals found in the biological system.

Free radicals are highly reactive and are capable of oxidizing lipids, proteins, carbohydrates and nucleic acids. Oxidation of lipids leads to the damage of the membrane; oxidation of proteins leads to the alteration or loss of protein function; and oxidation of the bases of nucleic acid DNA can cause mutation of genes leading to serious diseases like cancer.

The polyunsaturated fatty acids are especially susceptible for oxidation by the free radicals, leading to the formation of hydroperoxides. This oxidation of lipids, especially membrane lipids, called lipid peroxidation chain reaction can be terminated by vitamin E. Hence, it is an example for chain breaking antioxidant.

Vitamin E has anti-sterility function in rodents. It also has selenium sparing effect. Slenium is present along with the enzyme glutathione peroxidase. It catalyzes reaction by which the free radicals are detoxified. Since both glutathione peroxidase and vitamin E have antioxidant action, they have sparing action on each other.

The antioxidant action of vitamin E protects the –SH (thiol) groups of the proteins, prevents the damage to mitochondrial membrane and helps in cellular respiration, and prevents the oxidative damage, which can be caused by the atmospheric pollutants.

Deficiency

Deficiency occurs due to the malabsorption, as seen in the case of obstructive jaundice, pancreatic diseases, or diseases of the intestine. In experimental animals, it has been demonstrated that the deficiency of vitamin E causes sterility. In female rats, pregnancy is terminated because of the death and resorption of foetus. The deficiency of vitamin E can cause muscle pain due to muscular dystrophy. In infants fed with vitamin E deficient diet, increased fragility of RBCs and hemolysis may give rise to anaemia.

Hypervitaminosis E

When an excess of vitamin E is consumed, the process of blood clotting becomes abnormal.

Free Radical

Molecules containing one or more unpaired electrons in their outer orbital are called free radicals. Free radicals are highly reactive and have short lifespan. They oxidize lipid, protein etc., present in the cell and damage various tissues.

Free radicals are generated constantly during metabolism of nutrients. When oxygen is partially reduced in the body it becomes highly reactive and is called reactive oxygen species (ROS).

e.g., Superoxide anion
Hydrogenperoxide
Hydroperoxyl radical
Hydroxyl radical, etc.

Antioxidants

Free radicals are scavenged by various substances and enzymes. They are called antioxidants.

The enzymes involved in trapping and detoxifying free radicals:

1. Superoxide dismutase (SOD)
2. Glutathione peroxidase
3. Catalase

Vitamins detoxify free radicals

1. Vitamin E
2. Vitamin C
3. Provitamin form of Vitamin A Beta carotene

Other substances involved in scavenging free radicals

1. Caffeine
2. Dimethyl thio urea
3. Allopurinol
4. Uric acid, etc.

VITAMIN K

Vitamin K is a fat-soluble vitamin, which is required to promote the coagulation of blood. It is a naptoquinone derivative containing polyisoprenoid side chain. There are different forms of vitamin K:

1. Vitamin K_1 (phylloquinone)
2. Vitamin K_2 (menaquinone)
3. Vitamin K_3 (menadione)

Vitamin K_1 and vitamin K_2 are the naturally occurring forms. Vitamin K_1 is present in plants, whereas vitamin K_2 is synthesized by the bacteria. It is synthesized by the normal bacterial flora of the intestine. Vitamin K_3 is a synthetic analog of the vitamin.

Sources

Liver, green leafy vegetables, cauliflower, spinach, cabbage, soya bean, etc. are good sources of vitamin K. It can be synthesized by the normal bacterial flora of the intestine.

RDA

- Adults: 70 to 140 μg/d.

Absorption

Absorption of vitamin K takes place in the intestine with the help of bile salt. It is transported to liver along with the chylomicron.

Functions

Vitamin K is required for the post-translational modification of prothrombin and coagulation factors ii, vii, ix and x. It is required for the gamma carboxylation of the glutamate residues of these proteins. This modification of the clotting factors is required for their further activation. The gamma carboxylation of the glutamate residues of prothrombin is required to make this protein capable of binding to calcium. Vitamin K is also involved in the gamma carboxylation of glutamate residue of different proteins like osteocalcin of bone involved in calcium binding.

Deficiency

In adults, vitamin K deficiency is rare, since it can be synthesized by the bacterial flora and is also supplied in the daily diet. Deficiency can be seen if the gut is sterilized and simultaneously dietary content of vitamin K is less. When there is malabsorption of lipids, the absorption of vitamin K becomes defective, which leads to deficiency of this vitamin.

The gut of newborns is sterile. The mother's milk may not be able to provide the quantity of vitamin required during the initial period of life. Recommendation has been made to give intramuscular injection of vitamin K to all newborns as a prophylactic measure to prevent haemorrhagic disease.

Vitamin K is required for gamma carboxylation of the glutamate residues of different proteins. During the deficiency of vitamin K, there is a defect in the coagulation of blood. Hence, prothrombin time is increased.

Antivitamins

Dicumarol and warfarin are antagonists of vitamin K.

Hypervitaminosis K

Hypervitaminosis K is rare. It can be seen if a large dose of vitamin K is injected. In hypervitaminosis, haemolytic anaemia and hyperbilirubinemia are seen.

WATER-SOLUBLE VITAMINS

Riboflavin (Vitamin B_2)

Riboflavin is a water-soluble vitamin. Its active form is flavin mononucleotide (FMN) and flavin adenine dinucleotide (FAD). These active forms act as coenzymes. In the coenzyme form, riboflavin has several functions.

Chemistry

Riboflavin contains an isoalloxazine ring with methyl groups at 6th and 7th position. D-ribitol is attached to the 10th nitrogen of this ring structure (Fig. 20.4).

Sources

Liver, dried yeast, whole and skimmed milk powder are the richest natural sources of vitamin B_2. Good sources are cereals, milk, egg, green leafy vegetables, fish, meat, etc.

RDA

It has been recommended that 0.55 mg per 1000 kcal as the requirement. Men sedentary workers require about 1.3 mg/day. Men who are engaged in heavy workers require 2.2 mg/day.

Women sedentary workers require 1.0 mg/day. Women engaged in heavy work require 1.7 mg/day. During pregnancy, an additional 0.2 mg/day is required. During lactation, an additional 0.4 mg/day is required.

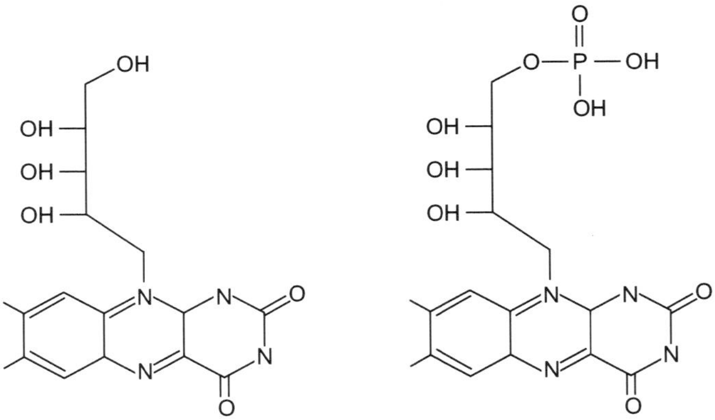

Fig. 20.4 Riboflavin.

Absorption

Vitamin B2 is absorbed in the small intestine and transported to the liver via blood. It is not stored in the body. Excess of vitamin is excreted in urine.

Functions

$$\text{RIBOFLAVIN} \xrightarrow{\text{Flavokinase}} \text{FMN} \xrightarrow{\text{FAD Synthase}} \text{FAD}$$

Riboflavin acts as a coenzyme. The two coenzyme forms of riboflavin are flavin mononucleotide (FMN) and flavin adenine dinucleotide (FAD). Synthesis of these coenzyme forms takes place in the liver and intestinal mucosa and other cells. Protein enzymes that use flavin coenzymes are called flavoproteins.

Enzymes Requiring FMN as Coenzyme

1. L-Amino Acid Oxidase

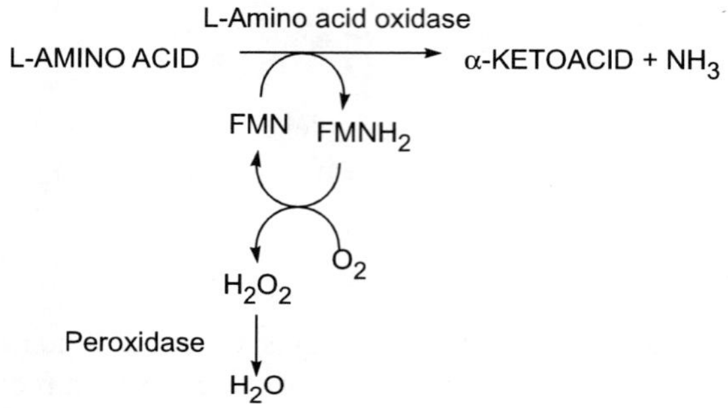

2. NADH Dehydrogenase of Electron Transport Chain
3. Acyl-CoA Dehydrogenase

Enzymes Requiring FAD as Coenzymes

1. Succinate Dehydrogenase

Succinate Dehydrogenase.

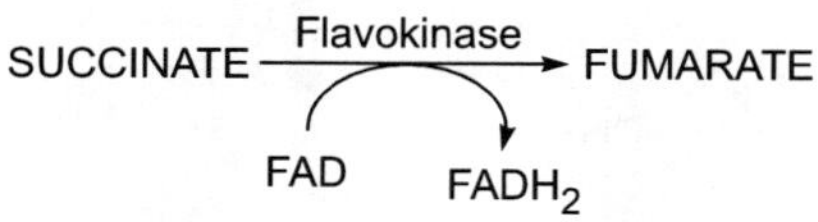

2. D-Amino Acid Oxidase

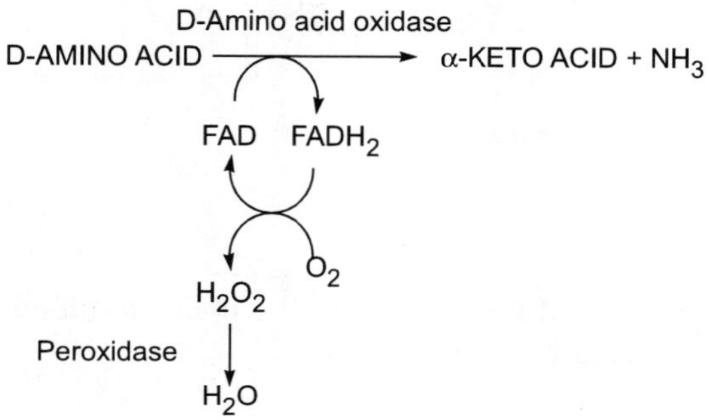

Deficiency

The deficiency of riboflavin is not known to be the primary cause of major human diseases. uncomplicated human riboflavin deficiency is characterized by:

1. Magenta-coloured tongue;
2. Fissuring at the corners of the mouth and lips (cheilosis);
3. Sebboric dermatitis; and
4. Corneal Vascularization, etc.

These signs and symptoms can also occur during the deficiency of niacin and iron. So, it may be difficult to state as to what extent riboflavin deficiency occurs in humans.

Antivitamin

Galactoflavin.

NIACIN (VITAMIN B_3)

Nicotinic Acid

It is a water-soluble vitamin. It is known as pellagra preventive factor of goldberg. Niacin is pyridine derivative. It is pyridine-3-carboxylic acid. The amide form of niacin is nicotinamide. Nicotinamide is highly soluble in water. The active form of nicotinamide is its coenzyme form—nicotinamide adenine dinucleotide (NAD^+), and nicotinamide adenine dinucleotide phosphate ($NADP^+$).

Sources

Cereals, pulses, and liver are rich in niacin. In addition to this, milk, egg, vegetables, and fish also contain niacin.

RDA

15 to 20 mg/day.

Absorption

Niacin is absorbed in the small intestine and transported to the liver via blood. It can be synthesized in the liver from the amino acid tryptophan. To synthesize 1 mg of niacin, about 60 mg of tryptophan is required. For the synthesis of niacin from tryptophan, another vitamin, pyridoxine (B_6) is required.

Functions

Niacin is converted to its active forms, NAD^+ and NAD^+. These active forms are the coenzyme forms of the vitamin. They catalyze the oxido-reduction reaction.

Nad$^+$ is mainly required for enzymes catalyzing the catabolic reactions, for example, reactions of TCA cycle, fatty acid oxidation, etc. Examples of enzymes, which require Nad$^+$ as the coenzyme, are isocitrate dehydrogenase, malate dehydrogenase, etc.

Nadp$^+$ is required as coenzyme for biosynthetic reactions, for example, fatty acid synthetic pathway, cholesterol biosynthesis, and reactions of HMP shunt pathway etc. Example of the enzyme, which requires Nadp$^+$ as the coenzyme, is glucose-6-phosphate dehydrogenase.

1. Examples of NAD$^+$ requiring reactions:

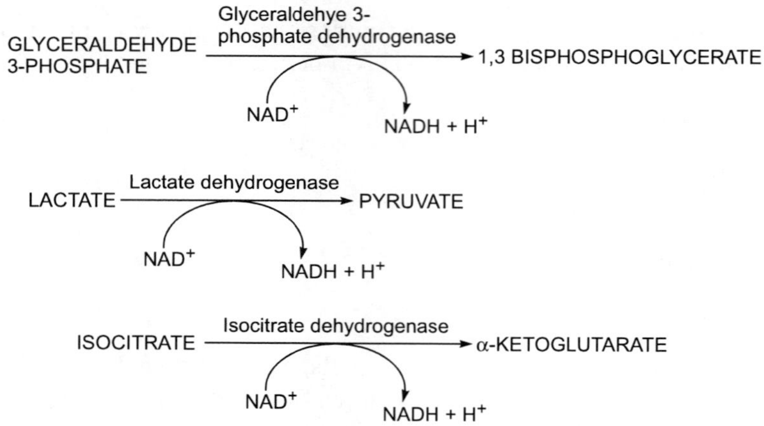

2. Example for NADP+ requiring reaction

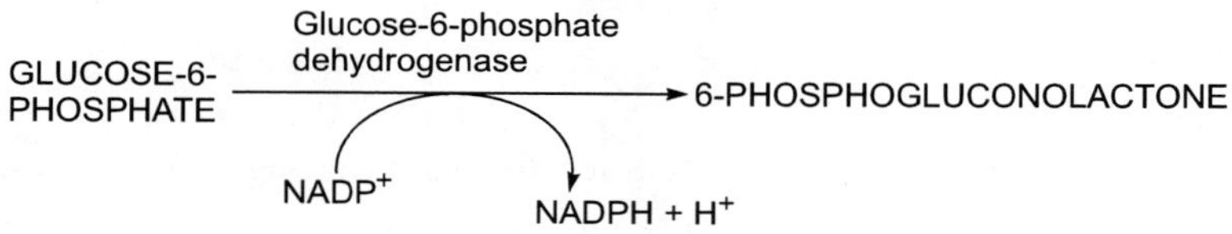

Deficiency

Several pathological conditions (e.g., hartnup disease and malignant carcinoid syndrome) as well as certain drug therapies (e.g., isoniazid) can lead to niacin deficiency.

In hartnup disease, tryptophan absorption is impaired, and in malignant carcinoid syndrome, tryptophan metabolism is altered, resulting in excess serotonin synthesis.

Isoniazid (the hydrazide derivative of isonicotinic acid) is the primary drug for the chemotherapy of tuberculosis.

THIAMINE (VITAMIN B_1)

Anti-Beriberi Factor; Anti-Neuritic Factor

Thiamine is a sulphur containing vitamin. It has a thiazole ring, which is connected to pyrimidine ring by methylene bridge. Thiamine as such is not biologically active.

In our body, thiamine is converted to its active form called thiamine pyrophosphate (TPP).

Source

Thiamine requirement depends upon the calorie intake. The outer layers of cereal grains are rich in thiamine. Hand pounded rice is a good source of the vitamin. In animal tissues it is present as phosphate ester.

RDA

1 to 1.5 mg/day

Absorption

When food is cooked at a temperature above 100°C, large amount of the vitamin is lost.

Phosphate esters of thiamine present in animal tissues are hydrolyzed by phosphatase in the intestine. The free vitamin, thiamine is absorbed availability of thiamine present in the ingested food is affected by the enzyme, thiaminase, present in certain fish, microorganisms present in the colon, etc. Thiaminase destroys thiamine.

Activation

Thiamine is converted to thiamine pyrophosphate. Thiamine pyrophosphate acts as a coenzyme for various biochemical reactions.

Thiamine is phosphorylated mainly in the liver and to a lesser extent in other tissues like muscle, brain reticulocyte, etc. Leading to formation of thiamine pyrophosphate.

Functions

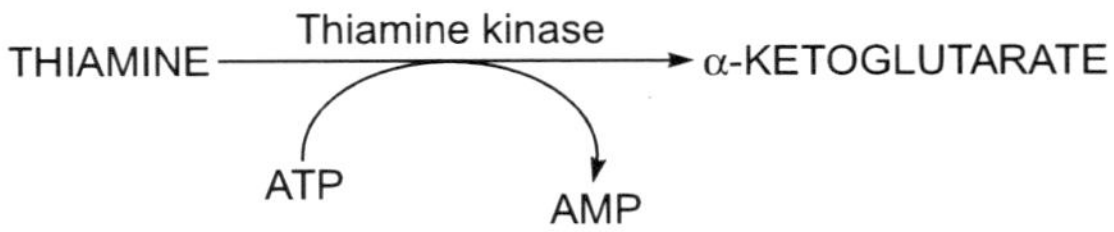

TPP is required as a coenzyme for the enzymes catalyzing the oxidative decarboxylation reaction, for example, pyruvate dehydrogenase, are α-ketoglutarate dehydrogenase. TPP is also required for transketolase enzyme.

Deficiency

The deficiency is mainly seen in alcoholics or under special clinical situations like haemodialysis, feeding after starvation, etc. In developing countries, deficiency is commonly due to the consumption of polished rice, or foods containing thiaminase or antithiamine agents.

Deficiency of thiamine gives rise to beriberi. The signs and symptoms are loss of appetite, muscle pain, palpitation, dyspnea, cardiac failure, edema, and polyneuritis.

In chronic deficiency, especially in alcoholics, paralysis and impaired mental functions are seen (Wernicke Korsakoff Syndrome).

Different types of beriberi

- Dry Beriberi: In dry beriberi, mainly the nervous system is involved and there is no oedema.
- Wet Beriberi: In wet beriberi, cardiovascular system is predominantly affected and oedema is seen.
- Infantile Beriberi: Beriberi seen in infants.

Transketolase activity of RBCs can be used to assess the vitamin status. In thiamine deficiency, transketolase activity is reduced.

Antivitamins

- Oxythiamin and pyrithiamin.

BIOTIN

Anti-Egg White Injury Factor

It is a water-soluble vitamin. It is readily soluble in hot water. Biotin acts as a coenzyme for various enzymes. It is bound to the enzyme by forming a covalent linkage between the e amino group of lysine residue of the enzyme and the carboxyl group attached to biotin.

Sources

Egg yolk, tomatoes, liver, and animal tissues are good sources of biotin. It is synthesized in the intestine by the bacteria flora.

RDA

100 to 300 mg/day.

Functions

It is required as a coenzyme in the carboxylation reaction. During carboxylation, reaction activated carbon dioxide is carried by the biotin. Later it is transferred to the substrate, which undergoes carboxylation.

Carboxylation reactions

$$\text{PYRUVATE} \xrightarrow[\text{Biotin; ATP; } Mg^{2+}]{\text{Pyruvate carboxylase}} \text{OXALOACETATE}$$

$$\text{ACETYL-CoA} \xrightarrow[\text{Biotin; ATP; } Mg^{2+}]{\text{Acetyl-CoA carboxylase}} \text{Malonyl-CoA}$$

Deficiency

Since the vitamin is synthesized in the intestine, its deficiency is rare. Deficiency manifestations can be seen if antibiotic or anti-bacterial drugs are used for a long time. Bacteria flora is destroyed and deficiency manifestation can be seen. If large quantity of avidine is ingested, it binds to biotin and retards the absorption of the vitamin.

In infants suffering from persistent diarrhoea on breast milk can suffer from deficiency of biotin because of poor absorption and low biotin content of milk. It is called Leiner's disease. In biotin deficiency, atropic glossitis, nausea, loss of appetite, muscle pain, etc., is seen.

PANTOTHENIC ACID

Pantothenic acid is a water-soluble vitamin. It is also known as vitamin B_5. Pantothenic acid is a complex of pantoic acid and B-alanine.

Sources

It is synthesized by the intestinal bacteria flora. It is present both in plants and animals. Good sources are liver, wheat germ, kidney, yeast, eggs, and leafy vegetables, etc.

RDA

10 mg/day.

Functions

Pantothenic acid is converted into coenzyme A. Pantothenic acid undergoes phosphorylation and combines with cysteine. Later it gets decarboxylated to form 4′ phosphopantetheine. The functional group of 4′phosphopantetheine is sulfhydryl (–Sh) group of the cysteine.

The 4′phosphopantetheine is present in the acyl carrier protein of fatty acid synthase complex and is required for the synthesis of fatty acids. The 4′phosphopantetheine is later linked to AMP, and the 3rd carbon of the ribose is phosphorylated to form coenzyme A. The functional group of coenzyme a is sulfhydryl group (–Sh) of the cysteine. Coenzyme a is abbreviated as CoASH, where CoA stands for coenzyme A and SH stands for sulfhydryl group. It acts as a coenzyme in variety of biochemical reactions.

Examples for Reactions Requiring CoA SH:

$$\text{PYRUVATE} \xrightarrow[\text{CoA SH}]{\text{Pyruvate dehydrogenase}} \text{ACETYL-COA}$$

$$\alpha\text{-KETOGLUTARATE} \xrightarrow[\text{CoA SH}]{\alpha\text{-ketoglutarate dehydrogenase}} \text{SUCCINYL-COA}$$

$$\text{FATTY ACID} \xrightarrow[\text{CoA SH}]{\text{Thiokinase}} \text{ACYL-COA}$$

Deficiency

Its deficiency has not been reported in human beings. Experimentally produced deficiency causes burning feet, gastrointestinal symptoms, fatigue, etc.

PYRIDOXINE (VITAMIN B_6)

Pyridoxal, pyridoxamine, and pyridoxine are collectively known as vitamin B_6. The functional group attached to the 4th carbon atom of the pyridine ring is the primary alcohol in case of pyridoxine, aldehyde in case of pyridoxal, and amine in case of pyridoxamine.

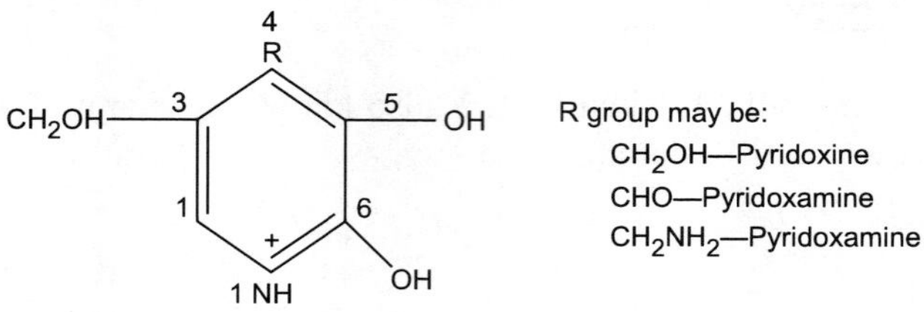

Sources

Vitamin B_6 present in food is mainly in the form of pyridoxine. Good sources are milk, meat, egg, green leafy vegetables, wheat, corn, etc.

RDA

1.5 to 2 mg/day.

Functions

All three different forms of vitamin B_6 are converted to their active form, pyridoxal phosphate (Plp).

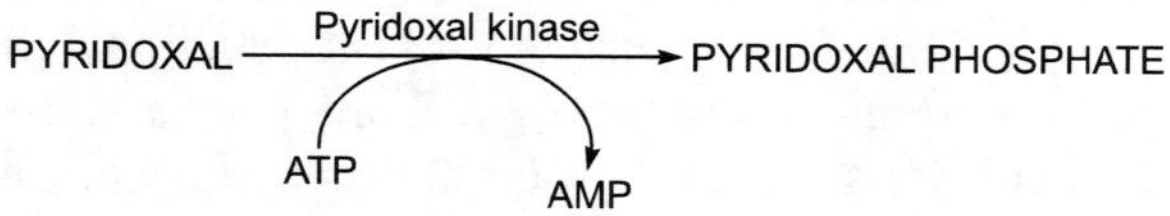

Pyridoxal phosphate acts as a coenzyme for various biochemical reactions.

Transamination reactions

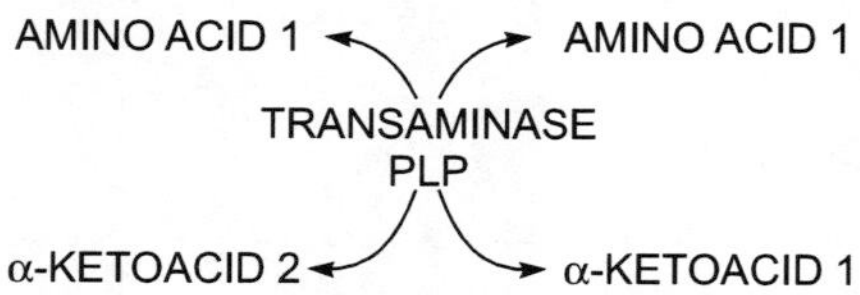

Non-oxidative deamination

$$\text{SERINE} \xrightarrow[\text{NH}_3]{\text{Serine dehydratase; PLP}} \text{PYRUVATE}$$

Decarboxylation reactions

$$\text{DOPA} \xrightarrow[\text{CO}_2]{\text{Dopa decarboxylase; PLP}} \text{DOPAMINE}$$

Deficiency

Pyridoxal phosphate is required as a coenzyme for the formation of various neurotransmitters. When there is a deficiency of pyridoxine, neurological manifestations like irritability, depression, convulsions and peripheral neuritis can be seen. Children with pyridoxine deficiency may have convulsions.

Other manifestations of pyridoxine deficiency are anaemia, pellagra (conversion of tryptophan to niacin requires PLP).

FOLIC ACID

Folic acid is pteroylglutamate, which is made up of pteridine, para-aminobenzoic acid, and glutamic acid. Based upon the number of glutamic acid present, it is called pteroylmonoglutamate, and pteroylpolyglutamate. It is heat stable and sensitive to light. Human beings cannot synthesize folate because of their inability to synthesize pteridine.

Sources

Green leafy vegetables, liver, yeast, kidney, meat fish, etc. Folic acid is also synthesized by the intestinal bacteria.

RDA

Adults require 0.1 mg/day, and in pregnancy, requirement is increased to 0.3 mg/day.

Functions

Tetrahydrofolate is the coenzyme form, folic acid is reduced to dihydrofolate. Dihydrofolate is further reduced to tetrahydrofolate by folic acid reductase. Tetrahydrofolate acts as a carrier of one carbon compounds. One carbon compounds are attached to N^5 and N^{10} positions of the tetrahydrofolate.

One Carbon Compounds

1. Methylene (= Ch_2)

2. Methenyl (= Ch)
3. Formimino (–CH = NH)
4. Formyl (–CHO)
5. Formate (–HCOO–)

One carbon compounds are formed during the metabolism of tryptophan, serine, and glycine. Tetrahydrofolate carrying the one carbon compounds are as follows:

(a) N^5-Methyl Tetrahydrofolate.
(b) N^5 N^{10}-Methylene Tetrahydrofolate.
(c) N^{10}-Formyl Tetrahydrofolate.
(d) N^5-Formimino Tetrahydrofolate, etc.

These one carbon complexes can be interconverted in liver and other tissues.

One carbon compounds carried by the tetrahydrofolate are utilized for various biosynthetic reactions

For example:

(a) Synthesis of purine ring (C^2 and C^8).
(b) Synthesis of serine from glycine.
(c) Synthesis of methionine from homocysteine.
(d) Methylation of dUMP to form TMP.
(e) Synthesis of glycine from CO_2 and NH_3.
(f) Synthesis of N-formyl Methionine containing tRNA.

Deficiency

The deficiency of folic acid causes megaloblastic (macrocytic) anaemia. Folic acid deficiency may be produced because of the dietary deficiency, use of folate antagonists, and defect or interference in the absorption of the vitamin.

The use of antibiotics can kill the bacteria flora, and its synthesis in the intestine is decreased. If diet is deficient or the need for the folic acid is increased, as in case of pregnancy, it also causes deficiency manifestations. Folate antagonists like aminopterin and amethopterin inhibit the enzyme dihydrofolate reductase competitively, and inhibit the formation of tetrahydrofolate. Intestinal diseases that interfere with the absorption of nutrients can produce folic acid deficiency.

The deficiency of folic acid can be detected by the estimation of formiminoglutamate (Figlu) in the urine. Figlu is formed during the catabolism of histidine. Formyl group is transferred to tetrahydrofolic acid. If folic acid is deficient, Figlu is excreted in the urine.

VITAMIN B_{12}

Cobalamine; Castle's Extrensic Factor; Antipernicious Anaemia Factor

Vitamin B_{12} is a water-soluble vitamin. It contains corrin ring. Corrin ring consists of four substituted and reduced pyrrole rings, which are connected to one another.

The nitrogens of the pyrrole rings form coordination complex with cobalt atom present at the centre of the corrin ring system. Cobalt can also form coordination bond with 5,6 benzimidazole and hydroxyl, cyanide, methyl or 5´deoxy adenosyl group. If the cobalt is linked to hydroxyl group, vitamin B_{12} is called hydroxycobalamin; if it is linked to cyanide, it is called cyanocobalamine; if it is linked to methyl group, it is called methyl cobalamine; and if it is linked to 5′deoxy adenosyl group, it is called deoxyadenosylcobalamine.

Sources

- Vitamin B_{12} is not synthesized by the plants or animals, but by the bacterial flora of the intestine.
- Milk, liver, meat, fish, kidney, eggs etc. are good sources of vitamin B_{12}.

RDA

Adults require 1 mg/day. During pregnancy and lactation, requirement is 1.5 mg/day.

Absorption

Cobalamine-protein complex from animal tissues is digested and vitamin B_{12} is released. Vitamin B_{12} binds to the intrinsic factor secreted by the stomach at the acidic ph. Intrinsic factor protects the vitamin and carries it to the site of absorption.

Vitamin B_{12} is actively absorbed by the intestinal cells. After absorption, it is transferred to plasma where it binds to transport protein called transcobalamine I and transcobalamine II.

Transcobalamine II-cobalamine complex binds to the cell surface receptors on various tissues. Vitamin B_{12} enters the cell by endocytosis and is released in the cytoplasm as hydroxycobalamine.

Vitamin B_{12} is stored in the liver as deoxyadenosylcobalamine. In the circulation, vitamin B_{12} is present mainly as methylcobalamine.

Functions

The two coenzyme forms of vitamin B_{12} are:

1. Methylcobalamin; and
2. Deoxyadenosylcobalamin.

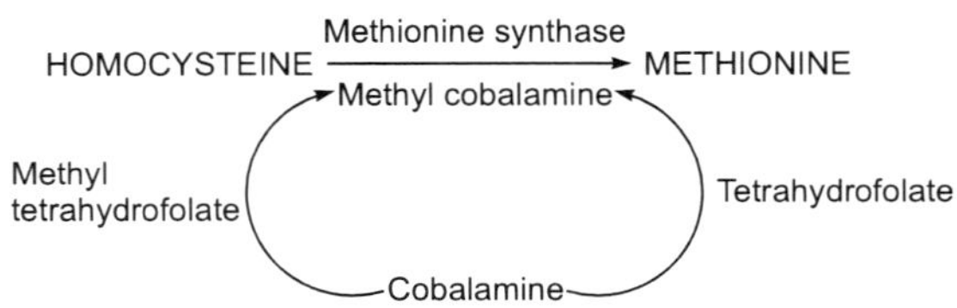

In vitamin B_{12} Deficiency, the methyl group from methyltetrahydrofolate cannot be transferred to cobalamine to form tetrahydrofolate. Methyl group and folic acid are trapped as methyltetrahydrofolate. It is called "Methyl Trap" or "Folate Trap".

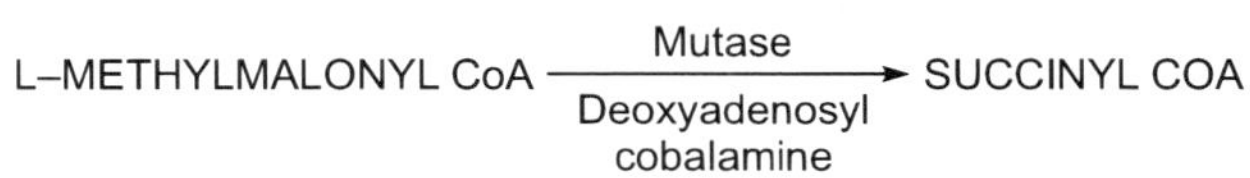

Deficiency

Vitamin B_{12} deficiency may be due to:

- Lack of intrinsic factor; and
- Defect in the intestinal absorption.

In vitamin B_{12} deficiency, folate is trapped as methyltetrahydrofolate. Tetrahydrofolate is required for the formation of RBC and in the bone marrow. It gives rise to megaloblastic anaemia.

Other manifestations of vitamin B_{12} deficiency are glossitis, neurological symptoms like parasthesia, and subacute degeneration of the spinal cord. In vitamin B_{12} deficiency, methylmalonyl CoA gets accumulated and is excreted in the urine as methylmalonic acid. The condition is called methylmalonic aciduria.

VITAMIN C

Ascorbic Acid; Antiscorbutic Factor

Vitamin C is a water-soluble vitamin. It is a hexuronic acid and serves as a reducing agent. Vitamin C is destroyed by the heat and atmospheric oxygen. It can exist as L and D isomeric form. L-ascorbic acid possesses the vitamin activity. The enzyme, which is required to form ascorbic acid from l-gulonolactone oxidase, is absent in guinea pigs and primates. Other than guinea pigs and primates, all other animals and plants are able to synthesize vitamin C.

Sources

Guava, Indian gooseberry (Amla), coriander leaves, and citrus fruits like orange, lemon, pineapple, etc. are rich sources of vitamin C.

RDA

- Adults require 60 mg/day.
- During pregnancy and lactation, requirement is 80 mg/day.

Functions

- Vitamin C is required for the synthesis of mature collagen, which is rich in hydroxyproline and hydroxylysine. Hydroxylation of proline and lysine requires vitamin C. Collagen is present in the extracellular matrix and capillary basement membrane, matrix of the bone, etc. In vitamin C deficiency, hydroxylation of proline and lysine is defective. It leads to the formation of defective collagen, and those tissues, which contain collagen, become fragile. So, bleeding from the gums on pressure, petichial haemorrhage (small haemorrhagic spots on the skin), increased fragility of the bone, and even internal bleeding in severe cases can be seen.
- Vitamin C is required for the absorption of iron. Iron is absorbed in the ferrous form. Since vitamin C has reducing property, it reduces the ferric form of iron to ferrous and facilitates its absorption.

- Vitamin C acts as an antioxidant in the aqueous phase.
- Adrenal gland contains large amount of vitamin C, which may be required for the synthesis of adrenal hormones.
- Bile acid synthesis requires vitamin C.
- Catabolism of the amino acid tyrosine requires vitamin C.
- Carnitine is a biomolecule, which is required to transport long chain fatty acids across the inner mitochondrial membrane. The synthesis of carnitine requires vitamin C.

Deficiency

- Vitamin C deficiency gives rise to scurvy. In scurvy, the clinical manifestations seen are swollen and bleeding gums due to increased fragility of the capillaries, delayed wound healing, petechial haemorrhages, increased fragility of the bones, and increased susceptibility to infection.

QUESTIONS

Long Essay

1. Discuss in detail about the chemistry, sources, RDA, functions, and deficiency manifestations of vitamin A.
2. Discuss in detail about the chemistry, sources and synthesis, RDA, functions and deficiency manifestations of vitamin D.
3. Discuss in detail about the chemistry, sources, RDA, functions, and deficiency manifestations of vitamin C.

Short Essay

1. Write some of the functions of vitamin C.
2. Discuss the chemistry, sources, RDA, functions, and deficiency manifestations of vitamin E.
3. Discuss the chemistry, sources, RDA, functions, and deficiency manifestations of vitamin K.
4. What are the coenzyme forms of niacin? Write any four reactions where the coenzyme forms of niacin are required.
5. What are the coenzyme forms of riboflavin? Write any four reactions where these coenzymes are required.
6. What is the coenzyme form of thiamine? Write any three reactions where the coenzyme form of thiamine is required.
7. Mention the coenzyme form of vitamin B_{12}. Give any two examples where these coenzymes are required.
8. Discuss the functions of folic acid.
9. Write about the deficiency manifestations of vitamin A.

Short Answer

1. Write short notes on the following:
   (a) Folate Trap
   (b) Pellagra
   (c) Beriberi
   (d) Scurvy
   (e) Rickets
   (f) Osteomalasia
2. Name the provitamin form of vitamin D.
3. Give any four good sources of vitamin A, and mention their provitamin forms.
4. What do you mean by retinol equivalents? mention the RDA of vitamin A.

21

Mineral Metabolism

INTRODUCTION

Adequate amount of minerals in the diet is important for normal health, growth and reproduction. Based on the daily requirement of the minerals, they can be classified into two groups

1. Microelements (daily requirement is more than 100 mg), e.g., calcium, phosphorus, magnesium, sodium, fluoride, chloride, etc.
2. Microelements (daily requirement is less than 100 mg), e.g., iron, copper, magnesium, zinc, iodine, molybdenum, selenium, etc.

CALCIUM

Calcium is the most abundant mineral in the body. The total calcium in the body is about 1.2 kilograms. Most of the calcium present in the body is concentrated in the bones and teeth (about 99% of the total body calcium). It is also present in the extracellular fluid (about 1% of the total body calcium). Normal blood level of calcium is 8.5 to 10.5 mg/dL.

Absorption

Calcium is absorbed from the upper part of the small intestine. Absorption of calcium is an active process; calcium is absorbed against the concentration gradient using energy. It requires calcium binding proteins.

Calcium absorption is regulated by vitamin D (Calcitriol) and paratharmone

1. Vitamin D (calcitriol) increases the absorption by increasing the synthesis of calcium binding proteins (calbindin)
2. Paratharmone increases the conversion of 25 hydroxy cholecalciferol to 1, 25 dihydroxy cholecalciferol (calcitriol)

Calcium absorption is affected by various factors:

(a) **Factors increasing the absorption of calcium**
Acidity, amino acids (lysine and arginine), high dietary lactate and citrate.

(b) **Factors decreasing the absorption of calcium**
High dietary phytate, phosphate and oxalate. Excess alkali intake decreases the absorption of calcium. Malabsorption of the fatty acids leads to formation of insoluble calcium soaps with calcium and decreases the absorption.

Sources

Milk and milk products are the richest source for calcium. Other sources are egg, fish, cereals, vegetables, etc.

Daily Requirement

Adults about 500 mg
Pregnancy and lactation requirement increases to about 1500 mg
Children about 1200 mg

Functions

1. Required for formation of bone and teeth
2. Regulation of transmission of nerve impulse
3. Activation of enzymes
4. Required for muscle contraction
5. Calcium acts as second messenger
6. Required for blood clotting
7. Cardiac conduction
8. Maintenance of cell membrane permeability
9. Decrease of neuromuscular excitability.

Excretion

Most of the calcium is excreted in faeces and about 200 mg of calcium is excreted in urine.

Plasma Calcium

Normal blood level of calcium is 8.5 to 10.5 mg/dL. Calcium present in the blood is present in different forms.

(a) Ionized calcium (50%)

It is the active form of the calcium which can diffuse from the blood to tissue. If ionized calcium is decreased it gives rise to tetany.

(b) Non-ionized (50%)

- Protein bound calcium (45%) – Mainly bound to albumin
- Complex with organic ions (5%)

Normally the calcium to phosphate ratio must be maintained. If the plasma phosphate level is increased as in case of renal failure, plasma calcium decreases to keep the ratio constant.

Ionization of calcium occurs at normal blood pH 7.4. When blood pH increases (alkalosis) calcium will bind to protein and protein bound fraction increases and ionized calcium level decreases. In case of acidosis ionized calcium level increases.

When the total protein level of the plasma decreases, total calcium level also decreases due to decrease in the protein bound form of calcium.

Regulation of Plasma Calcium (Calcium Homeostasis)

The plasma level of calcium is regulated by three hormones:

1. Parathyroid hormone (PTH)
2. Calcitriol
3. Calcitonin

Parathyroid hormone

Parathyroid hormone is secreted in response to decreased blood level of calcium. It mobilizes calcium from the bones (resorption of bone). Parathyroid hormone increases the absorption of calcium from intestine by enhancing the conversion of vitamin D to calcitriol. In kidney parathyroid hormone decreases the excretion of calcium and increases excretion phosphate.

Calcitriol (1,25 dihydroxycholecalciferol)

Calcitriol is the active form of vitamin D. It increases calcium absorption from the intestine and reabsorption of calcium by the renal tubules. It also mobilizes the calcium from bone.

Calcitonin

Calcitonin is a hormone secreted by the parafollicular or 'C' cells of the thyroid gland. It is secreted in response to increase blood calcium level and helps to decrease the blood level of calcium by inhibition of calcium mobilization from bones or by increasing the deposition of calcium in bones.

Alteration of Plasma Calcium

Hypercalcemia

Increase in plasma calcium is called hypercalcium.

Causes

(a) Primary hyperparathyroidism

(b) Ectopic cells producing parathyroid hormone as in case of some malignancies
(c) Milk alkali syndrome:- Hypercalcium present in patients who received excessive absorbable alkalis and milk for long periods as in case of treatment of peptic ulcer.
(d) Drugs like thiazide diuretics, etc.

Hypocalcemia

Decrease in plasma calcium is called hypocalcemia. Hypocalcemia causes tetany with carpopedal spasm. Chronic deficiency of calcium in children leads to rickets and adults osteomalcia.

Causes

(a) Hypoparathyroidism
(b) Alkalosis
(c) Kidney disease where activation of vitamin D is inhibited

PHOSPHATE

Phosphate is present mainly inside the cells. About 80% of the phosphorus is seen in bone and teeth, 20% are present in the muscles and other tissues.

Absorption

Phosphate is absorbed by an active transport mechanism in the mid jejunum. Absorption is regulated by calcitriol. Factors affecting the calcium absorption affect the absorption of phosphate in the same manner.

Sources

Milk and milk products, fish, meat, liver, kidney, leafy vegetables and egg.

Daily Requirement

About 500 mg in adults

Functions

1. Required for bone and teeth formation
2. It acts as buffer (phosphate buffer). Since the pK value of phosphate buffer is close to normal pH of blood, it is a more effective buffer.
3. It is present in nucleic acid DNA and RNA.
4. Phospholipids contain phosphate.
5. Phosphoproteins contain phosphate a.
6. High energy compounds like ATP and GTP creatine phosphate contain phosphate.
7. Phosphate containing cAMP and cGMP acts as second messengers.

8. Phosphate esters of sugars are intermediates in the metabolism
9. Coenzyme NADP contains phosphate

Blood Phosphate

Normal plasma inorganic phosphorus is 3 to 5 mg/dL in adults and 5 to 6 mg/dL in children.

Regulation of Blood Level of Phosphate

PTH decreases blood level of phosphate by stimulating its excretion. Hypophosphatemia (decreased blood level of phosphate) stimulates the renal activation of vitamin D (calcitriol). Calcitriol increases the absorption of phosphate in the intestine, renal reabsorption of phosphate and mobilize phosphate from the bone.

Alteration of Blood Phosphate

Hyperphosphatemia

Increase in blood level of phosphate is called hyperphosphatemia.

Causes

(a) Hypoparathyroidism
(b) Acidosis
(c) Hypervitaminosis D
(d) Hemolysis

Hypophosphatemia

Decrease in blood level of phosphate is called hypophosphatemia.

(a) Hyperparathyroidism
(b) Vitamin D deficiency
(c) Renal tubular disease

MAGNESIUM

The total body magnesium is about 25 g. About 60 to 70% of the magnesium is present in bone complexed with calcium. Magnesium is mainly present in intracellular fluid.

Absorption

Absorbed from the upper part of small intestine

Daily Requirement

Adult male requires about 400 mg

Functions

1. Activates many enzymes, e.g., hexokinase, phosphofructokinase, etc.
2. Required for the active transport of various cations like sodium, potassium and calcium where ATP is used as source of energy.
3. Lowers neuromuscular irritability
4. Glucose tolerance is improved by magnesium supplementation (magnesium deficiency decreases the insulin dependent uptake of glucose)

SULPHUR

Sulphur containing amino acids are the source of sulphur. Sulphur present in the body is oxidized to sulphate in the liver and excreted in the urine. Total amount of sulphur in urine is about 1 gm/day.

Sulphur present is the urine in three different forms:

1. **Organic sulphate (ethereal sulphate)** constitutes 10% of total sulphur excreted in urine
2. **Inorganic sulphate** constitutes 80% of total sulphur excreted in urine
3. **Neutral sulphur** (unoxidized sulphur) constitutes 10% of total sulphur excreted in urine

Functions

1. Many enzymes contain –SH group in their active site
2. Chondroitin sulphate is present in cartilage and bone
3. Thiamine, lipoic acid, biotin, Coenzyme A, contain –SH group.
4. Sulphur is required for detoxification reaction
5. Active sulphate present in the body is phosphoadenosine phosphosulphate (PAPS)
6. Keratin present in hair and nail are rich in sulphur

IRON

Total body iron content ranges from 2.5 to 3.5 g in normal individuals. Most of the iron is present in the blood associated with haemoglobin (about 75% of total body iron) and remaining is present in liver bone marrow and muscle and almost all tissues (associated with heme).

Ferritin contains about 15% and myoglobin contains about 5% of the total iron.

Sources

Liver, heart, kidney, spleen fish, sugar cane syrup (molasses) dates, egg yolk, jaggery, leafy vegetables, pulses, cereals. Milk is very poor source of iron.

Requirement of Iron

Daily requirement of iron for an adult Indian is 20 mg of iron. Children from 13 to 15 years of age require about 20 to 30 mg of iron per day and pregnant women require about 40 mg of iron.

Functions of Iron

(Heme contains porphyrin ring and iron)

1. Heme of haemoglobin contains iron. It is required for the transport of oxygen
2. Myoglobin contains iron
3. Cytochrome P450 contains heme which is required for detoxification of drugs
4. Various enzymes of the body contains heme, e.g., catalase, peroxidase, tryptophan oxygenase, etc.

Absorption

Absorption of iron takes place from duodenum and proximal part of jejumum.

Factors Affecting Iron Absorption

Factors increasing iron absorption

1. Cooking of food and gastric HCl
   Helps to liberate the iron from the organic compounds.
2. Various reducing substances
   Vitamin C and cysteine of dietary protein help the reduction of ferric irons into the absorbable ferrous iron.
3. Body need
   Absorption depends upon the need of the body.

Factors decreasing iron absorption

1. High dietary phytates and phosphates, calcium, copper
2. Steatroohea
3. Alkalis and tea

Regulation of Iron Absorption (Mucosal Block Theory)

Iron is absorbed in the ferrous state and within the mucosal cells. It is oxidized to ferric form and combines with apoprotein to form ferritin. Depending on the need of the body, iron is released to the circulation. When the iron content of the body is optimum all the apoferritin molecules become saturated with iron and absorption of iron is blocked.

Transport and Uptake of Iron

Absorbed iron enters the portal blood in ferrous state. In the plasma, iron is rapidly oxidized to ferric state. In the plasma, copper containing protein called ceruloplasmin catalyzes the oxidation of ferrous iron to ferric iron. Ferric iron binds to iron transporting protein called transferrin.

Transferrin is a glycoprotein. It has got half life of 7 to 10 days. One molecule of transferrin can bind two iron molecules. Normal plasma level is about 250 mg/100 mL. In iron deficiency the blood level of transferrin increases.

The total capacity of transferrin to bind iron is about 250 to 400 μg/100 mL (Maximum amount of iron that can be carried by transferrin per decilitre is called Total Iron Binding Capacity [TIBC]). Only one-third of the transferrin is saturated with iron (120 μg/100 mL). In iron deficiency anaemia plasma iron is decreased. Liver synthesizes more transferrin so the TIBC is increased.

Cells of our body especially heme synthesizing cells have got transferrin receptor. Transferrin receptor binds two molecules of transferrin and iron is transported inside the cell.

Iron is stored in the cell as ferritin. Apoferritin is the protein part of ferritin which binds up to 4,000 atoms of iron. In case of iron deficiency anaemia ferritin level is decreased.

When the body content of iron is more than the capacity of the apoprotein to store them, iron is found in granules deposited in tissues called **hemosiderin**. These granules are composed of iron, protein and polysaccharides. Hemosiderin may be a degraded ferritin.

Excess iron accumulation in the body is called **hemochromatosis (hemosiderosis).** When iron accumulates in the body it damages various tissues like liver, pancreas and heart, etc. **Pigmentation of the skin** and **diabetes mellitus** if frequently associated with the excess iron deposition in the body, it is called **bronze diabetes**.

COPPER

Copper is mainly distributed in muscles, liver, kidney, bone marrow, heart and hair. Total copper content of the body is about 100 mg.

Sources

Liver, kidney, dried legumes, and nuts. Milk is a poor source of copper.

Requirement

Daily about 1.3 to 3 mg of copper is required for a healthy adult.

Functions

1. Copper is essential constituent of several enzymes, e.g., cytochrome oxidase, tyrosinase, superoxide dismutase (SOD), dopamine hydroxylase, etc.
2. Copper is essential for haemoglobin synthesis, bone formation and maintenance of myelin of nerves.
3. It is required for the absorption of iron and incorporation of iron into haemoglobin.

Absorption

Absorbed from the upper small intestine

Excretion

Mainly through bile
Normally urine does not contain copper

Blood and Plasma Level of Copper

Normal blood level of copper is 100 microgram/dL. Normal serum level of copper is 25 to 50 microgram/dL. In the RBCs copper is associated with superoxide dismutase (erythrocuprein). In plasma copper is mainly associated with ceruloplasmin. Ceruloplasmin is a copper binding protein. Each molecule can bind 6 atoms of copper. It acts as ferrooxidase enzyme (oxidizes ferrous iron to ferric iron) during iron metabolism. Ceruloplasmin binds 90% of the plasma copper. The remaining 10% of the plasma copper is loosely bound to albumin.

Disorders of Copper Metabolism

Wilson's disease

In Wilson's disease copper excretion is defective leading to copper accumulation of copper in liver, brain, kidney and red blood cells. It is also called hepatolenticular degeneration. **Genetic defect in the copper binding ATPase** in cells leads to defective incorporation of copper into apoceruloplasmin and defective excretion of copper into the bile. Another cause is aceruloplasminemia which is a genetic disorder. Because of lack of ceruloplasmin copper gets accumulated in the body.

In Wilson's disease

- Cirrhosis of the liver,
- Degeneration of the lenticular nucleus of the brain,
- Greenish brown discolouration of the margin of cornea due to deposition of copper (Kayser Fleisher ring)
- Renal damage are seen

Anaemia due to copper deficiency

Microcytic normochromic anaemia is seen in copper deficiency. Copper is required for ALA synthase key enzyme in the heme synthesis. Cerulopsamin containing copper helps in transport of iron.

Menke's kinky hair syndrome

In Menke's kinky hair disease copper absorbed from GIT, but due to defective copper binding ATPase it cannot be transported in the blood. Copper deficiency in the tissues leads to defective cross-linking of connective tissue. It is X-linked disorder affect only male child.

Copper Toxicity

Copper toxicity gives rise to diarrhoea blue-green discolouration of the saliva, haemolysis, haemoglobinuria, proteinuria and renal failure.

ZINC

The adult human body contains about 2 g of zinc. Zinc is found in skeletal muscle, bone, hippocampus area of brain and prostate, etc.

Sources

Meat, grains, beans, shellfish, meat, milk, etc.

Requirement

In adults 10 to 20 mg/day.

Functions

1. Zinc is required for many enzymes
   e.g., Alkaline phosphates
   Carbonic enhydrase
   Superoxide dismutase
   Carboxypeptidase
   RNA polymerase
2. Zinc forms complex with insulin in the β cells of pancreas. It helps to stabilize insulin.

Absorption

Absorbed from small intestine mainly from duodenum.
Iron inhibits the absorption of zinc. Copper and zinc inhibit each other's absorption.

Zinc Deficiency

Zinc deficiency gives rise to

1. Dermatitis
2. Alopecia
3. Delayed wound healing
4. Confusion, apathy and depression
5. Impaired spermatogenesis
6. Reduction in the number of T and B lymphocytes

Zinc Toxicity

Zinc toxicity may produce nausea, vomiting, headache, gastric ulcer, anaemia, pancreatitis, pulmonary fibrosis. It is usually seen due to inhalation of zinc fumes by welders and in accidental ingestion of rat poison containing zinc.

IODINE

Iodine is required for the thyroid hormone synthesis. Total body iodine content is 25 to 50 mg. It is mainly concentrated in the thyroid gland, body fluids and other tissues.

Sources

Fish, seafood, vegetables grown near the seaboard, iodinated table salt, etc.

Requirement

150 to 200 μg/day

Functions

The known function of iodine is the formation of thyroid hormone.

Deficiency

Hypothyroidism and goitre in adult and cretinism in children.

SELENIUM

Selenium is required for the activity of the enzyme glutathione peroxidase. Selenium containing amino acid selenocysteine is present in this protein.

Glutathione peroxidase is required for the detoxification of hydrogen peroxide. Selenium and vitamin E has got complementary action. Vitamin E reduces the selenium requirement (selenium sparing effect of vitamin E).

Daily requirement of selenium is 50–100 μg/day. Selenium availability depends upon the selenium content of the soil in which food crops are grown.

Deficiency

Deficiency of selenium in the soil seen in China (Keshan Province) is responsible for Keshan disease. In Keshan disease multifocal myocardial necrosis, cardiac arrhythmias and cardiac enlargement are seen. Selenium deficiency seen elsewhere causes cardiomyopathy, liver cirrhosis and muscular dystrophy.

Selenium Toxicity

Hair loss, diarrhoea, weight loss are seen in selenium toxicity.

FLUORIDE

It increases the hardness of the bone and teeth and prevents caries. Fluorine content of drinking water is more it can give rise to fluorosis. In fluorosis blood level of fluoride is increased to about 50 microgram/100 mL. In fluorosis osteoporosis and osteosclerosis with brittle bones is seen.

COBALT

Cobalt is a component of vitamin B_{12}. It is necessary for the blood cell formation. Cobalt gives vitamin B_{12} its red colour.

Enzymes requiring cobalt

1. Methylmalonyl CoA
2. Homocysteine methyltransferase
3. Ribonucleotide reductase

QUESTIONS

Long Essay

1. Write the normal Serum total calcium level. What are the different fractions of total calcium?
2. Explain the blood calcium and phosphate homeostasis in human body.
3. Write the sources requirement functions and absorption of iron.

Short Essay

Write short notes on the following:

1. Magnesium
2. Copper
3. Fluorine
4. Fluorsis
5. Functions of zinc and magnesium.
6. Functions of iodine and selenium.

22

History of Nutrition; Nutritional Problems in India; National Nutrition Policy; Factors Affecting the Nutritional Status

INTRODUCTION

Every living cell requires energy. All living cells have got their own mechanism to procure energy. Cells which are not able to derive energy on their own, have got the ability to harness energy from their surroundings. Human beings are multicellular organisms. We depend mainly on plants as source of energy. Plants use sunlight, carbon dioxide and water to produce carbohydrates which is stored in the form of starch. When we consume this starch, by various mechanical and chemical processes taking place in our body, energy is released. It is used for various processes, e.g., muscle contraction, transport, etc.

Human body apparently is a mass of proteins. Proteins are built up of amino acids. Some of the amino acids are synthesized in the body. Which cannot be synthesised need to be supplied either from the plant or animal sources. Human body is protected by lipids; they also act as storage form of energy. Even though we can synthesize various lipids, some lipids have to be supplied from external animal or vegetable sources. Vitamins and minerals in addition to sufficient amount of water are also required to keep us healthy and fit.

Everyday we eat and drink at regular intervals to keep us alive, active and healthy. Substances that we eat is called **food**. Most of the time food is eaten after some culinary process like frying, baking, boiling, grinding, etc. Now food becomes **diet**. The culinary process used depends upon the culture of the family, society and to some extent on religion.

CLASSIFICATION OF FOOD

Food is Classified Based on Different Criteria

(a) Depending upon its source, food is classified into

1. **Foods derived from plant source**
   - **Cereal grains** (Rice, wheat, barley, corn, maize)
   - **Legumes/pulses** (Soya beans, peas, peanuts)
   - **Vegetables** (Potatoes, sugar beet, sugar cane, horseradish, cauliflower, carrot, tomato, spinach, bamboo shoots)

   Fruits, nuts and seeds
2. **Foods derived from animal source**

   Meat, liver, egg, milk, fish, shellfish, honey, etc.

(b) Depending on the function, food is classified into

1. **Body building food (rich in protein)**
   Milk and milk products, meat, egg, fish, pulses and dry legumes
2. **Energy yielding food (Rich in Carbohydrate)**
   Cereals, roots and tubers, sugar, etc.
3. **Protective food**
   Green-leafy vegetables, fruits etc.

History of Nutrition

Man cultivates a variety of vegetables, grains, fruits and rears birds and animals to use them as food. Thought and foresight had been exhibited in selecting the food best suited for promoting good health. In early days, selection of food was done by trial and error method. The knowledge of nutrition is accumulated by the observations made by several pioneers. They studied

1. The chemical nature of food
2. Energy output from the food
3. Importance of proteins in nutrition
4. Effect of modifying the composition of diet in treatment of certain diseases.

Some of the observations made by several pioneers are given below:

The Chemical Nature of Food

A.Fr.de Fourcroy on the basis of nitrogen content, had distinguished three kinds of animal products which laid a basis for the study of protein. He described about cholesterol and identified adipose tissue. **Marggraf in 1747** identified the milk sugar. **Prout in 1806** described malt sugar. **Chevreul in 1814** discovered that fats are composed of fatty acid and glycerol. Glucose in blood and glycogen in liver were discovered by **Claude Bernard and Baumann in 1895** discovered iodine in thyroid gland.

Energy Output from Food

Lavoisier and later Leibig's work had shown that carbohydrates, fats and proteins are oxidized in the body to CO_2 and water and produce heat. **Regnault in 1849** determined the respiratory quotient (RQ). **Carl Voit** by using the human respiration calorimeter determined the energy requirement in 1862. **Atwater** found out the energy requirement of the human beings using a human respiration calorimeter and found the physiological **calorific value of one gram each of carbohydrate, fat and protein as 4 kcal, 9 kcal and 4 kcal respectively.**

Importance of Proteins in Nutrition

Carl Voite and associates determined the protein requirement of adult man. **Chittenden** by his experiments found out that **protein requirement of adult men is about 35 to 45 g/day**. **Folin** distinguished two types of protein metabolism endogenous and exogenous protein metabolism. Later it was found out that lack of certain amino acids in the diet will retard growth.

Effects of modifying the composition of diet in treatment of certain diseases

Eijkman in 1897 found that addition of rice polishing extracts in the chicken feed cure polyneuritis in chickens. **Pekelharing** described the importance of small amount of milk instead of synthetic diet in promoting the growth of mice. **Jamos Lind in 1753** established that scurvy could be prevented by including fresh fruits or vegetables in the diet. British Navy introduced lime juice as part of the diet of their sailors. In **1907, Holst and Frolich** by feeding a diet based on oats produced scurvy in guinea pigs.

Hopkinsin (1906) by his experiments postulated that some unknown growth factors are present in milk and he called them accessory factors of the diet (now known as vitamins). Later **Funk in 1912** gave the name 'vitamin' to these accessory food factors. **McCollum and Davis in 1913** showed that butter fat, egg yolk, cod liver oil added to the diet promote growth. **McCollum and Davis** designated the factor as fat soluble vitamin A. In **1916 McCollum** and associates showed addition of whey of milk to synthetic diet helps to promote good growth of the rats. They called it 'water soluble vitamin B'.

In **1915 Gold Berger** found that addition of 200 g of meat and 30 g of brewers' yeast can prevent pellagra. In **1935 Cicely Williams** had shown that Kwashiorkor seen in children is caused by deficiency of protein and can be cured by feeding milk.

NUTRITION

Nutrition is the study of food starting from its ingestion including its digestion, absorption, transport and metabolism until the disposal of waste products formed from it. In simple words, it is a branch of science which helps to understand the fate of the food in the body.

Dietetics deals with the study of nutrition in healthy individuals as well as in disease conditions. It involves planning as well as prescription of diet to healthy individuals and in disease conditions.

Nutrients are different chemical components of food (e.g. carbohydrates, proteins, fat, etc). Nutrients present in the food determine its quality. They are required in proper quality and proportion to maintain health, growth and reproduction.

Functions of Nutrients

1. Provide energy required to sustain life and for day-to-day activities
2. Supply essential molecules that cannot be synthesized in the human body or synthesized in amounts that are not sufficient to support growth and maintenance. E.g: essential amino acids, essential fatty acids, vitamins and minerals.

Food which we ingest every day contains carbohydrates, proteins, lipids, vitamins and minerals. Food doesn't mean only proteins, fats and carbohydrate but much more than this. It is a part of civilization and security of the society. Absolute or relative excess or deficiency of one or more nutrients gives rise to a pathological state called **malnutrition**. Malnutrition includes undernutrition, overnutrition, imbalance and specific deficiency of nutrients. **Undernutrition** is due to insufficient food ingested over a period of time. **Overnutrition** is due to consumption of excess quantity of food over a period of time. Disproportion in the essential nutrient content of the food also leads to malnutritition.

Concepts

Nutrition was the topic confined to the laboratory and the subject of interest only to physiologist and physicians. It was only the health sector which was responsible for nutritional problems of the society. Now science of nutrition is not confined to the laboratory but, it is also applied to the community. It gives rise to new concept in nutrition called **nutritional epidemiology**.

Nutritional epidemiology includes epidemiological assessment of nutritional status of the communities, nutritional surveillance, nutritional and growth monitoring and rehabilitation, etc. These epidemiological methods

(1) Help to identify the cause, risk factors of the disease
(2) Planning and evaluation of the nutritional programmes.

During the past years, significant advances have been made in the field of nutrition. Attention has been drawn on association of nutrition with immunity, fertility, maternal, child and family health. Research is being conducted to know the association of food with diseases like coronary heart disease, diabetes, metabolic syndrome, cancer, etc.

During the recent years, it has been recognized that nutrition is the indicator of the socioeconomic development of the nation. Nutrition was one of the eight elements of the campaign for health for all. Wherever possible great importance is given to integrate nutrition with the primary health care system.

NUTRITIONAL PROBLEMS IN INDIA

India got independence in 1947. Before independence, famine and starvation were commonly seen. India had scarcity of food for about two decades after independence. Now India has become self sufficient in food production and has got surplus stock of food grains. In spite of excess food production, scarcity of food is seen in some parts of the country. It may be due to abject poverty in the population or a result of uneven distribution of food due to weak public distribution system. The **Green Revolution** has helped to increase the food grain production and helped to keep pace with the growing population. Increase in the food production also helped to decrease the incidence of protein energy malnutrition and other forms of deficiency disease.

Malnutrition specifically undernutrition is a major problem in India. It is commonly seen in rural areas and in preschool children. Many of the mild or moderate cases go unnoticed. Incidence of malnutrition varies widely between different states, social groups, age and gender.

Poor **maternal nutrition** is frequently seen in India. It is considered to be one of the risk factors for low birth weight (weight less than 2500 g immediately after birth). **Low birth weight** has been considered nutritional indicator to assess improvement in health status of the population.

Vitamin A deficiency (xeropthalmia) if untreated by supplementing vitamin A in diet or by injection can lead to blindness. It is called **nutritional blindness**. It is a major nutritional problem in India commonly seen in preschool children. According to the recent data, prevalence of vitamin A deficiency is more common in the southern states of India. It may be related to dietary factor.

Iron, folate and vitamin B_{12} deficiency causes **nutritional anaemia**. Most common cause of nutritional anaemia is iron deficiency anaemia. Incidence of iron deficiency anaemia is seen more in women, young children and in rural areas.

Deficiency of iodine causes **goitre**. It can also give rise to a wide spectrum of disorders starting from womb till tomb. Iodine deficiency is a health problem not only in the "Himalayan goitre belt," but also in Madhya Pradesh, Gujarat and Maharashtra.

Fluorosis is caused due to excess fluorine in drinking water. Endemic fluorosis is also a major health problem in certain regions of the Andhra Pradesh, Punjab and Karnataka.

Lathyrism is due to consumption of **"Khesari dal"** in large quantity over a period of time. It is due to the toxin **β–oxalyl amino alanine**. Neurolathyrism is prevalent in parts of Madhya Pradesh, Uttar Pradesh and Bihar.

NATIONAL NUTRITIONAL POLICY

The national nutritional policy (NNP) was adopted in the year 1993 by Government of India. NNP aims

(1) To find group of people who need immediate action to improve their nutritional status
(2) To consider nutritional aspects in various policies of the government.

It stresses the importance of direct nutritional intervention such as short-term measures for groups of people who are vulnerable.

The goals of National Nutritional Policy

To decrease:

- The moderate and severe malnutrition in pre-schoolchildren by 50%
- Chronic undernutrition and stunted growth in children
- Incidence of low birth weight by <10%
- Prevent blindness due to vitamin A deficiency
- Decrease the incidence of iron deficiency anaemia in pregnant women by 25%

- Reduce the incidence of iodine deficiency disorder to 10% by iodization of salt

To increase

- The annual production of food grains

To improve

- The appropriate diet and healthy lifestyle
- Household food security by poverty alleviation programme
- Geriatric nutrition

NNP has identified different areas like food production, education, food supply, rural development, healthcare, women and child development, people with need for special nutritional care and monitoring.

1. Direct short-term intervention is to provide nutritional care to vulnerable group by
   - Expanding the network of organizations involved in nutritional intervention, e.g., Integrated Child Development Scheme, Oral Rehydration Therapy, etc. so as to improve the child survival and malnutrition.
   - Providing nutrition and health education
   - Providing through ICDS all support to the girls in adolescence and pregnant women. Improving the growth monitoring of the child (0 to 3 years) by involvement of the mother.

Direct short-term intervention also includes fortification of essential foods, control of deficiency of micronutrients among the vulnerable groups and making low cost nutritional food popular.

2. Indirect long-term intervention for institutional and structural changes includes improving
   - The production of nutritionally rich foods such as pulses, oilseeds, vegetables, fruits, meat, fish, etc. This will give food security and improve the pattern of diet.
   - The purchasing power by actively implementing the poverty alleviation programmes.
   - The public distribution system for equal distribution of food in all seasons.

Indirect long-term intervention also includes:

- Implementing land reforms
- Providing basic health and nutritional knowledge, improving the status of the women and ensuring the community participation
- Evolving an effective system for nutritional surveillance, information, communication and preventing food adulteration
- Monitoring various nutritional programme
- Implementation of minimum wages
- Research about various aspects of nutrition

National Plan of Action in Nutrition (NPAN) was prepared in 1995. It mainly aims to decrease the various forms of malnutrition and to achieve an optimal state of nutrition for the people. Government of India has initiated several programmes in nutrition to control and prevent major nutritional problems.

Some of the programmes designed to improve the overall nutritional status of the individual are

1. Applied nutritional programme
2. Supplementary feeding programme
3. Midday meal programme for schoolchildren

Programme to overcome the specific deficiency diseases are:

1. National goitre control programme
2. Vitamin A prophylaxis programme
3. Iron and folate distribution to expectant mothers

Programmes aimed to improve the nutrition of the population are:

1. Integrated child development scheme
2. India population project.

Applied Nutritional Programme

It was started in 1963 with an aim to increase the production of nutritious foods such as fruits, vegetables, eggs, fish, meat, etc. It also aims at increasing the consumption of nutritious food by the vulnerable group (expectant and nursing mothers and children).

Supplementary Feeding Programme

Supplementary feeding programme benefitted the children between 6 months and 3 years of age. Supplementary feeding programme includes on the spot feeding cooked food and once a week take home ready to eat preparations. On the spot feeding programme benefit the children because targeted population gets the additional nutrition. Take home ready to eat preparations are usually shared by all the family members and the targeted population gets less benefit.

Midday Meal Programme

Midday meal programme, also called **school lunch programme,** is operational since 1961. This programme attracts more children to get admitted to the school and decreases the school dropouts. This helps to improve the literacy rate.

Midday meal programme is under the Ministry of Education. It should supply at least one-third of the total energy requirement of the child. Midday meal should not be a substitute to the home diet but supplementary to it. Locally available food when used, it reduces the cost and is more acceptable. Menu need to be changed frequently to avoid monotony.

The National Institute of Nutrition, Hyderabad has prepared recipes for the preparation of meals acceptable by children in north as well as south India.

Integrated Child Development Scheme (ICDS)

It was started in 1975. The beneficiaries are pre-schoolchildren below the age of 6 years, adolescent girls between 11 and 18 years, pregnant and lactating mothers. It includes supplementary nutrition, vitamin A prophylaxis, iron and folic acid distribution.

At the village level, the services are delivered to the target population by Anganwadi workers. Anganwadi unit covers a population of about 1000. The Anganwadi workers are helped by a network of Mahila Mandals. Field supervision of the Anganwadi workers is done by the Child Development Project Officer (CDPO).

National Goitre Control Programme

In 1962, the Government of India launched the national goitre control programme in the Himalayan region. It was launched with the objective of (1) identifying the goitre endemic area (2) to supply iodinated salt and to assess the impact of this iodine supplementation over a period of time.

The problem of iodine deficiency disorder and goitre are more widespread. Nearly 145 million people live in the known goitre endemic areas. The programme called **iodine deficiency disorder control programme** was launched in 1986 to replace the edible salt with iodized salt by 1992.

Vitamin A Prophylaxis Programme

Vitamin A prophylaxis programme was launched in 1970 by the Ministry of Health and Family Welfare on the basis of technology developed by National Institute of Nutrition (NIN), Hyderabad.

A single massive dose of vitamin A (2,00,000 IU or 110 mg of retinol palmitate) is given orally to all pre-schoolchildren in the community once every six months. Evaluation of the programme has shown that there is significant reduction in the vitamin A deficiency.

Iron and Folate Distribution to Expectant Mothers

Government of India during the fourth five year plan launched a National Programme to prevent nutritional anaemia. To prevent nutritional anaemia, iron and folic acid tablets are distributed to all pregnant women and young children between 1 and 12 years. Mother and Child Health Centres in urban areas and Primary Health Centres in rural areas are involved in the implementation of this programme.

VARIOUS FACTORS THAT AFFECT THE NUTRITIONAL STATUS

Culture of the family and the society have great influence on the nutritional status. Food habits of the population are mainly influenced by the culture. Influence of the culture on food and nutrition of the population may vary from region to region and country to country. Food habits of the population are mainly shaped at the family level and passes from one generation to another. It is the most deep rooted and oldest aspect of the culture. In the population, customs and beliefs mainly affect the nutrition of the vulnerable groups (infants, expectant and lactating women, etc.)

Socioeconomic status of the family has great impact on the nutritional status of the family members. Malnutrition is the outcome of **poverty, ignorance, insufficient education lack of knowledge about nutrition, large family size, inadequate hygiene and sanitary environment**.

Religion also has great influence on the food habits. The religious compulsion prevents people from consuming some nutritious food even when they are easily available. **Cooking practices** and **child rearing practices** vary widely from region to region and influence the nutritional status.

FOOD FADDISM

Optimum nutrition is absolutely necessary for the all-round development of an individual. The food habits of an individual are affected by the social customs, educational, economical and dietary habits of the family. Lack of knowledge and ignorance about nutrition, false beliefs regarding the food which affect the nutrition and in turn the health of the individual. The traditional beliefs, social customs, personal likes and dislikes of an individual play an important role in nutrition and it is called 'food fads'. Following are some of the examples.

1. Feeding colostrum (yellow liquid secreted in small amount rich in vitamin A) to new-born is avoided thinking that it is not clean.
2. Some families avoid feeding pulses to child during weaning thinking that it produces gas.
3. Some think that use of garlic helps to increase the flow of milk.
4. In India, food is described as hot and cold food. Hot foods (meat, fish, papaya, etc.) are considered suitable during rainy seasons. Curds, buttermilk, fruit juice, etc. are considered cold food suitable for summer season.
5. In some parts of the country, people think that feeding children with buffalo milk will affect their intelligence.
6. Some people believe that eating fish and milk together in a meal will lead to development of white spots on the face and chest.

PRODUCTION OF FOOD

To increase the availability of food to all, food production should be increased. It also requires an effective system of distribution of food in the society to help prevent malnutrition. Scarcity of food as a factor responsible for malnutrition may be true at the family level. Scarcity of food may not be a factor for malnutrition in some of the countries where malnutrition is common. It is the result of uneven distribution of the food. By using the best available techniques, food production can be increased.

QUESTIONS

Long Essay

1. Discuss the nutritional problems in India. Describe the national programmes implemented to tackle problems of malnutrition among our population.
2. Discuss national nutritional policy.

Short Essay

Write short notes on the following:

1. Midday meal programme.
2. Vitamin A prophylaxis schedule.
3. Iron and folate distribution to the expectant mothers.
4. Various factors that affect the nutritional status.
5. Food faddism.
6. Medicinal value of food.
7. ICDS.

23

Toxins in Food; Medical Value of Food; Food Adulteration; Food Standards; Cookery Rules and Preservation of Nutrients

TOXINS IN FOOD

Food which we consume may not be safe. It might contain various toxic substances producing deleterious effects on health. Food consumed especially, of plant origin, might naturally contain certain toxic substances (anti-nutritional principles)

For example, Goitrogens found in cabbage, mustard, etc.
Lathyrogen found in *Lathyrus sativus*
Protease inhibitor present in legumes

When these foods are consumed in excess for a long period of time, it produces toxic effects and is responsible for deterioration of health.

(a) Some toxic substances can enter the food as contaminants, e.g., various seeds containing toxic substances can get access to grains.

(b) Various bacteria and moulds can contaminate the food if there is high moisture content. They may cause food-borne diseases like typhoid, cholera, diarrhoea, food poisoning, etc. Bacteria can grow in food and produce toxins which are harmful when consumed. Some other bacteria after entering the gastrointestinal tract along with the food will grow there and produce diseases. Some of the fungus or moulds which grow on food might produce toxins which prove to be carcinogenic if consumed for a long period of time.

Food might get contaminated with various parasites, e.g., amoeba, roundworm, tapeworm or liver fluke. Even drinking water gets contaminated with hepatitis virus and can cause viral hepatitis in a person who consumes contaminated water.

(c) Various chemicals and metals contaminate the food and can have toxic effect on the health. Metals such as arsenic, mercury, lead, antimony, cadmium can get access to food and in high concentration produce toxic effects.

(d) Some of the packing materials can contaminate the food, e.g., metal, plastic or paper pieces. If the food is acidic in nature like certain fruit juices, it can dissolve metal and their level will increase.

(e) Residues of pesticides and chemicals used during cultivation or processing of the food can contaminate it, e.g., feeds of the animals might contain antibiotics which might be present in the meat or milk derived from it. Pesticides used to control the pests during farming or chemicals used during processing or storage of food can contaminate it.

(f) During roasting polycyclic, hydrocarbons can be produced from fatty acids. Nitrosamines are formed during fermentation. Unpermitted processing technique like artificial ripening using calcium carbide, lime, etc. can add toxic substances to the food.

(g) Pollution of the environment by the automobile exhaust or industrial waste contaminate the water and the food.

Lathyrism

It is prevalent in those who consume large amount of **lathyrus sativa.** It is commonly seen in Uttar Pradesh, Bihar, Madhya Pradesh and Rajasthan. Lathyrism is characterized by spastic paralysis of the lower limbs which develops gradually due to consumption of diet containing lathyrus sativus. If the diet rich in (more than 30%) lathyrus sativus is consumed for six months, it can cause the disease. Toxic effects of β- N- oxalye amino –L- alanine (BOAA) present in lathyrus sativa is responsible for the disease.

The toxin can be removed from the lathyrus seed

(a) By soaking it in boiling water and draining the water

(b) By parboiling: Seeds are first soaked in luke warm water and later steaming it for 15 minutes.

MEDICINAL VALUE OF FOOD

Many plants and plant products which are used as spices and condiments in our food have got medicinal value. **Coriander (Dhania)** has got a role in reducing blood cholesterol level, it also helps in digestion and it is a good diuretic. Coriander seeds check excessive menstrual flow.

Clove when fried and mixed with honey, helps to control vomiting. It also helps to reduce toothache and has got anaesthetic effect on the stomach and gullet. Chewing of clove with the common salt reduces expectoration.

Cumin seed (Jeera) is used against indigestion, diarrhoea, flatulence and morning sickness. It has a role in reliving sleeplessness. It is also useful in common cold and fever.

Cinnamon (Dalchini) is used in sore throat, vomiting, diarrhoea, etc. It helps to relieve flatulence and also indigestion. It can be used in the treatment of black heads and acne.

Asafoetida (Hing) reduces distension of abdomen by expelling wind. Relieves toothache. It is used in painful and excessive menstruation and leucorrhoea.

Cardamom (Chhoti Elaichi) can be used to relieve gas and heartburn. It is used in indigestion and to relieve hiccups.

FOOD ADULTERATION

Food adulteration refers to mixing of toxic or injurious substance with the food, concealing the quality of the food, sale of decomposed food, giving false label, etc. to get more profit. Food adulteration is an age old problem and different parts of the country have got different practices to adulterate food. Some common adulterated foods are:

1. Cheese is adulterated with vanaspathi/animal fat
2. Cereals like rice and wheat are adulterated with stones, sand, etc.
3. Milk is adulterated by addition of water, starch, paper pulp, etc. or by removal of cream
4. Toor dal is adulterated with Khesari dal
5. Black pepper is adulterated with dried papaya seeds
6. Chilli powder is adulterated with sawdust, brick powder
7. Coffee powder is adulterated with date husk, tamarind husk etc.
8. Tea powder is adulterated with used tea powder, black gram husk, tamarind seed powder
9. Haldi powder (turmeric powder) adulterated with lead chromate powder

Food adulteration is a social evil. To control and to prevent food adulteration, consumers should be alert and should rise up against the people involved. To prevent food adulteration the Government of India passed the Prevention of Food Adulteration Act, 1954.

Prevention of Food Adulteration Act, 1954

Prevention of Food Adulteration Act was amended in 1964, 1976 and later in 1986.

The Central Committee for Food Standards has framed the rules,. The State Government is authorized to appoint food inspectors and analysts. A chain of food laboratories are there to test the food samples drawn from the suspected food. Any food that is not conforming to the minimum standards is considered adulterated.

For a proven case of adulteration under this act punishment not less than 6 months of imprisonment with a minimum fine of Rs. 1000 is imposed. If adulteration leads to injury to the health of the consumer severity of punishment also increases. It may go up to life imprisonment and a fine minimum of Rs. 5000.

Food Standards

Food with the stated nutritional value which is hygienic, free from toxins, contaminants is the one which helps to maintain or promote healthy growth and reproduction. To be healthy food there should be certain guidelines to monitor the quality of the food. **Food standards** defined by various regulating bodies help to maintain the quality of the food.

Prevention of Food Adulteration (PFA) Standards

Under the Prevention of Food Adulteration Act, 1954, certain guidelines have been established to maintain minimum quality of the food attainable under Indian conditions. The guidelines of this Act are to prevent adulteration, or contamination of the food and to ensure that the standard of the food is suitable for safe consumption.

It is mandatory to conform to the standards mentioned in this act and any aberration from these standards will invite punishment and fine.

Codex Alimentarius Commission

It was created in 1963 by WHO and FAO to develop food standards. It develops various guidelines, standards, codes of practice and other recommendations to protect the consumer from consumption of unsafe food. It also ensures fair food trade, coordination of all food standard works undertaken by various organizations. Based on the Codex Alimentarius, food standards in India are derived.

AGMARK Standards

AGMARK is an acronym for Agricultural Marketing. Standards are set by the Directorate of Marketing and Inspection of the Government of India. This organization certifies the food products for their quality. The certification of the food products by AGMARK is voluntary, but it is made compulsory for certain food items which are exported.

Indian Standard Institutions (ISI) Standards

Bureau of Indian Standards has defined the standards which the food needs to meet. Certification by ISI standards is voluntary. Certification by ISI or AGMARK shows the excellent quality of the food and consumer is assured of good quality of food.

FOOD PRESERVATION

Food is liable to get contaminated and spoiled at any point from its production till consumption. Deterioration of the food might lead to accumulation of toxic substances and loss of nutritive value making food useless or harmful for consumption.

Human beings by way of long-term experience and research have developed various techniques to process and preserve the food for longer time. Food processing is done mainly to remove all harmful, toxic, indigestible hard portions of the harvest by milling, peeling, polishing, filtration, pneumatic separation, etc. Food processing also helps to separate the commodities with unusual appearance or infected with spoilage organisms. It includes various processes used to improve the taste, appearance, texture, flavour and nutritive value of food.

Ultimately by various processes like dehydration, pasteurization, fermentation, salting, irradiation and freezing shelf life of many perishable food is improved.

Food preservation helps to

- Keep food in good shape so that decay, spoilage of food can be prevented
- Nutritive value of the food is retained
- Lifespan of the food is prolonged
- Make seasonal food available throughout the year
- Provide food when there is shortage of food
- Avoid any wastage and save time in procurement

Principles of Preservation of Food

(a) Airtight packing can be used to keep organisms out from the food (asepsis).
(b) Certain enzymes present in the food can be inactivated by mild heat treatment (blanching)
(c) Rancidity of oils and fats can be prevented by using antioxidants (prevent oxidation)
(d) Liquid foods can be filtered through bacteria-proof filters (filtration)
(e) UV rays can be used to sterilize the food, knives, etc. Ionizing radiation is also used to preserve fruits and vegetables (irradiation).

Methods of Preservation of Food

Methods of preservation are of two types

1. Bacteriostatic method
2. Bactericidal method

1. BACTERIOSTATIC METHODS

Methods of preservation of food by preventing the growth of the microorganisms is called **bacteriostatic method.**

(a) Dehydrating

- Microorganisms require moisture for their growth. When moisture is removed by dehydrating the food, growth of microorganism can be retarded.

(b) Salting

- Salt can be used as a preservative it absorbs moisture and prevents the growth of microorganisms, e.g. dried fish.

(c) Topping foods with a layer of oil

- Topping certain foods with a layer of oil will prevent the growth of moulds, microorganisms and yeast, e.g., layer of oil placed on pickle.

(d) Coating

- Coating of certain foods with specific chemical, helps to preserve it, e.g., coating of egg with silicate closes the pores and preserves the egg.

(e) Chemicals

- Chemicals can be used as preservatives, e.g., benzoic acid, vinegar, etc.

(f) Chilling

- By chilling the food at lower temperature food can be stored. When temperature is lowered the activity of the enzymes as well as the growth of the microorganisms decreases. It helps to preserve the food for a long time.

2. BACTERIOCIDAL METHOD

Method of preservation of the food by killing the microorganisms is called **bactericidal method**.

(a) Heating

- Heating will kill almost all the microorganisms and denaturation of various enzyme proteins help to preserve food for a long time.
  Milk is made safe for consumption and can be preserved over a period of time in household by pasteurization. **Pasteurization** is a process of preservation of milk by heating the milk at a particular temperature for specific period of time to destroy the pathogens present in it. It preserves the colour, smell, flavour, taste and nutritive value of the milk.

Pasteurization can be done by

1. **Holder method**

   Milk is heated to 65°C (145 to 150°F). It is kept at that temperature for 30 minutes and suddenly cooled to a temperature below 5°C. It is called Holding process which is slow and suitable for rural communities. It is a British Method.

2. **High temperature and short time process (HTST) method**

   Milk is heated to 72°C and is maintained at that temperature for 15 seconds and later cooled to less than 5°C. It is a rapid method, widely used and recommended for urban communities. It is an American method.

3. **Ultra high temperature (UHT) method**

   Milk is heated to 88°C under normal pressure. Then temperature is raised to 125°C under pressure for few seconds and rapidly cooled and packed quickly.

(b) Canning

- Food which is heated, is placed in sterile cans and sealed to make it airtight.

(c) Irradiation

- Irradiation of the food with ultraviolet rays and even with ionizing radiation helps to preserve it.

FOOD ADDITIVES

Food additives are non-food substances added to food in small quantities to improve the flavour, storage property, appearance or texture, or any substance which gets incorporated into the food during the course of packing, storage and transportation which is legally permitted.

Food Additives can be Categorized into Two Different Groups

1. *Direct additives or intentional additives*

They are deliberately added to food

(a) To maintain palatability
Benzoates, sulphites, citric acid, butylated hydroxyl toluene, etc.

(b) Artificial flavours
Constitute the largest single group of food additives. Monosodium glutamate, following agents like vanilla essence used in most of the Asian foods. Monosodium glutamate was in news because of the symptoms it caused in about 1 to 2% of the population called **Chinese restaurant syndrome**. These sensitive individuals would have burning sensation in the chest and facial flushing or throbbing headache.

(c) Nutrients as additives
Nutrients may be added as antioxidants (vitamin C and vitamin E) or as colouring agents (β – carotene)

(d) To stabilize the food, e.g., starch, gum, dextrin, etc.

(e) To increase the palatability, e.g., benzoic acid, citric acid

(f) To increase the sweetness, e.g., aspartame

2. *Indirect food additives or incidental additives*

Substances which find their way to food, during the process of harvesting, processing, production or packing.

Examples for indirect food additives are small bits of plastic, paper, tin, etc. added during packing. Use of bleached papers for packing, paper plates, milk cartoons, etc. will add dioxins (compound formed when wood pulp is treated with chlorine).

The hormones, bovine growth hormone produced artificially is given to cows to increase the yield of meat and milk. Traces of bovine growth hormone remains in food derived from cows treated with this hormone. But it was shown that animal growth hormone does not have any effect in human body since structurally it is different from human growth hormone.

Growth hormone treated cows are more prone for infection of the udder. To treat this, a lot of antibiotics may be used. These antibiotics make their way to milk or organisms which have developed antibiotic resistance can find their way to the milk and meat.

When radiation is used to sterilize certain foods, it can add certain radiolytic products to the food. Probably many of the radiolytic products may not be hazardous to the health.

COOKING

Most of the food we eat is cooked. Even though there is no clear evidence as to when cooking was invented, some of the primatologists suggest that it was as far as 1.8 to 2.3 million years ago. But there is no consensus among the researchers regarding when cooking was invented.

Cooking is a process by which heat is applied to food and various ingredients are combined together so that flavour, texture, appearance and chemical properties of the ingredients can change. It makes the food more acceptable, palatable and easily digestible. The technique of cooking and various ingredients used vary widely across the globe. It is mainly influenced by the cultural, environmental and economical factors.

Effects of Cooking

1. Makes food soft. It can be easily chewed and swallowed.
2. Makes the food more appetising.
3. Cooking of meat increases its digestibility but overcooking may cause shrinkage of coagulated protein and decrease its digestibility. Boiling of egg will inactivate harmful avidin present in it. Collagen present in meat is converted to gelatin-like protein with high digestibility.
4. Cooking of vegetables increases its water content as well as digestibility. Cooking breaks the cellulose wall and release starch which can be easily digested. The trypsin inhibitor present in soya bean is inactivated and its biological value is increased.
5. Increases the texture, flavour, colour and taste of the food.
6. Helps to make variety of dishes and variety in the meal.
7. Helps to preserve the food longer by killing the spoilage organisms. It helps to make the food safe and sterile.
8. Vitamins are lost in the process of cooking. Specifically, thiamine and vitamin C is destroyed. When food is fried in open pan, vitamin A is lost. Addition of tamarind (acidic material) helps to preserve vitamins present in the vegetables.
9. Repeated washing and cooking in excess of water and discarding the excess water will lead to loss of vitamins and minerals.

Different Methods of Cooking

Foods can be cooked by different methods as follows:

1. Cooking by moist heat
2. Cooking by dry heat
3. Cooking by frying in oil, ghee, etc.

Cooking in moist heat

Food is cooked in boiling water, heated water or in the steam. Cooking by moist heat may be in three different ways.

(a) **Boiling:** Food is covered with adequate amount of water and is heated to boiling. Before we cook food by boiling we need to wash the food thoroughly. Water should be boiled first and then add the food to boiling water. Food should be covered with sufficient water. Cover the mouth of the vessel used to cook with a lid and avoid the drying of water. Food is cooked fast when lid is placed in position. It is not advisable to boil the food for longer than required.

Water used for cooking contains various nutrients. To preserve these nutrients, use this water to prepare some other dish or use it in the preparation of gravy.

(b) Simmering: Cooking by gentle heating for long time using little water and low flame once the boiling starts. Suitable for cooking hard foods like dal, meat, etc. While cooking by this method we need not constantly monitor the process and there is less chance that the food gets burnt.

Most of the foods are boiled before simmering. The disadvantage of this method is that it is slow and time-consuming. Because food is exposed to heat for long nutrients like vitamin C may be lost.

(c) Steaming: Food cooked by using water vapour with or without pressure is called steamed food and the cooking process is called steaming. It is a healthy cooking technique. Steaming reduces the time required for cooking. Food does not come in direct contact with the water, so all the water soluble vitamins and various nutrients are retained in the food. Food becomes easily digestible and nutritious. When food has to be cooked in bulk, steaming is not a suitable method.

Steam can be used under pressure by using pressure cookers. Food is cooked in short duration at high temperature. It is suitable to cook large quantity of food. By this method of cooking, all bacteria present in the food are killed and the food becomes safer for consumption. Hard food can be easily made soft and easily digestible. Disadvantages of pressure cooking is that food cooked is not as tasty as steamed food and little carelessness can lead to overcooking.

Cooking by dry heat

Hot air is used to cook food. There are three different methods of cooking food by dry heat.

(a) Baking: Food is kept inside a closed box called oven. Air present inside the box is heated by fire or electricity. Food kept inside the oven is cooked by the hot air which circulates around the food. Bread, cake, biscuits, etc. are cooked by this method. Baking is a slow method of cooking.

(b) Roasting: Food is kept directly over the tava, hot sand or fire and cooked

(c) Grilling: Food is cooked by keeping it directly over the fire supported on an iron grill or on the electrically heated grills. Food comes in direct contact with the flame.

Few advantages of cooking by hot air

- Large quantity of food can be cooked uniformly.
- Food becomes tasty and light and digestible.

Disadvantage is that a few of the vitamins are lost as food comes in direct contact with the heat.

Frying

Frying is a quick method of cooking food by using ghee or oil. Frying the food makes it more appetizing and long lasting. While frying, many pieces of food should not be put at the same time to avoid lowering of temperature. Food should be fried until it becomes golden brown in colour on either side.

Frying may be

(a) Shallow frying by using little oil or fat.
(b) Deep frying by immersing the food in oil or ghee

Braising is cooking by searing or browning in fat or oil and then immersing in liquid (e.g., water) on low heat in a covered vessel. It is a way to cook economically tough cuts of meat to make it tender and tasty.

Cooking by using Microwave: It is a new method of cooking by using microwave radiation. Microwave radiation generates heat in the food by rapidly vibrating water in it. Food cooked by this method retains its flavour. Disadvantage is that the equipment is costly and the food looks unattractive since it does not brown the food.

Dredging is cooking by coating wet or moist food with dry ingredients like flour before cooking, e.g., frying fish after covering it with rava or rice flour.

Marinating is a method of cooking to make food (especially meat) flavourful and tender by soaking in seasoned liquid.

Methods to Conserve Nutrients

Nutrients present in food can be preserved by the following guidelines.

1. By using under milled or hand pound rice.
2. By sprouting the pulses
3. By keeping the food in clean, dry and airtight containers
4. By washing the food using minimum water
5. By cutting vegetables just before cooking into large pieces preserves vitamins and minerals
6. Boil water before you cook the food.
7. Do not use baking soda since vitamins are lost at alkaline pH
8. Cook potatoes and sweet potatoes in pressure cooker without peeling.
9. Boil vegetables for not more than 15 minutes.
10. Cook eggs at temperature just below the boiling point of water.
11. Add salt at the end to avoid loss of vitamins.
12. Addition of little weak acidic ingredients such as tamarind, vinegar, lime juice, etc. helps to preserve the nutrients.
13. Steaming is preferable to boiling.
14. Frying of food in oil leads to loss of nutrients.
15. Avoid repeated heating at frequent intervals.
16. Eat food when it is hot.

FOOD FORTIFICATION

Process of maintaining or improving the quality of food of a group, a population or a community, by addition of small quantity of nutrients is called food fortification. Food fortification helps to control or to prevent some of the disorders due to nutritional deficiencies.

Food which is consumed regularly by the population is used as vehicle for fortification. The fortification should not increase the cost of the food beyond the reach of the population. Nutrients added must be able to provide their requirement. By adding the nutrients there should not be any change in the taste, colour, smell, etc. of the food. The effectiveness of the fortification on the population should be constantly monitored.

Fortification commonly done are:

1. Fortification of common salt with potassium iodide.
2. Milk is fortified with vitamins A and D
3. Vegetable fat fortified with vitamins A and D
4. Fortification of common salt with iron salt to combat nutritional anaemia
5. Fluorination of water to prevent dental caries

QUESTIONS

Long Essay

1. Explain various methods of food preservation.

Short Essay

Write short notes on the following:

1. Lathyrism
2. Food toxins
3. Food adulteration
4. Prevention of Food Adulteration Act
5. Food standards
6. Various methods of food storage
7. Pasteurization
8. Food additives and its principles
9. Effect of cooking on food
10. Methods of cooking
11. Cooking by moist heat
12. Cooking by dry heat
13. Food fortification

24

Carbohydrates, Lipids, Proteins and Malnutrition

INTRODUCTION

Cells present in human body require energy to sustain life. They are not primary producers of carbohydrate. They depend upon the carbohydrates produced by the plants as a major source of energy. Fats are concentrated form of energy stored in the adipose tissue. Tissues constantly undergo turnover, i.e., continuous synthesis and breakdown. So the constant supply of basic building materials such as monosaccharides, fatty acids and amino acids are required. In addition, cells also require accessory food factors to utilize the nutrients and minerals. The requirement of these substances varies depending upon the activity of the individual and different phases of life.

CARBOHYDRATES

Plants are primary producers of carbohydrate. Diet contains a large amount of a variety of carbohydrates. The major dietary carbohydrate is starch. Diet also contains oligosaccharides (sucrose, lactose, etc.) and monosaccharides (glucose, fructose, etc.).

Carbohydrates present in the diet (starch, lactose, sucrose, maltose, etc) are digested in the gastrointestinal tract (GIT) leading to formation of monosaccharides (glucose, galactose, fructose). These monosaccharides are absorbed. Absorbed carbohydrate is mainly used for energy purpose and excess amount is stored as glycogen or converted to fat, cholesterol and other biomolecules (e.g., non-essential amino acids).

Carbohydrates have to provide about 70 to 75% of the total calories required by an individual. 55 to 65% of the total calorie requirement should come from carbohydrates which are easily digestible polysaccharides like starch. About 10 to 15% of the calorie requirement can be derived from monosaccharides and disaccharides. Diet should contain more than 40 g of undigestible carbohydrates called dietary fibres. Starch is the major carbohydrate present in food. Cereals, pulses and tubers are rich in starch. Starch is digested by an amylase present in saliva, pancreatic juice and by various other enzymes present on the lining epithelial cells of the small intestine.

Dietary restriction of carbohydrate adversely affects the lipid and protein metabolism even when the calorie intake is adequate. Body then derives energy from lipids. When diet deficient in carbohydrate is taken, fat (triglyceride) stored in the adipose tissue is mobilized. Blood level of fatty acid is increased and proportionate increase in the oxidation of fatty acid in the liver. End product of beta oxidation of fatty acid is acetyl-CoA.

When excess amount of acetyl-CoA is formed in the mitochondria of liver cell, it enters into ketogenic pathway. This leads to formation of ketone bodies (acetoacetate, beta hydroxybutyrate and acetone). Under this condition fatty acid synthesis is inhibited.

When the diet is rich in carbohydrate body derives energy mainly from oxidation of carbohydrates. Excess of carbohydrates are **converted to glycogen** and stored in the body. When there is surplus amount of carbohydrates (e.g., fructose, glucose, etc.) in the diet they can be converted to fatty acids. Fatty acids are esterified to glycerol to form triacylglycerol (triglyceride, fat). Triacylglycerol is stored in the adipocytes of adipose tissue. Excess carbohydrate ingestion can increase the adipose tissue mass and cause **obesity.**

The **intermediates of carbohydrate metabolism** (e.g., α-ketoglutarate, oxaloacetate and pyruvate) are **converted to nonessential amino acids** by transamination reaction. Sufficient carbohydrate supply in the diet decreases the breakdown of proteins for energy purpose. This is called the **protein sparing action of carbohydrate**.

The **vitamin thiamine** is required for the metabolism of carbohydrates. As the carbohydrate metabolism is increased, the requirement of thiamine will also increase.

Dietary Fibres

Carbohydrates which are not digested in the gastrointestinal tract constitute **dietary fibres**, e.g., cellulose, hemicelluloses, pectin, etc.

1. Dietary fibres increase the bulk of the intestinal content and retain water. This helps to **increase the bowel motility and prevent constipation**.
2. Dietary fibres help to **reduce the blood level of cholesterol and improve the glucose tolerance.**
3. It reduces the risk of **colonal cancer.**
4. It has **got satiety value** since it increases the bulk of the food consumed.
5. It helps to **reduce weight.**

PROTEIN

Proteins are polymers of amino acids. Nitrogen is the characteristic element present in protein which is present to an extent of 16%.

Nitrogen Balance

When the dietary intake of protein nitrogen is equal to the daily loss of nitrogen from the body individual is said to be in **nitrogen balance**. If excretion of nitrogen exceeds intake, individual is said to be in a state of **negative nitrogen balance**.

Negative nitrogen balance is seen in various debilitating diseases, protein deficiency or deficiency of essential amino acids in the dietary protein, prolonged starvation, etc.

When intake of protein is more than the excretion the individual is said to be in a state of **positive nitrogen balance**. Positive nitrogen balance is seen during active growth, pregnancy, convalescence, etc.

Amino acid composition of protein and its digestibility determines the quality of the protein. Proteins containing balanced amount of all the essential amino acids and which is easily digestible is a good quality protein. If one or more essential amino acids are deficient in the dietary protein, body cannot efficiently use the amino acids of these proteins. These amino acids which are lacking in the dietary protein are called **limiting amino acids** (because they limit the efficient use of the other amino acids of the dietary protein).

Lysine is the limiting amino acid in wheat and tryptophan is the limiting amino acid in sorghum.

The problem of limiting amino acids in the dietary proteins can be solved by taking mixture of proteins, e.g., cereals like rice, wheat, etc. are deficient in lysine, on the other hand, pulses are deficient in methionine. Pulses and cereals should supply proteins in the ratio of 1:4 to 1:5 ratios. Mixing these two in the diet helps to improve the nutrition of the individual by providing adequate amount of all essential amino acids

Proteins which are easily digestible and contain adequate amount of all the essential amino acids are called **complete proteins** or **"first class" proteins**. Whole egg, milk protein especially lactalbumin has got highest amount of essential amino acids in proper proportion. The next group of proteins with high nutritive value are meat, fish, yeast and soya bean.

Proteins present in cereals, legumes, etc. are poor in one or more amino acids, so they are **incomplete proteins** and have poor nutritive value.

Tyrosine is produced from phenylalanine and cysteine is synthesized from methionine. If tyrosine and cysteine content of the dietary proteins are less, requirement of phenylalnine and methionine increases.

Quality of the Protein can be Measured by Different Indicators Like

1. Digestibility coefficient of the protein
2. Protein efficiency ratio
3. Chemical score (index of essential amino acid content of the protein)
4. Biological value of the protein
5. Net protein utilization

Digestibility coefficient

Digestibility coefficient gives an idea about the percentage of the food absorbed. Normally about 92% of the protein is absorbed from the food. Cooking of the protein increases its digestibility coefficient because it breaks the cell wall so that proteolytic enzymes get access to the proteins present inside the cell.

Proteins present in the legumes and seeds have got lower digestibility than meat, egg and milk proteins. Digestibility coefficient (DC) is calculated by the formula

$$DC = \frac{\text{N present in the food} - (\text{Faecal N} - {}^*\text{Metabolic N present in the faeces})}{\text{Nitrogen present in the food}}$$

*Metabolic nitrogen = Faecal N estimated after the animal is kept on protein free diet

Protein efficiency ratio

Protein efficiency ratio is the weight gain of the animal per gram of protein consumed. It is given by the ratio of weight increase to gram of protein consumed.

$$\text{Protein efficiency ratio} = \frac{\text{Weight increase (in gram)}}{\text{Gram of protein consumed}}$$

Chemical score

The essential amino acid content of a protein can be expressed as chemical score. Egg protein contains all essential amino acids and is considered reference protein and is given chemical score 100.

Chemical score is the ratio between the amount of limiting amino acid of the protein to the amount of the same amino acid in egg protein expressed as percentage.

E.g. Chemical score of

Egg – 100 ; Milk – 65; Meat – 70; Fish – 70; Rice – 60; Wheat – 42

Biological value (BV) of protein

Biological value is the percentage of the absorbed protein retained in the body. It is the ratio of the protein nitrogen retained in the body to the protein nitrogen absorbed over a specific time interval.

$$BV = \frac{\text{Nitrogen retained in the body}}{\text{Nitrogen absorbed}} \times 100$$

$$BV = \frac{N_2 \text{ present in the food} - (\text{Faecal } N_2 - {}^*\text{Metabolic } N_2 \text{ present in the faeces}) - (\text{Urine } N_2 - \text{Endogenous } N_2 \text{ present in the urine})}{N_2 \text{ present in the food} - (\text{Faecal } N_2 - \text{Metabolic } N_2 \text{ in the faeces})} \times 100$$

E.g. Biological value of Egg – 94; Mutton – 60; Fish– 76; Rice – 86; Wheat – 67.

Net protein utilization (NPU)

Net protein utilization is the percentage of food nitrogen retained in the body. NPU gives an idea about the quantity and quality of dietary protein.

$$\text{Net protein utilization} = \frac{\text{Retained nitrogen}}{\text{Protein intake}} \times 100$$

NPU is a better indicator of quality of protein than BV.

E.g. Net Protein Utilization of Egg – 91; Milk – 84; Meat – 76; Fish – 72; Rice – 57; Wheat – 47.

Carbohydrate and Fat are Protein Sparers

Dietary requirement of protein is reduced by the dietary carbohydrate and fat. The oxidation of carbohydrate and fat produce energy in various tissues which spares protein from catabolism. Intermediates of carbohydrate metabolism, e.g., pyruvate, α-ketoglutarate, etc. can be converted to amino acids. Dietary carbohydrate increases the blood glucose level and the necessity of gluconeogenesis from proteins is reduced. Carbohydrates are better than fat as protein sparers.

Requirement of Protein

Adults require about 1 gram of protein/kg body weight/day. An adult women requires about 45 g, adult man requires about 55 g, pregnant women needs 60 g and lactating women require 70 g of protein per day. Infants and children require 1.5 to 2.0 g/kg body weight. Protein requirement is increased if the dietary protein has low digestibility and biological value.

Of the daily protein supply about one-third (about 30%) to one half (about 50%) should be derived from animal proteins, e.g., egg, milk, meat, etc.

Protein Deficiency

Low protein diet over a period of time causes anaemia, hypoalbuminaemia, and oedema. In children protein deficiency causes 'Kwashiorkor'. (Refer page 265)

FAT

Dietary fat is mainly oxidized to provide energy. Fat has got higher calorific value than carbohydrate and protein. Neutral fat constitutes the largest fraction of the food lipids. Fats containing medium chain, short chain fatty acids and unsaturated fatty acids are well digested and absorbed when compared to long chain fatty acids and saturated fatty acids. Fat present in the diet is also responsible for satiety.

Under normal conditions fat provides 20 to 25% of calories of the diet. It should not exceed 30% of the total calories. When diet content of fat gives more than 30% of the total calories it may elevate the blood level of cholesterol especially in sedentary and moderately active individuals.

Fat present in the diet may be visible fat or invisible fat. Fats consumed as such are called **visible fats**, e.g., ghee, butter, vegetable oil, etc. Visible fat content of the diet should be about 5 to 15% of the total calories. Fat present in different components of the diet is called **invisible fat**, e.g., fat present in cereals, legumes, milk, meat, etc. Invisible fat should not exceed 10 to 20% of the total calories.

Dietary fat should contain adequate amount of polyunsaturated fatty acids. The polyunsaturated fatty acid content should be equal to about 3% of the total calories. Essential fatty acid requirement during lactation increases (about 6% of the total calories). Sunflower oil, safflower oil, groundnut oil, etc. are rich in essential fatty acids.

Linoleic acid, linolenic acid and arachidonic acid are essential fatty acids. Arachidonic acid can be produced from linolenic acid in human body. Essential fatty acids decrease the serum cholesterol. ω-3 polyunsaturated fatty acid (e.g. timnodonate and linoleate) are more useful in decreasing the serum cholesterol than ω-6 polyunsaturated fatty acids (e.g., linoleic acid and arachidonic acid)

ω-3 polyunsaturated fatty acid produces PGI3 which helps to prevent the platelet aggregation and reduce the risk of developing atherosclerosis. Essential fatty acid deficiency gives rise to **phrynoderma.**

MALNUTRITION

Undernutrition is the problem in developing countries especially affecting children. It may be due to

1. Insufficient intake of calories, or inadequate intake of essential nutrients like proteins
2. Inefficient absorption of nutrients (malabsorption)
3. Increase in the requirement or increased destruction or excretion of nutrients secondary to disease.

Classification

Malnutrition can classified into the following types

1. Kwashiorkor
2. Marasmic Kwashiorkor
3. Marasmus
4. Nutritional dwarfism
5. Underweight child

Kwashiorkor

Diet deficient in protein, with adequate calories supplied for over a period of time gives rise to **Kwashiorkor**. Kwashiorkor is seen in older children (one to five years of age). Kwashiorkor is seen especially in those infants who are weaned from mother's milk and fed with starchy diet (protein poor diet). In addition to diet poor in protein, other predisposing factors like respiratory infection, acute diarrhoea, etc. act as precipitating factors.

Serum albumin is less than 2 g/dL. Because of hypoalbuminaemia, colloidal osmotic pressure of the blood is decreased because of which oedema of the face and lower limbs can be seen in Kwashiorkor. Skin of the leg, buttocks and perineum shows patches of hypopigmentation, desquamation and ulceration. Hair is dry, and brown or reddish yellow in colour. Sometimes fatty liver is seen in Kwashiorkor.

The body weight of the child is between 80% and 60% compared to the standard weight for the age, and deficiency in weight for height is seen.

Marasmus

Marasmus is due to the deficiency of calories and protein over a period of time. It is usually seen in exclusively breast-fed infant of a malnourished mother due to the grossly reduced milk supply or inadequate supplementation with other foods. It can be seen in babies on artificial feeds insufficient in protein and calories. Marasmus is seen in babies less than one year of age. Diarrhoea is almost always present.

Child is grossly emaciated and there is growth retardation. Skin becomes thin and dry. Subcutaneous fat disappears and muscle wasting causes thinning of the limbs. Intestinal mucosa and pancreatic acini may be atrophied. This causes lack of intestinal and pancreatic enzymes, which causes impaired digestion giving rise to flatulence and diarrhoea. Body temperature is lowered. Oedema is not seen in marasmus.

The body weight of the child is less than 60% compared to the standard weight for the age, and gross deficiency in weight for height is seen.

Marasmic Kwashiorkor

Symptoms of Kwashiorkor and marasmus are seen depending upon the relative degrees of the calorie and protein deficiency.

The body weight of the child is less than 60% compared to the standard weight for the age, and gross deficiency in weight for height is seen. Oedema is present.

In nutritional dwarfism the body weight of the child is less than 60% when compared to the standard weight for the age. There is no oedema and deficiency in weight for height.

Underweight child has body weight 80% to 60% compared to the standard weight for the age. There is no oedema and deficiency in weight for height.

Obesity (Overnutrition)

It is a major problem found in developed countries. It is the disorder of excess accumulation of fat in the body. Overeating accompanied with reduced physical activity (energy intake is more than energy expenditure) usually seen in affluent society results in sustained deposition of fat and results in obesity. Obesity has become a global problem. It is difficult to measure the amount of fat present in the body directly. Body fat is determined indirectly by calculating body mass index using height and weight of the individual.

Body mass index is calculated from the height and weight of the individual. It is calculated by the formula

$$\textbf{BMI} = \textbf{(Weight in kilogram)/(Height in metres)}^2$$

Body mass index is indirect measure of body fat in most of the individuals (exceptions are athletes having large amount of muscle mass). **Normal BMI** is 19.5 – 25. Individuals who are **overweight** will have BMI >25 and up to 29.9. If BMI is more than 30 then the individual is said to be **obese**.

Obesity is a risk factor for various diseases. Health risk due to obesity depends on anatomic distribution of body fat. Accumulation of excess fat in the central abdominal area of the body is called **upper body ("android"; "apple-shaped") obesity**. This type of obesity is associated with higher risk for hypertension, insulin resistance; diabetes etc. If fat is accumulated in lower extremities, i.e., around hip and gluteal region is called **lower body ("gynoid"; "pear-shaped") obesity**. This type of obesity is usually seen in females and is relatively benign.

In obesity the number and size of the adipocytes in adipose tissue is increased. Obesity is often observed clustered in family. Genetic mechanism plays a role in obesity. When both parents are obese there is a greater chance (70 to 80%) of the child becoming obese.

Adipocytes store fat and it **acts as an endocrine cell** by secreting leptin, adeponectin and resistin. Leptin secretion is decreased by depletion of fat in the adipose tissue and is increased by deposition of fat. **Leptin increases the rate of metabolism and decreases the appetite.**

Obesity especially upper body obesity is associated with various metabolic abnormalities like insulin resistance, glucose intolerance, dyslipidemia and hypertension.

NUTRITION SURVEILLANCE

Nutrition surveillance helps to keep watch on the nutritional status of the population and helps to take necessary action to improve the nutrition. Nutrition surveillance helps to get correct information on the prevalence and geographic distribution of nutrition related problems. It also helps to identify the individual or the population which need assistance to improve their nutritional status. Unless the information about these facts are available it is difficult or not possible to formulate nutrition policies or programmes. Nutrition surveillance also helps to evaluate the effectiveness of programmes designed to improve the nutrition by various agencies.

Assessment of the nutritional status can be done by the following methods

1. Clinical Examination

By physical examination various signs and symptoms which are specific for or non-specific with malnutrition will help to assess the nutritional status. Based on the physical signs malnutrition cannot be quantified. Some of the deficiency disorders may not be associated with physical signs

2. Anthropometry

Measurements such as height, weight, skinfold thickness and midarm circumference indicate nutritional status. In infants and young children, head circumference and chest circumference are also used to assess the nutritional status. When anthropometric measures are made over a period of time at frequent intervals it helps to study the pattern of growth and to find out any deviation of growth from the normal pattern.

3. Laboratory Tests

Estimation of haemoglobin is of importance in the assessment of nutritional status. It is an indicator of the overall status of the nutrition. Faecal examination for parasites and urine examination for sugar and albumin also has got value in the nutrition surveillance. Estimation of concentration of various nutrients, e.g., serum iron, serum retinol, serum folate, serum vitamin B_{12}, etc helps to assess the nutritional status. Detection of abnormal concentration of various intermediates of metabolism as a part of load tests helps to assess the nutritional status. Estimation of the enzymes which require certain vitamins as coenzymes helps detect the deficiency of these vitamins.

Biochemical tests cannot be used on a large population considered for the study of nutritional status because it is time-consuming and costly. Biochemical tests give an idea about the present status of the nutrition and mild deficiency of certain nutrients can be detected.

4. Functional Indicators

Like erythrocyte fragility, leucocyte chemotaxis, prothrombin time, sperm count, heart rate, dark adaptation, etc. is used as indicators to assess the nutritional status

5. Assessment of Dietary Intake

Requires dietary survey. It may be by weighing the food which is going to be cooked and consumed as well as wasted or discarded over a period of 1 to 21 days. Usually, it is carried out for a period of 7 days (one dietary cycle). This type of dietary survey is popular in India. Dietary survey can also be done by weighing the cooked food or by oral questionnaires. When oral questionnaires are used, details regarding the food consumed during the past 24 to 48 hours is collected. If efficiently done, results of this method are reliable. Questionnaire method is convenient to carry out survey over a short period of time on a large population.

6. Vital Statistics

Like mortality and morbidity data indicates the nutritional status of the population. Certain vital statistics indicating the nutritional status of the society are:

(a) Infant mortality rate
(b) Rate of low birth weight
(c) Life expectancy
(d) Morbidity data (e.g., hospital based) related to protein energy malnutrition, vitamin deficiencies, goitre, diarrhoea, parasite infestation, etc.

Agencies Working Towards Food and Nutrition Care

In the year 1945, Cooperative for Assistance and Relief Everywhere (CARE) was founded in North America. It is world's largest non-profit, independent organization providing emergency aid and long-term developmental assistance. The operation of CARE in India began in 1950. Until 1980, CARE-India was involved in providing food for the children. Now, it is helping in various projects like

1. Integrated nutrition and health project
2. Better health and nutrition project
3. Anaemia control project
4. Improving women's health project
5. Improved healthcare for adolescent girls project
6. Child survival project, etc.

It is working with Government of India, State governments, NGOs, etc.

FAO

In 1945, Food and Agriculture Organization (FAO) was formed with its headquarters in Rome. The main aim of FAO is to increase the food production and to keep pace with the growing world population and to ensure that food is consumed by people who need it, in sufficient quantity and proper proportion. It was the first United Nations Organization. It was started with several objectives.

1. To help nations to improve the standard of living
2. To improve the nutrition of the population
3. To improve the farming, fisheries and forest
4. To improve the living conditions of people living in rural areas

In 1960, it organized worldwide campaign called Freedom from Hunger Campaign (FFHC) to fight against malnutrition.

CFTRI

Central Food Technology and Research Institute (CFTRI) was started in 1950. The aim of this research institute is to generate and to apply the knowledge to conserve and to effectively utilize the food resources of the country. It also has got different objectives like using the research and development efficiently, use of agriculture resources, contribute in the sustained development, food security and food safety of the nation, setting standard for the food quality of the nation, to integrate the food supply chain from the cultivator to the consumer which will benefit both cultivator and consumer. There are several achievements by the CFTRI, which gave worldwide recognition to India with respect to its food technology.

NIN

Sir Robert McCarrison in the year 1918 started a 'Beri-Beri' Enquiry Unit in a single room laboratory at the Pasteur Institute, Coonoor, Tamil Nadu. Over a short period it became 'Deficiency Disease Enquiry' unit and later in (1928) it became "Nutrition Research Laboratories" (NRL). In the year 1958, it was shifted to Hyderabad. In the year 1969 it was renamed as National Institute of Nutrition (NIN).

It is involved in the study of various nutritional problems, to monitor diet and nutrition situation in the country. It is also involved in the development of effective method to manage nutritional problems and conducts research connected with planning and implementation of national nutrition programme.

QUESTIONS

Long Essay

1. Discuss carbohydrates in the following headings:
   (a) Classification
   (b) Functions
   (c) Dietary sources and requirements.
   (d) Deficiency and over consumption.

Short Essay

Write short notes on the following:

1. Dietary fibres.
2. Carbohydrates in daily diet.
3. Nitrogen balance.
4. Limiting amino acids.
5. Quality of the protein.
6. Proteins in diet.
7. Fat in diet.
8. Biological value of proteins and its importance in our body.
9. Kwashiorkor and Marasmus.

10. Drying and dehydration.
11. Effect of over consumption of fat in the diet.
12. Protein energy malnutrition.
13. Assessment of Nutritional status.
14. Care.

25

Food Groups; Energy Requirement and Balanced Diet; Dietary Principles in Diseases Condition; Nutritional Education

INTRODUCTION

The components of food needed to be supplied in daily diet for normal health, growth and reproduction are divided into 6 main classes

1. Carbohydrates
2. Proteins
3. Fats
4. Vitamins
5. Minerals
6. Water

Each of these components have got their own function in the body. The diet which we consume must contain adequate amount of nutrients so as to supply all the essential nutrients to the body for optimum growth, health and reproduction.

Nutrients required by all individuals are same but the amount of these nutrients required varies depending upon the physical activity, age and sex. Different natural foods contain varying amounts of essential nutrients and they constitute basic food groups. Food is divided into different groups

They are

1. Protein rich foods
   (a) Milk and milk products
   (b) Meat, fish, egg, pulses and dry legumes, etc.
2. Cereals, roots and tubers
3. Green leafy vegetables, fruits

4. Other vegetables
5. High calorie foods

Milk and Milk Products

Milk contains protein, carbohydrate, fat, calcium, phosphorus, vitamin A and some of the vitamin B complex. Milk proteins are mainly lactalbumin, lactglobulin and casein (phosphoprotein). Lactalbumin and lactglobulin have a very high biological value. Lactalbumin has got highest digestibility.

Casein is a phosphoprotein, Casein content is high in cow's, buffalo's and goat's milk when compared to other milk proteins. Boiling of milk increases the digestibility of casein. Milk proteins have adequate amount of tryptophan. Since 60 mg of tryptophan gives 1 mg of niacin, it compensates the low niacin content of milk.

The fat content of cow's milk is higher (about 5.0 g%) when compared to human milk (about 4.0 g%). Milk fat is palatable and digestible fat. It contains even carbon saturated fatty acids. Unsaturated fatty acids like oleic acid, linoleic acid, linolenic acid and arachidonic acid are also present in milk. Unsaturated fatty acid content in human milk is more when compared to cow's milk.

Carbohydrate present in the milk is lactose (milk sugar). It is a reducing disaccharide made up of glucose and galactose.

Milk is rich in calcium and phosphorus. About 100 ml of cow's milk contains 120 mg of calcium and 90 mg of phosphorus. Calcium and phosphorus content of buffalo milk is higher than the cow's milk. Human milk content of calcium and phosphorus is lower than the cow's milk. Milk is rich in sodium, potassium, chloride and zinc. Milk is poor in iron and copper.

Milk is a good source of vitamin A. It also contains some of the water-soluble vitamins. Milk is a poor source of vitamin D. Usually, milk is fortified with vitamin D to supplement vitamin D to growing children to prevent rickets. Vitamin C content of milk is very low.

Cow's milk is diluted with water to lower its protein content and glucose, lactose or sucrose is added to raise its sugar content to that of human milk. This process is called **humanization of cow's milk.**

Meat

Meat is rich in proteins and poor in carbohydrates. Proteins of meat are of high biological value. Fish is also rich in proteins of higher biological value. Oily fish, fish liver and viscera are sources of vitamin A and vitamin D. Sundried fish are very good source of vitamin K. Fish bone if chewed and swallowed, it will supply good amount of calcium and phosphorus.

Because of the flavour and taste, fish and meat are used in daily diet by most of the individuals. But due to the high cost, all sections of the society cannot afford to include meat and fish in their daily diet.

Egg

Egg is highly nutritious and supplies easily digestible proteins of high biological value. Egg white and egg yolk both contain proteins of high biological value. Egg white contains mainly

ovalbumin and glycoprotein called **avidin**. Avidin tightly binds to water-soluble vitamin biotin. When egg is boiled avidin is denatured and it cannot bind biotin. This increases the bioavailability of biotin.

Egg yolk (yellow part) contains **ovoitellin**. It is also a protein of high biological value. Egg yolk contains large amount of cholesterol (160 mg/100 g), vitamin A, thiamin, biotin, pantothenic acid, calcium, phosphorus, iron, etc.

Egg yolk contains about 30% phospholipids, mainly cephalin and lecithin. Fat present in the egg yolk is readily digested and absorbed. It also contains unsaturated fatty acids.

Pulses

Pulses contain protein, carbohydrate (starch) and some vitamin. **Proteins present in the pulses are rich in lysine but deficient in methionine**. Pulses are rich in protein and cheaper than meat so they are called **"poor man's meat"**. Pulses are rich sources of vitamins like thiamine, riboflavin and niacin. Germination of pulses increases their vitamin content. Nuts have got low carbohydrate and high fat content. They are also rich in proteins.

"Khesari dal" used in certain parts of the India have got a toxin called **b-N-oxalylamino-1-amine.** It can produce spastic paralysis of the lower limbs, a condition called **'lathyrism'**.

Cereals

Cereals are rich in carbohydrates. Protein content is less when compared to pulses. Major proteins present in cereals are glutelin and gliadin. Rice and wheat are commonly consumed in our country. Protein present in rice is deficient in lysine. Unpolished, parboiled rice is rich in water soluble vitamin thiamine. Polishing of the rice will lead to loss of this vitamin. Parboiling will help to preserve thiamine content even during milling. Parboiled rice has got more thiamine content than raw rice.

Wheat has got more protein content than rice. Wheat protein lacks lysine. Wheat contains starch and it is rich in phosphorus, calcium, iron, vitamin B_1, B_2, B_6, nicotinamide and vitamin E.

Roots and Tubers

Commonly used tubers, potato, carrot, sweet potato, colocasia, etc. also come under this group. They are rich in carbohydrates (starch). Potatoes contain good quantities of vitamin C and iron. Carrot is rich in carotene which is a precursor of vitamin A.

Green Leafy Vegetables

Leafy vegetables are good source of carotene, B-complex vitamins like riboflavin, folic acid, vitamin C, iron and calcium.

Cooking of the vegetables leads to loss of vitamin C and B-complex vitamins, hence uncooked vegetables consumed as salads have great nutritive value. Raw vegetables also provide dietary fibres which enhances the peristaltic movement of intestine by improving the bulk of the intestinal content.

Fruits

Energy value of fruits is twice that of vegetables due to high content of sugar and starch. Fruits have got low fat and protein content. Fruits are good sources of B-complex vitamins, vitamin C and vitamin A. Orange, lemon, guava, sweet lime, etc. are rich in vitamin C.

High Calorie Foods

They are rich in sugar or fat. Butter, ghee, vegetable oils, jaggery, sugar, etc. are high calorie foods.

PLANNING MENU

Everyday, we eat at regular intervals. We eat different types of food so as to get sufficient energy and building blocks to keep us healthy. Planning a suitable menu for the family members so as to maintain the physical, mental and social well-being is an art which requires scientific knowledge about food and nutrition.

When we plan a menu it should provide adequate nutrition. Planning the daily food need of the family is to give nutritionally balanced, appetizing food within the budget of the family. During planning the menu, it is necessary to consider the following points.

- Taste, desire of the family.
- Seasonal availability of the food.
- Occasion for which food is being prepared should be considered.
- Budget of the family.
- From all basic food groups, food should be included every day. At least three of the five basic food groups should be included in each meal.
- Quality and texture of the food depending on the preference of the consumer.
- Variety with respect to colour, texture, flavour, appearance; variety with respect to the combination of carbohydrate, protein and fat rich diet; variety with respect to the method of cooking should be considered.
- Number of servings and serving size needs to be kept in mind.
- Personal preferences, religious and social customs should be considered.
- Amount of energy and time the person involved in cooking requires also needs to be considered.

Energy Yield

Energy yield is measured in **calories.** One calorie is the amount of heat required to raise the temperature of 1 gram of water from 15°C to 16°C (i.e. by 1°C). One kcal is 1000 calories. **Joule** has been adopted recently as the universal unit for all types of energy including food energy. One joule is the energy required to move a mass of 1 kg through 1 metre by a force of 1 Newton acting on it.

One kcal = 4.184 kilo Joule [kJ]

1 kJ = 0.239 kcal

Calorific Value

Energy yield of the nutrients is usually expressed as kilocalorie (kcal or Cal [C]). Calorific value is the energy yield of the nutrient per unit weight.

The average calorific value of different nutrients in the body

1. **Carbohydrates** **4 kcal/g**
2. **Proteins** **4 kcal/g**
3. **Fat** **9 kcal/g**

ENERGY REQUIREMENT

Energy balance should be maintained (energy input should be equal to the energy output). When energy input is more than energy output it leads to obesity. If energy output is more than input it leads to weight loss.

Energy required by an individual is the amount of energy exhausted by that individual. Energy expenditure can be calculated by the measurement of amount of energy spent. Energy is mainly spent for

(a) Basal Metabolic Rate
(b) Specific Dynamic Action (Diet induced thermogenesis)
(c) Physical Activity

Basal Metabolic Rate (BMR)

Basal metabolic rate is the energy required by an individual who is awake and at a state of mental and physical rest, 12 to 18 hours after meal in a comfortable condition (i.e., temperature, pressure and humidity) to maintain life.

Basal metabolic rate can be determined by different methods. Accurate measurement can be done by using **Tissot method** and **Douglas method**. It is cumbersome and BMR cannot be measured accurately. Basal metabolic rate can be measured with reliable accuracy by using **Benedict-Roth basal metabolism apparatus (Indirect calorimetry; closed circuit method).**

Indirect Calorimetry

BMR is calculated based on the volume of oxygen consumed by the individual at the nonprotein RQ.

Benedict-roth basal metabolism apparatus

Individual resting in a comfortable surrounding breathes oxygen from the cylinder of the apparatus. Carbon dioxide exhaled is absorbed by the soda lime in the apparatus. The volume of oxygen consumed can be recorded on the drum. The subject is asked to breathe for 6 minutes.

Oxygen consumed is multiplied by 10 to convert it to oxygen consumption per hour. The energy produced when one litre of oxygen is consumed is calculated as 4.82 kcal. The oxygen consumed per hour when multiplied by 4.82 it gives the energy produced in kcal per hour under basal conditions. The energy produced (kcal/hour) is divided by the surface area of the body since **basal metabolic rate is expressed as energy produced/square metre of the body surface area/hour under basal conditions.**

Surface area of the body can be calculated by **Du Bois formula** as below:

$$A = H^{0.725} \times W^{0.425} \times 71.84$$

A = Surface area

H = Height in centimetre

W = Weight in Kilogram

The surface area thus obtained in square centimeter is converted to square metre by dividing 10,000.

Normal range for BMR: (a) in male is about 34 to 37 kcal/sq.meter/hour, (b) in adult females 30 to 35 kcal/sq.metre/hour. For the purpose of calculation of energy requirement of an adult individual, BMR can be considered 24 kcal/kg body weight/day.

Factors affecting BMR

1. **Age:** As the age advances BMR decreases.
2. **Sex:** Infants of both sexes have almost identical BMR. In adults, BMR is higher in males than in females.
3. **Nutritional status:** Starvation will reduce BMR
4. **Lean body mass:** Diminishes the BMR
5. **Fasting and starvation:** Decreases BMR
6. **Climate:** People who live in warm climate have lower BMR than who live in cold climate
7. **Hormones:** BMR is increased by thyroid hormone and also by the growth hormone. In hypothyroidism, BMR is decreased and in hyperthyroidism BMR is elevated
8. **Pregnancy:** BMR increases.
9. **During growth of the child:** BMR is increased
10. **Stress:** Increases BMR
11. **Fever:** BMR is increased
12. **Smoking:** Increases BMR

For the convenience of calculation BMR can be considered 24 kcal/kg body weight.

Specific Dynamic Action (SDA)

Extra amount of heat generated than what is expected by catabolism of the food ingested is called **specific dynamic action (SDA)** or dietary thermogenesis.

1. Carbohydrate is 5 to 6%,
2. Protein is 30%
3. Fats 4%
4. Mixed diet 10 to 12%

The exact reason for SDA is not known. It may be energy used to prepare the non-nitrogenous part of the biomolecule for storage. When energy requirement of an individual is calculated, an additional 10% of calories (for mixed diet) should be added to account for the energy lost as SDA.

Respiratory quotient (RQ)

Respiratory quotient is the ratio of the volume of CO_2 produced to the O_2 consumed during the given time.

$$RQ = \frac{\text{Volume of } CO_2 \text{ produced}}{\text{Volume of } O_2 \text{ consumed}}$$

RQ value for different components of the diet is different. When carbohydrates are metabolized in the body amount of CO_2 produced is same as that of O_2 consumed.

$$C_6H_{12}O_6 + 6O_2 \rightarrow 6CO_2 + 6H_2O$$

$$\text{RQ of carbohydrate} = \frac{6\,CO_2}{6\,O_2}$$

RQ of carbohydrate = 1
RQ value of protein is about 0.8
RQ value of fat is about 0.7

Diet which is a mixture of varying proportion of carbohydrate, protein and fat (mixed diet) is about 0.85.

Physical Activity

The amount of energy required by an individual mainly depends upon the age, sex, climatic condition in addition to basal metabolic rate and physical activity. Physical activity requires a large amount of energy.

Physical activity is one of the factors on which the energy requirement of an individual depends. Depending upon the level of physical activity, work done by the individuals can be divided into

1. Sedentary work (Teacher, Priest, Executive, Retired person)
2. Moderate work (Weavers, Drivers, Fitters, Carpenters)
3. Heavy work (Blacksmith, Mine workers, Labourers, Stone cutters)

BALANCED DIET

The diet containing adequate amount of energy, building blocks, accessory food factors, minerals and water so as to support health, growth and reproduction is called a **balanced diet**. Depending upon the age, sex, physical activity as well as physiological state (pregnancy, lactation) amount of different food groups used in the diet will also change.

When planning suitable diet following points should be kept in mind.

(a) It should provide all the required amount of energy (calorie), protein (of good quantity and quality), fat, vitamin and mineral.
(b) It should have proper proportion of carbohydrate, protein and fat.
(c) It should be locally available and within the reach of the individual or family.
(d) It should adhere to the local food habits.

(e) It should be digestible and palatable.
(f) It should contain enough dietary fibres.
(g) It should be convenient to prepare.

Energy requirement according to ICMR recommendation considering the reference man and women with body weight of 60 kilogram and 50 kilogram respectively

	Adult man	Adult woman
1. Sedentary worker	2425 kcal/day	1875 kcal/day
2. Moderate worker	2875 kcal/day	2225 kcal/day
3. Heavy worker	3800 kcal/day	2925 kcal/day

Infants of 6 to 12 months require 98 kcal/kg per day. Children above one year require 1200 kcal to 2000 kcal/day at different ages.

Calorie need of an individual or family can be expressed as average calorie need of a reference man. It is called **adult consumption unit (ACU).** The energy requirement of an adult man who is a **sedentary worker is about 2400 kcal/day and it is accepted as 1 ACU.**

	Adult man	Adult woman
1. Sedentary worker	1.0 ACU	0.8 ACU
2. Moderate worker	1.2 ACU	0.9 ACU
3. Heavy worker	1.6 ACU	1.2 ACU

The ACU of a family containing three members can be calculated as follows:

Adult man Moderate worker = 1.2 ACU
Adult woman Moderate worker = 0.9 ACU
Child of 5 years = 0.5 ACU
Total = 2.6 ACU
= 2.6 × 2400 = 6240 kcal

In general, calorie requirement of an individual with ideal body weight doing sedentary or moderate work can be considered approximately **40 kcal/kg body weight.**

A 60 kg man with sedentary lifestyle requires 60 × 40 = 2400 kcal of energy per day. He requires 60 grams of protein. One gram of protein gives 4 kcal of energy so 60 grams of protein gives 60 × 4 = 240 kcal of energy.

2400 – 240 = 2160 kcal of energy should come from carbohydrate and fat. Fat content of the diet should provide 20% of the total energy requirement and the remaining should come from carbohydrate.

By knowing the nutritive value of the foodstuffs, diet can be designed to provide efficient nutrition. Foods with the same nutrient composition can be grouped together, e.g., 2 chapattis, 2 slice of bread, 1 cup of rice, 2 idlis give almost the same amount of energy and protein. Food items having almost the same nutritive value can be grouped together to provide variety in the food to avoid monotony.

Total quantity of food items calculated should be divided into three or four parts and given three to four times a day depending on the food habits of the individual.

1. Nutrition of Healthy Adult Males and Females

Healthy adult male and female (sedentary or moderate worker) require approximately 40 kcal/kg body weight.

The energy yielding nutrients should be supplied in proper proportion. Carbohydrate should provide 60 to 70% of total energy requirement mainly in the form of starch. Protein should provide about 10 to 12% and fat about 20 to 25% of the total energy requirement. Protein requirement is about 1 gram per kg body weight in adults. Fat or oil supplied should contain 1/3 part each of saturated fat, polyunsaturated fat and monounsaturated fat.

Fresh vegetables and fruits of different variety and green leafy vegetables should be included in the diet. Seasonal fruits should be taken daily. About 100 gram of fruit should be consumed every day.

Calcium, phosphorus, vitamin D and iron should be supplied in sufficient amount. To replace the loss of calcium and phosphorus from the body, about 0.5 grams should be supplied in the diet. Rich sources of calcium and phosphorus like milk, carrot, bony fish, ragi can be included in the diet. Daily about 100 IU of vitamin D should be included in the diet. Daily dietary iron requirement is about 10 mg for adult man and 18 mg for adult women. Daily supply of vitamin A should be about 750 mg per day.

2. Nutrition in Pregnancy

Pregnancy is a physiologic state through which most of the women pass through during their reproductive period. During pregnancy, foetus consisting of millions of different types of cells is produced from one single cell called zygote. In addition to formation of foetus various other physiological changes take place in the prospective mother. Her uterus and mammary gland grows in size; placenta and amniotic fluids are formed. All these changes help to protect, nourish and to deliver a healthy baby.

As a woman, she requires various nutrients to maintain her health and growth. During pregnancy, requirement of nutrients increases to cope up with the nutrients used by the growing foetus, the organs of the reproductive system as well as to prepare the mother for lactation.

Maternal nutrition has got great effect on the overall development of the infant. Poor maternal nutrition may be responsible for low birth weight, infection, and infant mortality. It can also be responsible for maternal mortality.

ICMR nutrition expert group has recommended an extra energy of 300 kcal/day to the pregnant lady. Extra calories requirement is mainly during the second and third trimester. Additional protein requirement is about 15 g/day in the last two trimesters. For example, a woman who is 50 kg, her protein requirement is about 50 grams/day. In the second and third trimester of her pregnancy, protein requirement becomes 65 grams/day. Enhanced protein requirement is to provide amino acids for foetal tissues, maternal reproductive organs and for the synthesis of maternal plasma proteins to increase the blood volume. At least 1/4th of the protein should be supplied by proteins of animal origin having high biological value.

The daily requirement of calcium, phosphorus, iron and vitamins is also increased. The extra calcium is required to deposit in the foetal tissue. So a total daily allowance of about 1 gram of calcium and phosphorus each and about 200 IU of vitamin D is safe. Iron requirement during the entire period of pregnancy is increased. Daily about 38 mg of iron should be provided throughout the pregnancy. About 0.3 and 1.5 μg of folic acid and vitamin B12 is required in the diet to help in the nucleic acid synthesis and blood cells in the mother and foetus. Vitamin K (3IU) needs to be supplemented especially during the last trimester to provide vitamin K to the foetus. Additional thiamine, riboflavin and nicotinamide (0.2, 0.2 and 2 mg respectively) are provided in the diet.

Extra nutrients required can be provided by adding extra cereals, sugar, jaggery, pulses, milk, vegetables, meat, fish and enough fibres in the food to meet this extra requirement.

3. Nutrition in Lactating Mother

Lactation is a process of production and secretion of milk. It is a physiological phase in mother's life through which she passes after parturition (delivery of the baby). According to ICMR nutrition expert group, milk output is about 600 mL/day. So the lactating mother requires an additional energy of 700 kcal/day.

Protein requirement is increased by about 25 grams in lactation. An additional requirement of calcium of about 500 to 600 mg/day should be provided. During lactation, iron supplementation is not recommended by ICMR nutrition expert group. Sufficient iron should be provided in the diet to meet the normal requirement. Care should be taken to minimize the interference of phytates with the iron absorption.

Vitamin A is secreted in the milk. About 400 mg of additional vitamin A per day is required. Other vitamins like folic acid, vitamin C, vitamin D, thiamin rich foods should be supplied in the daily diet.

4. Nutrition of Elderly Individuals

In **elderly individuals** special attention should be given to their nutrition. Various physiological changes accompany the process of aging. Nutritional requirement of geriatric age group is different from that of adults.

Elderly individuals need food which is soft and easily digestible. It is better to steam or boil the food. Since many of them have sedentary lifestyle calorie requirement is reduced. If an individual is obese, calorie intake should be adjusted in such a way so as to reduce the excess weight. Protein supplementation is required. Food should contain sufficient fat, minerals and lot of water. Most of the elderly individuals suffer from constipation. To avoid constipation sufficient fibres should be included in the diet.

5. Nutrition of the Infants

Infants grow rapidly. They require supply of good quantity of carbohydrate, protein, fat, vitamins minerals and sufficient amount of water to support the rapid growth. Energy requirement of the infant is very high so, it requires good nutritious food.

Infants during the first three months of life require about 120 kcal per kg body weight, from three months to six months infants require 110 kcal/kg body weight, from six months to 12 months 100 kcal/kg body weight. Protein requirement of infants and children is increased. In the diet, about 0.5 to 0.6 gram of calcium is required for infants and 0.4 to 0.5 grams for children. 200 IU of vitamin D is required for infants and children. Requirement of vitamin A and C are increased in infancy and childhood. Every day about 12 to 20 mg of iron should be supplied in the diet.

Breastfeeding

Infants do not have teeth, hence diet should be liquid or semisolid. Mother's breast milk is the ideal food for infants. It contains all the nutrients required to support the growth of the infant. Human milk can be easily digested. Milk flow starts between third to fifth day. Breastfeeding should be started on the same day the baby is delivered. Infant should be put to the breast regularly. Infant need not drink much in first few days.

Breast milk is the best food for the newborn up to one year of age. Even during the second year breastfeeding has got importance in the nutrition of a child. Good nutrition during the first year of life helps the child to grow better in the later years and gives a good start to life.

Advantages of breastfeeding

- Mother can produce as much milk as required by the baby. If baby sucks more milk, the mother will produce more milk. A special mechanism of demand and supply exists between them.
- Mother's milk is clean since milk passes directly from breast to infant's mouth.
- Milk is available throughout the day ready with optimum temperature suitable for immediate consumption and free of cost.
- Breastfeeding helps to establish a close relationship between mother and infant. It gives a sense of security and love to the baby.
- Mothers who breastfeed do not menstruate immediately after delivery. This helps in spacing of children and mother can give more attention to the baby.
- Incidence of breast cancer is less in mothers who breastfeed their child.
- Helps in the retraction of the uterus and in the restoration of reproductive organs to normal state.
- Sodium content of human milk is low which is advantageous for infants. Infant's kidneys do not have the capacity to handle large amount of sodium.
- Milk contains antibodies which protect the baby against infections.

Sometimes, mother may not be able to feed the infant due to various reasons (e.g., mastitis, lack of breast milk, sore nipple, or death of the mother). It is advisable to find a wet-nurse (a lactating woman who is responsible and willing to feed the infant).

Cow's milk and buffalo milk has got high protein content compared to human milk. Fat content of the buffalo milk is more when compared to human milk. Carbohydrate content of the cow's milk is low compared to human milk.

Animal milk can be modified so that it can be made suitable for infants. It can be done by removing excess fat content especially in case of buffalo milk. It is essential to dilute animal milk with clean water add sugar. Utensils used to feed the infant have to be washed and boiled. Before preparing the food, wash hands thoroughly with soap.

Colostrum

Yellow viscous fluid secreted by the mammary gland during first three days after parturition. It is rich in vitamin A, D, E, calcium and phosphorus. Colostrum is poor in casein, fat and lactose. It is rich in immunoglobulin derived from the mother. These immunoglobulins are absorbed as such and give immunity to the newborn.

Weaning

Weaning is to detach from that to which one is strongly habituated. It is a process of shifting the baby from breast milk to semisolid food so that the baby takes more semisolid/solid food. It is advised to feed only breast milk for the first six months. About fifth month onwards, breast milk alone may not be enough to support the growth of the baby.

During the transition from breast milk to weaning food mother should be very supportive to the child. Transition should be gradual and start feeding with small quantity of semisolid food. Even when the child gets accustomed to semisolid food it is advisable to feed breast milk along with semisolid food.

To begin with porridge made of cereal flour which is very thin can be fed in small quantity (about one teaspoon) in addition to breast milk and gradually the quantity can be increased. Temperature of the food should not be too cold or hot.

It is advisable to feed the child same time every day at a place where all family members eat so as to make the child more comfortable. The thickness of the porridge of cereal flour can be increased as the baby grows. Cereals which are well cooked, smashed and made soft can be used. All utensils used including the hands of the individual preparing the food should be clean.

Once the child is accustomed to the semisolid food later cooked legumes, chopped and smashed vegetables, smashed egg white, smashed fruit can be given. Cooked and smashed cereal and legume combination makes a good feed for the growing child. By the end of one year, child is able to eat the food cooked for the family members. Priority should be given to the feeding of the child.

6. Nutrition of Pre-schoolchildren; Schoolchildren and Adolescents

Pre-schoolchildren require good nutritious food which helps them to thrive better. It gives a solid basis for their future health. This group is particularly vulnerable for malnutrition. There is hardly any difference in the nutritional requirement of boys and girls until about 9 years of age.

Nutrition of adolescents is equally important since maximum growth takes place during this period of time. Nutritional requirements differ between boys and girls during puberty.

Since growth is rapid and they are active, diet rich in protein and energy should be provided. In addition to this, all other essential nutrients like calcium, phosphorus, iron, various vitamins like vitamin A, vitamin D, vitamin C, and vitamin B-complex should be provided in the diet.

DIET IN DISEASE CONDITIONS

Diet in Diabetes

In the treatment of diabetes mellitus, diet control and exercise has got great importance. The main aim is to maintain the blood glucose level near normal. In diabetic individual who is not obese (sedentary worker) requires about 30 kcal/kg body weight. In obese diabetics energy requirement is calculated as 20 kcal/kg body weight (or a total of about 1000 kcal/day).

50% of the total calorie requirement should come from carbohydrate. 25 to 30% of the total calories should come from fat and protein should provide about 20% of the energy requirement (should not exceed 30%).

Protein provided in the diet should be proteins of high biological value. If there is renal impairment, then the protein intake can be reduced to about 0.6 gm/kg body weight.

Carbohydrate supplied in the daily diet should be polysaccharide (starch–rice, wheat, potatoes, etc). Monosaccharides and disaccharides should be avoided in the diet.

Fats supplied in the diet should be free of or contain minimum quantity of saturated fats. Unsaturated fatty acids (polyunsaturated and monounsaturated) are beneficial to the diabetics.

The food should be consumed at regular intervals. Taking total calorie requirement of the day in small quantities at frequent intervals benefits diabetics by keeping their postprandial blood glucose level low.

Sufficient raw vegetables and fresh fruits need to be included in the diet to provide vitamins and antioxidant activity.

Diet for Heart Disease

In person with **heart disease or hypertension,** diet should be free of saturated fatty acids. It should be rich in polyunsaturated fatty acids and some monounsaturated fatty acids. ω-3 fatty acid containing foods like marine fish should be included. Sufficient amount of dietary fibres should be included. To provide antioxidants sufficient leafy vegetables and other vegetables should be taken along with one fresh fruit every day.

Salt should be restricted and animal protein should be avoided. If renal impairment is present, protein intake should be minimized in the diet. Carbohydrate should provide 50% of the total energy requirement.

Diet in Renal Disease

In acute renal failure, fluid and salt intake should be restricted. About 2000 kcal of energy should be provided. During the oliguric phase protein intake should be minimized and during the diuretic phase gradually the protein intake is increased. When the urine output improves then potassium supplementation can be done by giving fruit juice and care should be taken to avoid hyperkalaemia. Concentrated forms of water-soluble vitamin can be given.

In chronic renal failure, adequate calories containing minimum amount of proteins of high biological value are provided in the diet. Water intake should depend on the water output (water input output chart should be maintained). Minimum salt intake is allowed and phosphate content of the diet should be kept very low.

Diet in Febrile Condition

Fever commonly results from infection. It is a defensive mechanism of the body to ward off the infectious agent. In febrile condition BMR is increased. Energy requirement may be slightly increased. Large portion of the energy should be supplied as carbohydrate. Proteins of high biological value and easily digestible should be supplied. Person with fever requires to drink a lot of water.

NUTRITIONAL EDUCATION

Educating the population regarding nutrition plays a great role in improving nutrition. Nutritional education can be given at various places like home, hospital, small gathering arranged in the rural areas, etc. It is important to understand the feeding habits of that population.

Mothers are keen to learn about nutrition and improve the health and nutrition of the family. They may be ignorant about this aspect. Ignorance about nutrition can be superadded by health related problems like unhygienic environment where they live and shortage of clean drinking water. Health workers can teach the individuals of the society regarding the importance of hygienic conditions and how to achieve it.

The cultural belief and superstitions about the food and feeding especially during pregnancy, lactation and feeding of the young ones of the family can be overcome by education of the family regarding nutrition.

Advantages of breastfeeding, knowledge about nutritive value of the locally available foods, cultivation of vegetables and fruits in the country yard, etc. can be imparted.

QUESTIONS

Long Essay

1. Explain the methods of meeting the nutritional needs of a pregnant women.
2. Define balanced diet. What are the factors you consider while planning the menu?
3. Define BMR. What are the factors affecting BMR?

Short Essay

Write short notes on the following:

1. Food groups.
2. Balanced diet.
3. Weaning
4. Energy requirement in lactation.
5. Dietary needs in fever.
6. BMR
7. Dietary management of Diabetes mellitus.
8. RDA for heavy worker female adult.
9. Balanced diet for toddler.
10. Diet for lactating mother.
11. Breast feeding.
12. Nutrition in preschool children.
13. Diet in fever.
14. Diet in hypertension.
15. Nutritional education.
16. Nutritional need of the elderly.

Appendices

Appendix-1

Recommended Dietary Allowances

(as per ICMR recommendation-dietary guidelines for Indians – a manual)

	Man	Women	Preg	Lactation	Infants	Children	Boys	Girls
Protein g/dl	60	50	+15	0 – 6 months + 25 6 – 12 months + 18	0 – 6 month 2.05/kg 0 – 12 months 1.65/kg	1 – 3 years 22 4 – 6 years 30 7 – 9 years 41	10 – 12 years 54 13 – 15 years 70 10 – 18 years 78	10 –12 years 57 13 – 15 years 65 16 – 18 years 63
Fat g/dl	20	20	30	45	–	25	22	22
Calcium Mg/day	400	400	1000	1000	500	400	10 – 15 years 600 16 – 18 years 500	10 – 15 years 600 10 – 18 years 500
Iron Mg/day	28	30	38	30	–	1 – 3 years 12 4 – 6 years 18 7 – 9 years 26	10 – 12 years 34 13 – 15 years 41 16 – 18 years 50	10 – 12 years 19 13 – 15 years 28 16 – 18 years 30
Vit A µg/d Retinol	600	600	600	950	350	1 – 6 years 400 7 – 9 years 600	600	–
β – Carotene Thiamine mg/dl.	2400	2400	2400	3800	1200	1 – 6 years 1600 7 – 9 years 2400	2400	2400

(Contd.)

	Man	Women	Preg	Lactation	Infants	Children	Boys	Girls
Thiamin µg/d	Sedentary 1.2 Moderate 1.4 Heavy 1.6	Sedentary 0.9 Moderate 1.1 Heavy 1.2	+ 0.2	0 – 6 months + 0.3 6 – 12 months + 0.2	0 – 6 months 55 µg/kg 6 – 12 months 50 µg/kg	1 – 6 years 1600 7 – 9 years 2400	10 – 12 1.1 13 – 15 1.2 16 – 18 1.3	1.0 1.0 1.0
Riboflavin µg/d	Sedentary 1.4 Moderate 1.6 Heavy 1.9	1.1 1.3 1.5	0.2	0 – 6 month + 0.3 6 – 12 months + 0.2	0 – 6 month 65 µg/kg 6 – 12 months 60 µg + 0.2	1 – 3 Years 0.7 4 – 6 Years 1. 0 7 – 9 Years 1.2	10 – 12 1.3 13 – 15 1.5 10 – 18 1.6	1.2 1.2 1.2
Nicotinic Acid µg/d	Sedative 16 Moderate 18 Heavy 21	12 14 16	+ 2	0 – 6 month + 4 6 – 12 month + 3	0 – 6 710 µg/kg 650 µg/kg	1 – 3 Years 8 4 – 6 Years 11 7 – 9 Years 13	10 – 12 15 13 – 15 16 16 – 18 Y 17	10 – 12 Y 13 13 – 15 Y 14 16 – 18 Y 14
Pyridoxins µg/d	0.2	0.2	2.5	2.5	0 – 6 month 0.1 6 – 12 month 0.4	1 – 6 Years 0.9 7 – 9 Years 1.6	10 – 12 Y 1.6 13 – 18 Y 2.0	10 – 12 Y 1.6 13 – 18 Y 2.0
Ascorbic acid µg/d	40	40	40	80	25	40	40	40
Folic Acid µg/d	100	100	400	150	25	1 – 3 Y 30 4 – 6 Y 40 7 – 9 Y 60	10 – 12 Y 70 13 – 18 Y 100	10 – 12 Y 70 13 – 18 Y 100
Vit B_{12} µg/d	1	1	1	1.5	0.2	0.2 – 1.0	0.2 – 1	0.2 – 1.0

Appendix-2

Balanced Diet for Adult Man

(as per ICMR recommendation)

Food groups	Sedentary	Moderate	Heavy
Cereals & Mellitus (g)	420	480	690
Pulses (g)	60	90	90
Roots & tubers (g)	200	200	200
Milk (ml)	300	300	300
Green leafy vegetables (g)	100	100	100
Other vegetables (g)	100	100	100
Fruits (g)	100	100	100
Sugars (g)	25	40	55
Fats & Oils (visible) (g)	20	35	55

One portion of pulse (30 g) can be substituted with one portion (50 g) of egg, meat, chicken or fish.

Appendix-3

Balanced Diet for Woman

(as per ICMR recommendation-dietary guidelines for Indians – a manual)

Food groups	Sedentary	Moderate	Heavy
Cereals & Mellitus (g)	300	360	480
Pulses (g)	60	75	90
Roots & tubers (g)	100	100	200
Milk (ml)	300	300	300
Green leafy vegetables (g)	100	100	100
Other vegetables (g)	100	100	100
Fruits (g)	100	100	100
Sugars (g)	20	25	45
Fats & Oils (visible) (g)	20	30	40

One portion of pulse (30 g) can be substituted with one portion (50 g) of egg, meat, chicken or fish.

Appendix-4

Balanced Diet for Infants and Children

(as per ICMR recommendation – dietary guidelines for Indians—a manual)

		Infants 6-12 months	1–9 years	10–12 years		13–18 years	
				Boys	Girls	Boys	Girls
1.	Cereals & Mellitus (g)	45	1-3 year 120 4-6 year 210 7-9 year 270	270	330	300	420
2.	Pulses (g)	15	1-3 year 30 4-6 year 45 7-9 year 60	60	60	60	60
3.	Milk (ml)	500	500	500	500	500	500
4.	Roots & tubers (g)	50	1-3 year 50 4-6 year 100	100	100	100	200
		Infants 6-12 months	1–9 years	10–12 years		13–18 years	
				Boys	Girls	Boys	Girls
5.	Green leafy vegetables (g)	25	1-6 year 50 7-9 year 100	100	100	100	100
6.	Other vegetables (g)	25	50	100	100	100	100

(Contd.)

7.	Fruits (g)	100	100	100	100	100	100
8.	Sugars (g)	25	1-3 year 20 4-9 year 25	30	35	30	35
9.	Fats/Oil (g)	10	1-3 year 20 4-9 year 25	25	25	25	25

One portion of pulse (30 g) can be exchanged with one portion (50 g) of egg, meat, chicken or fish. Egg, meat, chicken or fish can be introduced to infants by around 9 months.

Index